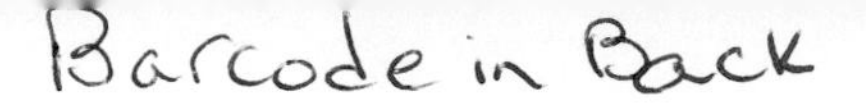

COLONIALISM AND RACISM IN CANADA

HISTORICAL TRACES AND CONTEMPORARY ISSUES

MARIA A. WALLIS
York University

LINA SUNSERI
Brescia University College, affiliated with University of Western Ontario

GRACE-EDWARD GALABUZI
Ryerson University

NELSON EDUCATION

NELSON EDUCATION

Colonialism and Racism in Canada: Historical Traces and Contemporary Issues
by Maria A. Wallis, Lina Sunseri, and Grace-Edward Galabuzi

Associate Vice President, Editorial Director:
Evelyn Veitch

Editor-in-Chief, Higher Education:
Anne Williams

Executive Editor:
Laura Macleod

Senior Marketing Manager:
David Tonen

Developmental Editor:
Lily Kalcevich

Permissions Coordinator:
Daniela Glass

Production Service:
Macmillan Publishing Solutions

Copy Editor:
Tanya Koehnke

Proofreader:
Dianne Fowlie

Indexer:
Maura Brown

Production Coordinator:
Ferial Suleman

Design Director:
Ken Phipps

Managing Designer:
Franca Amore

Interior Design:
Olena Sullivan

Cover Design:
Olena Sullivan

Cover Image:
© Spencer Grant/Photo Edit

Compositor:
Macmillan Publishing Solutions

Printer:
Webcom

Printed and bound in Canada
1 2 3 4 12 11 10 09

For more information contact Nelson Education Ltd., 1120 Birchmount Road, Toronto, Ontario, M1K 5G4. Or you can visit our Internet site at http://www.nelson.com

Library and Archives Canada Cataloguing in Publication

Wallis, Maria A. (Maria Antoinette), 1960-

Colonialism and racism in Canada: historical traces and contemporary issues / Maria A. Wallis, Lina Sunseri, Grace-Edward Galabuzi.

Includes bibliographical references and index.
ISBN 978-0-17-650066-5

1. Canada—Race relations—Textbooks. 2. Racism—Canada—Textbooks. 3. Canada—Social conditions—Textbooks. I. Sunseri, Lina II. Galabuzi, Grace-Edward III. Title.

FC104.W34 2009
305.800971 C2008-907998-1

ISBN-13: 978-0-17-650066-5
ISBN-10: 0-17-650066-9

CONTENTS

PART FIVE: Responses and Resistances 285

ACKNOWLEDGMENTS

We would like to express our appreciation to the many reviewers who offered very helpful comments and suggestions: Roger Albert, North Island College; Ron Bourgeault, University of Regina; Augie Fleras, University of Waterloo; Jeffery Klaehn, Wilfrid Laurier University; Cynthia Levine-Rasky, Queen's University; Roksana Nazneen, Concordia University; and Tracy Nielsen, Mount Royal College.

Maria A. Wallis
Lina Sunseri
Grace-Edward Galabuzi

My special thanks to Lina Sunseri for working so diligently on this book. Your company during this journey will always be remembered as an honour and a privilege. To Scott Couling, thank you for skillfully seeing this book through the initial developmental stages. Special thanks to Lily Kalcevich for guiding us through the book's review process with great empathy, patience, and intelligence. We appreciate the attention to detail and your passion for the work. Thank you to Tanya Koehnke, Susan Calvert, Gunjan Chandola, and Liisa Kelly for their diligence and expertise in copyediting and overall management of this project during the final stages. Thank you to the entire Nelson Education Ltd. team for making the publication of this book such an enjoyable experience for me.

Colonialism and racism are present in Canada today in new forms. My recent personal experiences—another future book—are a testimony to this fact. I arise from these challenges due to the solidarity and companionship of friends such as Grace-Edward Galabuzi; my comrades in CUPE 3903, representing contract faculty, teaching assistants, and research assistants at York University; and many others. I am supported daily by my family—my daughter Rose, and my chocolate Labrador retriever, Hershie.

I dedicate my work in this book, and in these struggles, to my daughter, Rose Mansi Wallis. Your namesake is Rosa Luxemberg, but your future will be yours to define. It will be built on the struggles of the seven generations that have come before you. To them, this book is a profound and deeply felt *thank you.* I am energized by the spirit of the past, and am constantly inspired by Rose's very being and possibilities.

Maria A. Wallis

My Longhouse name is Yeliwi:saks, Onyota'aka (Oneida) is my Indigenous nation, a?nowal (Turtle) is my clan, and I also have Italian heritage from my father's side. Oneida Nation of the Thames is my Indigenous community, and it is one of the three existing Oneida nations of North America. As an Ukwehu:wé (Original) person, I want to first give thanks to Shukwayatisu (Creator), all my relations, which includes all living things given to us by Creator, especially all my family, my partner Chris and his children, and all Onyota'aka people who have come before me. I also want to acknowledge all the support given to me by my colleagues at Brescia University College, to my friends/co-authors of this book, to Nelson Education Ltd. for giving us the opportunity to develop this project so that our words and those of the various scholars included herein can be shared with others interested in issues of racism and colonialism and who dream of a better world. I also appreciate the constructive criticisms provided to us by the reviewers so that the book could become stronger and clearer and to the detailed work done by the editing team. I would like to dedicate this book to the memory of my grandma Dorothy Day who now lives in the Spirit world and all Indigenous warriors who have struggled to maintain our cultures and teachings throughout the centuries of colonialism. May our generation continue to be warriors so that the next seven generations can have a better life here in Turtle Island.

Tane•to nyohtuhake ukwanikula (Our minds will stay now as one).

Lina Sunseri

I would like to celebrate my co-authors, Lina (Yeliwi:saks) and Maria for their patience, wisdom, encouragement, and strength throughout this project. Thanks to my family, especially Shelina, my life partner, and Makula and Sanyu, our children. The sacrifices they make, in quality time and attention, so that I can work at something I love is often taken for granted. To Auntie Zabeti, who delights in my achievements—may you live long! I extend my gratitude to my colleagues at Ryerson University for creating a comfortable space for me to work and grow. A special thank you to my former colleague, Janet Conway, for her friendship and support. Thanks also to my students through the years, who have inspired me and do so daily with their curiosity, enthusiasm, and friendship. You hold a special place in my heart. To the team at Nelson Education Ltd. through the project—thank you for supporting the vision and investing financially, intellectually, and through hard labour to make it happen. Finally, I want to pay special tribute to the many anti-racists and anti-colonial activists and scholars with whom I have shared the barricades from time to time and from whom I have learned so much! I continue to grow in my understanding of the complicated relationship between Indigenous peoples and racialized settlers in Canada. This project has opened my eyes wider still and I hope I stay true to the path of knowledge, self-reflection, and action.

Grace-Edward Galabuzi

Introduction

TRACKING COLONIAL AND CONTEMPORARY RACIALIZATION PATTERNS

Canada's emergence as a multicultural nation with a reputation as a global leader in "ethnic and racial diversity management" is steeped in a complex history and current configurations of structural hierarchies of racial inequality and attempts to erase its indigenous reality. Two key concepts are implicated in this construction of the contemporary Canadian reality: colonialism and racialization. Both colonialism and racialization exist in Canada's history and present an unacknowledged continuity that defines its dominant and structural social, economic, political, and cultural orders. The emergence of Canadian society was an act of colonial fiat and erasure that involved the rationalization of a new colonial social, political, and economic order imposed on the Indigenous nations whose territories were "legally" characterized as people-less by the British and French monarchies, their operatives, and the economic interests they represented. Invoking the legal notion of Terra Nullia—the idea that the land to be settled is empty and uninhabited—the British and French invaders "legally" disregarded the existence of Indigenous nations already on the land and set about to create a "new" Europe in the "new world." Canada thus engaged in a pattern of European colonization that would become generalized to the rest of the world outside Europe. This process, which similarly took place in other "white settler colonies" such as the United States, Australia, and New Zealand, was, as Razack points out, characterized by "the dispossession and near extermination of Indigenous populations by the conquering Europeans. As it evolved, white settler societies continued to be structured by a racial hierarchy. In the national mythologies of such societies, it is believed that white people came first and that it is they who principally developed the land."[1]

This colonial process of socio-political ordering also introduced the world's peoples to the power of structural racism and white supremacy, phenomena that would transcend geography and history. By white supremacy, we mean that in white settler colonies such as Canada, whiteness became a socially constructed category from which racial hierarchy is produced and maintained, ultimately leading to various forms of unearned privileges afforded to white peoples at the expense of other groups. Colonial consequences of wealth disparities and labour market demands resulted in the phenomenon of transnational migration to the lands of colonial powers—centres of Capital—where other racialized people would encounter socio-economic conditions similar to those of the Indigenous peoples whose land they now inhabited. The experience of colonization became a shared experience with its own specificities for Indigenous peoples in Canada as well as racialized immigrants to Canada through the internationalized historical processes of colonization and the social processes of racialization that resulted in forms of "internal colonization"—that being the existing process whereby Indigenous peoples of white settler societies like Canada still live in conditions where their political, social, and economic autonomy is undermined by an imposing State. This term also speaks of the oppressive and unequal conditions some racialized groups face, often leading to their relative lack of autonomy and power in what is often labelled a "postcolonial" society such as Canada. This reveals a stunning continuity in the global-local colonial order that transcends space and time.

The continuity of colonization and racialization manifests itself in the composition of many of Canada's institutions whose racially defined social hierarchies reflect the historical processes of colonization rooted in white-settler access to free land acquired either forcibly or by some form of unequal treaty process from Indigenous peoples. Cheap labour provided by racialized people also assured privileged access to power for European, white settlers. As a consequence, Canada's present and future are burdened by this settler-colonial beginning whose liberal, democratic, and capitalist national character is moored in the imposed racialized colonial order that sought to build an assumed exclusive European civilization and a political economic organization that effectively continues to devalue the contributions of non-European peoples and ultimately "sets up" or structures Indigenous peoples and settlers of colour to lower socio-economic status that many of the indicators in this volume show. The Euro-supremacist nature of the Canadian project is characterized by racial hierarchies that define the contours of the current neo-colonial order in Canadian society.

The same colonizing logic and racialized ideologies of European supremacy incubated in present-day North America were to become standard fare operating principles upon which was constructed an imperial world order that today imposes great burdens of economic, political, and social inequality on much of the global South. This colonizing logic operates on a continuum, from imperial world order to domestic internal colonization, to varied oppressions experienced in the communities and lives of Indigenous and racialized peoples of Canada.

By suggesting that colonization and racialization are intertwined and as Canadian as Canada itself, this book considers the legacies that history has left us, as responsible for unequal access to the labour market and its benefits; the disproportionate experience of poverty among Indigenous peoples and racialized people; and a set of disadvantages and burdens, now exacerbated by the neo-liberal character of the contemporary economy. This book is designed to introduce students to these historical and contemporary processes of colonization and racialization as continuous processes that play key roles as determinants of socio-political and economic outcomes that determine access to resources and life chances for Indigenous and racialized groups in Canada. Through the work of various authors across disciplines, we explore how the processes of colonization and racialization structure unequal citizenship for various groups in Canada. This interdisciplinary approach highlights how academic disciplines such as sociology, political science, and cultural studies are using different perspectives and lenses to explore multiple facets of current forms of colonization and structural racism. We present the persistence of racial hierarchies in a historical context, within which Canada's political economy took shape and its continuing impact on key Canadian institutions including the government, the labour market, immigration, education, the criminal justice system, health care, and social services.

By examining the interconnections between colonialism and racism, this book builds upon a growing scholarship of decolonizing anti-racism[2] that aims to critique the exclusion of Indigenous peoples' histories and perspectives in many anti-racist theories. This book also examines the ways in which a new dialogue can be fostered wherein Indigenous decolonization is at the forefront of anti-racist perspectives and movements. We do not minimize the differences of experiences and histories that have existed and continue to exist between Indigenous peoples and other non-white racialized groups. Indigenous peoples have a rich and unique spiritual connection to this land since the creation of Turtle Island. As well, they have been struggling against colonial domination, and their resistance is articulated with a discourse of self-determination that demands a nation-to-nation relationship with the State and Canadian society. Their demands are ultimately tied to land, sovereignty, and nationhood. As Lawrence and Dua point out,[3] such "Aboriginal activism against settler domination takes place without people of color as allies" and one must acknowledge (unwillingly and unconsciously as it might be) the complicity of people of colour in the ongoing project of colonization by living and owning "land that has been appropriated from Indigenous peoples and by having some rights and privileges that are denied to Aboriginal peoples collectively, and that are deployed to deny Aboriginal rights to self-government."[4]

We must recognize that the "Indigenous experience" was the first example of the racialization process in Canada; colonialism needed to construct Indigenous peoples as inferior racialized "Others" (the Indian "race") to justify the dispossession of their land and attacks on their cultures. As other people of colour came to settle on this land, they were also constructed and treated as inferior races, and although their experiences differed from those of Indigenous peoples, racism could be one point of juncture from which to build an honest dialogue and form solidarity. As we move toward this reality, however, Lawrence and Dua point out, "if they are truly progressive, antiracist theorists must begin to think about their personal stake in this struggle, and about where they are going to situate themselves." A progressive decolonizing antiracist approach does not shy away from differences but it works through/with these differences in order to commence a true and just coalition-building that examines how our various experiences and histories differ and are sometimes similar while recognizing that our demands might be different due to the locations we occupy in this land.

From a political-economic point of view, the context of continuity between colonial orders and contemporary racial inequality is accentuated by the emergence of neo-liberalism as the dominant ideology of our time. It represents a world view that seeks to delegitimize the historical claims for redress and equity by Indigenous and racialized peoples while entrenching the persistent social and economic hierarchies arising from the capitalist mode of economic organization. Neo-liberalism is a contemporary adaptation of a historical form of liberalism associated with the emergence of capitalism and a form of liberal democracy in the 17th and 18th centuries that discounted the citizenship of Indigenous and racialized peoples. As the dominant global ideology today, neo-liberalism has enormous impact on the lives of Indigenous and racialized peoples, providing a framework for policies aimed at effecting the retreat of the State from its traditional role as a mediator of social conflict. It emphasizes lower taxes, deregulation of the economy, and the privatization of public enterprises involved in such services as water, hydro, and transportation that were once considered the mainstay of the public service sector.

Broadly speaking, neo-liberalism advocates economic efficiency through free markets; a smaller role for governments in the economy and in the distribution of society's resources, consumer sovereignty, and trade liberalization; and a borderless world for capital investments that intensifies the vulnerabilities of the subjects of racialized hierarchies. Its results are the proliferation of free-trade agreements that disrupt local economies and empower elites in which Indigenous and racialized peoples are grossly under-represented while shifting economic decision-making toward more globalized institutions and away from the moderately accountable nation-state governments. These results call into question the role of the State in shaping social and economic policy and in acting as a refuge of last resort for those socially excluded by colonial and racialized structures. Neo-liberalism also calls into question the future of democratic accountability as States lose some of the autonomy they once held to make key decisions that affect their citizens. In that regard, it generates new challenges for the historically marginalized groups in society disadvantaged by the policies it inspires and it intensifies their exclusion. Its advocacy for unfettered exploitation of resources threatens to deplete natural resources that are still the subject of contestation between Indigenous peoples and settler populations, increasingly rendering their claims moot and denying Indigenous peoples their inheritance before they can build a new racially just order through self-determination and autonomy over their own nations.

Recent events demonstrate the colonial continuity that further entrenches the dispossession of Indigenous peoples of their inheritance. These events speak to the unyieldingly colonial vision of the Canadian federal government and its relationship with the Indigenous peoples of Canada. It is a vision that maintains a stubborn disregard for its constitutional obligations to recognize the self-determination of Indigenous peoples, negotiate self-government, settle land claims, and grant treaty rights. The federal Conservative Party's recent refusal to endorse the United Nations' Declaration of the Rights of Indigenous Peoples (See Appendix A), together with its withdrawal of the Kelowna Accord signed up by its predecessor designed to invest in long-term solutions to pressing deplorable conditions, and its silence about possible equitable ways to address issues facing Indigenous communities, represent a remarkable continuity in a relationship defined by the legacy of colonization and racialization. Even when it acted by making an apology to the survivors of the residential schools, to set up a Truth and Reconciliation Commission, and to speed up the claims disputes process, it was unclear whether these actions were motivated by the desire to build a new, post-colonial relationship, one that would define a changed space that the Indigenous peoples would occupy, and whether these actions represented a new understanding of Indigenous peoples' historical relationship with this land and with the Canadian State.[5] Instead, viewed in context of the government's other actions taken in regard to Indigenous matters, it seemed motivated by political and electoral considerations in a period of minority government.

The UN Declaration that the federal government rejected for fear of its "legal" implications has no legal force binding on its signatory members; however, it provides international norms for the recognition of Indigenous rights, including land ownership, protection of traditional language and culture, access to natural resources, self-determination, and overall commitment to address issues of poverty and discrimination. Although it is not binding, its symbolic significance cannot be dismissed and could be seen as a tool to make governments occupying Indigenous lands more accountable to Indigenous peoples so that no longer can matters that affect Indigenous peoples be decided without justly engaging with them. The rejection of the Declaration has been criticized by Indigenous peoples and non-Indigenous groups who argue that it signals a failure of commitment on the part of the contemporary Canadian State to transcend its history of Indigenous oppression and its socioeconomic impacts. As the First Nations Leadership Council states, "If Canada is serious about addressing the unacceptable levels of poverty on First Nations land and speeding the land claims process, it should accept the minimum standards in the Declaration." (August 2007, http://www.ubcic.bc.ca/News_Releases/). This criticism was echoed by The Council of Canadians and Amnesty International that called it an "appalling vote against a human rights declaration that clearly affirms the rights of Indigenous peoples."[6] Although not binding, the Declaration's parameters for recognizing Indigenous realities within their historical contexts and the rights that accrue from them, could help the member States, Canada included, move toward a better and healthier relationship with Indigenous peoples in the future. The Canadian government's explanation for rejecting the will of the overwhelming majority of the members of the United Nations (only the white settler nations of Australia, United States, and New Zealand, with the support of the United Kingdom, marched in lock-step with Canada's Government) was that it feared it could undermine the Charter of Rights, giving Indigenous peoples "special" rights compared to other groups, or jeopardizing past and ongoing negotiations—a clearly flawed explanation given that the Declaration has no legal force. The refusal, however, sent a clear message to Indigenous peoples: that your place and claims to this land to which your ancestors and you are intimately tied to is still mostly tenuous.

The federal government's disregard for international consensus was only consistent with its lukewarm response to the UN resolutions on racism arising out of the World Conference Against Racism, Racial Discrimination, Xenophobia, and Related Intolerance, held in Durban, South Africa in September 2001 and the subsequent rejection of any further participation in the ongoing processes to implement that agenda, including the upcoming WCAR II conference, also in Durban, South Africa.[7]

On January 23, 2008, the Canadian government of Stephen Harper announced that it would not be involved in the preparatory process for Durban Two in 2009. The government's Press Release adds, "By no longer participating in the Durban Review Conference, Canada will continue to focus its efforts on genuine anti-racism initiatives that make a difference," said Secretary of State Kenney. "Our government's decision to seek full membership on the Task Force for International Cooperation on Holocaust Education, Remembrance, and Research demonstrates that we remain committed to the fight against racism and the promotion of freedom, democracy, human rights, and the rule of law at home and around the world." (Press Release, Canadian Secretary of State for Multiculturalism, January 2008) This divisive strategy of focusing on Holocaust education and equating it with Canada's commitment to the fight against racism is conveniently silent on the lack of any substantive initiatives and results from Canada's Action Plan Against Racism announced in March 2005. The rhetoric of Canada's commitment to the fight against racism has not been matched by action. This will further the Canadian government's crisis of legitimacy as it continues to represent white-settler, systemic bias that privileges dominant groups as *the* democratic option for the good of all Canadians. This can only ring hollow for Indigenous communities facing the lack of federal recognition when Canada refuses to sign the UN Declaration of the Rights of Indigenous Peoples, when it withdraws from a significant international effort, Durban Two in 2009, to eliminate racism, and when Canadian citizens, specifically Indigenous Peoples and racialized individuals, disproportionately face poverty levels that drastically compromise their quality of life and, in some cases, life itself.

The context of colonization and racialization is also observable in several other current issues. It is highlighted dramatically through the actions of political leadership and the discourses of "limits of tolerance" and "reasonable accommodation" as some citizens have demanded the imposition of Anglo-Franco conformity in what they see as an unwieldy and overly accommodating multicultural society.[8] It is a vision that fits hand in glove with the clash of civilization dogma that threatens to turn neighbour against neighbour on the basis of suspicions of religious and cultural loyalties assumed to be inconsistent with Canadian nationalism. The differential impacts of national security, poverty, and human rights administration will be specifically highlighted in relation to Canada's racialized communities. In the post-9/11 era, concerns about national security betray a distinct neo-colonial tenor of the discourses about who the enemy is and who must be protected by the Canadian State. Canadian state security practices are increasingly informed by the "clash of civilizations" and its Orientalist overtones, especially as deployed in the "war against terror" and the derivative community security operations that enforce informal modes of racial profiling. The post-September 11, 2001 period has seen an intensified process of racialization relating to the religious character of certain groups in Canadian society—particularly those of Arab, other Middle Eastern, South Asian, and African descent, and the proliferation of Islamophobia in both State and civil societal processes.

In the midst of this "war against terror," Canada's poverty rate is undergoing a transformation of sorts. The Canadian economy went from a recession in the 1990s with a high poverty rate of 16.2 percent to an even higher rate of 19.7 percent in 1995. In the late 1990s, however, there was a period of economic growth. Canada's economy has doubled in size since 1981—it is now the world's ninth-richest nation. According to a Canadian Centre for Policy Alternatives report (2007), the income gap between rich and poor is at a 30-year high: some are getting richer while others are working longer to simply maintain their earnings levels.[9] Between 1999 and 2005, the median net worth of families in the top fifth of the wealth distribution increased by 19 percent while the net worth of the lowest fifth remained unchanged. So, unlike the period after the recession of the early 1980s when Canadians were able to "catch up" economically, Canada is now confronted with a serious persistence and depth of poverty. The Canadian Council on Social Development's Urban Poverty Report, 2007, noted: "Although the economic engine was operating full steam ahead by the end of the decade, Canada made no progress in reducing its comparatively high level of poverty during the 1990s. Indeed, the poverty gap widened and, in absolute terms, the number of poor Canadians increased…" (2007:4).[10] According to another CCPA report by Armine Yalnizyan, while the poorest 10 percent of families raising children—over 376,000 households—lived on less than $23,000 after taxes in 2004, half of these families lived on less than $17,500 a year.[11] The report suggests that the problem is not so much the creation of adequate levels of wealth by the Canadian economy, but that the wealth has not been equitably distributed. The simplest explanation in the dramatic changes in the economy and income distribution is the nature of work available and how the workplace relationships are organized. In the early 21st century, economic restructuring and demands for labour market flexibility have made precarious employment the fastest growing form of work that includes contracts; temporary, part-time, piece, and shift work; and self-employment. Characteristics of these types of employment include low pay, no job security, poor and often unsafe working conditions, intensive labour, excessive hours, and low or no benefits. This has combined with historical discrimination in employment to make certain groups more vulnerable in the Canadian economy.

This persistent and deepening poverty is having a disproportionate impact on Indigenous peoples and other racialized communities in Canada. In a Press Release, May 2005, the Assembly of First Nations highlighted the findings of the U.S. State Department's Report on Human Rights: "Aboriginals remained underrepresented in the workforce, over-represented in welfare rolls and in prison populations, and more susceptible to suicide and poverty than other population groups."[12] In other racialized communities, the impact of poverty is also greater than the national average. While the Canadian low income rate was 14.7 percent in 2001, low income rates for racialized groups ranged from 16 percent to as high as 43 percent.[13] The depth of poverty is seen in the details, in large urban centres such as Toronto where racialized communities are concentrated. According to a recent United Way of Greater Toronto report titled *Poverty by Postal Code*, in Toronto in 2001, racialized group members and immigrants were almost three times as likely to live in poverty whether they are employed or not. About 29.5 percent of them lived below the poverty line, 24 percent of immigrants compared to the overall average of 11.6 percent among non-racialized. The overall Toronto poverty rate at 19 percent was higher than the Canadian rate at 14.7 percent. While the poverty rate for the non-racialized population fell by 28 percent, poverty among racialized families rose by 361 percent between 1980 and 2000.[14] Michael Ornstein's analysis of the 2001 Census regarding ethno-racial inequality in the City of Toronto highlighted that despite their educational qualifications, unemployment rates for Africans, Blacks, and South Asians have skyrocketed. Among Ethiopians, Ghanaians, Somalis, and "other African nations," the overall unemployment rates were higher for the Pakistani and Bangladeshi, Sri Lankan, Tamil, and "Multiple South Asian" groups.[15] Within this context, the Conservative government implemented $725 million in tax cuts. The response was quick from the National Anti-Poverty Organization: "The Harper government's intended $725 million tax cut following a nearly $14 billion federal surplus last year cannot be justified in the wake of persistent poverty in Canada (September 28, 2007).[16] Also in 2007–2008, Canada's military spending will reach $18.24 billion per year, an increase of 9 percent over 2006–07 (Stephen Staples and Bill Robinson, Canadian Centre for Policy Alternatives, October 2007).[17] The federal government's priorities on tax cuts and military spending and its silence on poverty, given the impact on Indigenous and other racialized communities, is a process of non-action in a critical area of national responsibility that is resulting in the increasing racialization and feminization of poverty in Canada. This situation of growing racialized poverty is combined with the neo-colonial tenor of "security" and "war on terror" discourses to intensify the experience of social exclusion for Indigenous peoples and other racialized groups in Canada.

Given this current situation in Canada in 2008, this book concludes with a part on responses and resistances to the status quo, that is, dominant social relations that attempt to continue institutionalizing colonialism and racialization in Canada. The growing inequalities in national and world economies that coincide with neo-liberal policies, and the concerns around environmental degradation arising from the emphasis on unequal economic growth, have led to the emergence of new social movements and the re-energized existing movements such as the Indigenous rights movement and the anti-racism movement. Such social movements reveal the contours of contested terrains. In addition, these resistances, some of which are highlighted in Part Five, demonstrate the active and courageous efforts these communities are involved in to create *their* vision of Canada, a country that respects Indigenous right to self-determination, and the right of all citizens to a life of human dignity. By highlighting this vision of an alternative future, accompanied by a critical analysis of systemic oppression and social exclusion, this book aims to showcase and celebrate the hope, courage, and vision of Indigenous peoples and other racialized communities in Canada. These are invaluable contributions to Canada and to all Canadians.

ORGANIZATION OF THIS BOOK

This book is organized in five parts: 1) Historical Analysis, 2) The Canadian State, 3) Everyday and Cultural Racism, 4) Contemporary Issues, and 5) Responses and Resistances. The themes of colonialism, racism, and resistance are explored beginning with the premise that racism is not an aberration but a constitutive feature of Canadian society with racial classifications transformed into objective structures determining access to resources, opportunity, and power in the Canadian polity. Each of the parts opens with an introduction that connects the articles to the overall theme of the book and the specific part.

From a historical context, racist-colonial processes unfolded in the development of Canadian political economy, defining the nature and the role of the State in the construction of racial hierarchies, racialized regimes, and the racial character of key institutions such as Canadian immigration, labour markets, the education system, political systems, the media, and the criminal justice system. This book highlights, as opposed to dismissing, the "special" nature of Indigenous rights, beginning with an important history lesson of the nation-building project of Canada as a white-settler colony and linking it to the racist discourses and policies experienced by Indigenous peoples and, later on, by other racialized groups. Chapter One of Part One begins this examination with Mackey's "Settling Differences: Managing and Representing People and Land in the Canadian National Project," wherein she deconstructs the persistent myth of "tolerance" that greatly lingers in the Canadian national identity. This discourse is contrasted with examples of racist discriminatory practices that defined and ultimately treated non-whites as "different" and not part of the imaginary settler colony. The relationship between Indigenous peoples and the settlers is given particular attention, hence examining how, from the early days of settlement, Indigenous peoples had to be "managed," and this process of management has since undergone different stages, yet lingering into the present. One only needs to look at either the earlier examples given above of the recent federal government's stance on Indigenous issues, or at the death of Dudley George in 1995 in Ontario by the Ontario Provincial Police while he was protesting unarmed for the return of traditional territory of Ipperwash to his nation, or at the current land dispute happening at Caledonia between the Six Nations of the Grand River people and the developers and the provincial government to understand how Indigenous peoples have a particular relationship with the Canadian State and mainstream society.

In Chapter Two, Lawrence's article "Rewriting Histories of the Land: Colonization and Indigenous Resistance in Eastern Canada" rewrites the history of Eastern Canada from a much needed and insightful Indigenous perspective, looking at the role of racism in the colonization of Indigenous peoples, yet being mindful that such colonization has always and is still being resisted, so that we can indeed envision a better, just, and healthier Canada. But this cannot be done until we look at the racist history right in the eye and recognize the injustices that have been imposed on Indigenous peoples and other racialized groups.

Chapter Three looks at the experiences of a specific racialized group, Black Canadians, beginning with the article by Sylvia Hamilton, "Our Mothers Grand and Great: Black Women of Nova Scotia." Hamilton reviews the historical experiences of Black people, in particular women in Nova Scotia, and brings to the surface the ugly racism faced by these women while inspiring us to envision a better future by showing us the collective resistance of these women. Resistance and rebuilding are central themes that our book aims to make clear to the readers so that collectively we can decolonize this country and free it of its lingering injustices.

Part Two of the book explores the often contradictory role of the Canadian State in efforts to address racism, including State-mandated race relations, anti-racism, and multiculturalism. In Chapter Four, Himani Bannerji's "On the Dark Side of the Nation: The Politics of Multiculturalism and the State of 'Canada'" explores Canada's official multicultural policy, the idea of multiculturalism as a core Canadian value, and new and enduring questions about its effectiveness in addressing racial and gender oppression and ethnic diversity.

In Chapter Five, Patricia Monture-Angus's "Theoretical Foundations and Challenges of Aboriginal Rights" critiques European legal frameworks and discourses of liberal democracy and investigates possible paths for achieving independence and self-determination for First Nations. Chapter Six is Frances Henry and Carol Tator's "State Responses to Racism in Canada" that explores the conflicting role of the Canadian State in promoting and fighting racism. The chapter looks at three major case studies and introduces the concept of 'democratic racism'. In Chapter Seven, Sujit Choudhry "Protecting Equality in the Face of Terror: Ethnic and Racial Profiling and s. 15 of the Charter" discusses issues arising out of the tensions between Charter guarantees of equal treatment and the growing resort by the State to the use of ethnic, religious, and racial profiling as a tool of national security under the new "regime of terror."

Part Three addresses everyday and cultural forms of racism. Literature, media, and sport cultures are interesting places to critically examine how ideas of a racialized "other" are tied with criminalization of racialized bodies that can have negative impacts on the everyday lives of Indigenous groups and other racialized "others." They provide the ideological contexts under which forms of exclusion/inclusion are lived in everyday lives. It begins with Gamal Abdel-Shedid, who, in Chapter Eight, "Running Clean: Ben Johnston and the Unmaking of Canada," reviews sports within the framework of nationalism and exposes racism and racial marginalization through sports. In Chapter

Nine, "Literature, Image, and Societal Values," Janice Acoose examines themes of representation of Indigenous women in colonial discourses and mainstream literature. In Chapter Ten, Glenn Deer's "The New Yellow Peril: The Rhetorical Construction of Asian Canadian Identity and Cultural Anxiety in Richmond" examines emerging everyday racial discourses of Asian Canadians based on moral panics about the threat of "Asian invasion" of otherwise pristine British Columbia. Chapter Eleven, Angela Aujla's "Others in Their Own Land: Second Generation South Asian Women, Racism, and the Persistence of Colonial Discourse" laments how racialized groups never seem to be "Canadian" enough to belong and how these social exclusions are perpetuated in popular discourses.

Part Four examines contemporary issues that explore how colonization and racialization continue to thrive in a Canada that remains structurally a white-settler society. In Chapter Twelve, Yasmeen Abu-Laban and Tim Nieguth's "Reconsidering the Constitution, Minorities and Politics in Canada" discusses the place of Canadian minorities in the context of Canada's constitutional guarantees of equality and the gap between that ideal and the substantive reality of inequality compounded by immigration policy and globalization. In Chapter Thirteen, Grace-Edward Galabuzi's "Social Exclusion" examines the growing racialization of poverty in Canada and its implications for social determinants of health and well-being for the subject population. In Chapter Fourteen, "On Being *Not* Canadian: The Social Organization of 'Migrant Workers' in Canada," Nandita Sharma investigates the social organization of migrant labour and identifies the racist and nationalist ideologies embodied in the categories of "migrant workers" that make them un-free labour and therefore vulnerable to exploitation and abuse. In Chapter Fifteen, "Whose Transnationalism? Canada, 'Clash of Civilizations' Discourse and Arab and Muslim Canadians," Sedef Arat-Koc analyzes Arab and Muslim communities in Canada as they confront a condition of "siege" in the post-September 11, 2001 period that has intensified their vulnerability to religious and racial profiling in the name of national security.

Finally, Part Five of this book explores some of the efforts by Indigenous peoples and other racialized peoples to fight back and establish a space for validating difference in the face of Canadian Anglo-conformity. It also discusses the contemporary fascination with the concept of multiculturalism as a constitutive Canadian value and the increasingly contested role of the official policy of multiculturalism as a vehicle for negotiating diversity in a multicultural society. In Chapter Sixteen, Agnes Calliste's "Nurses and Porters: Racism, Sexist, and Resistance in Segmented Labour Markets" addresses the forms of resistance at intersections of race and gender manifest in the health professions in the 1980s and 1990s and draws parallels to the earlier struggles by African Canadian sleeping car porters in the 1950s. In Chapter Seventeen, "Indigenous Voice Matters: Claiming Our Space through Decolonizing Research" Lina Sunseri provides another form of resistance in the process of knowledge production. By illustrating how the term "research" is often linked to European imperialism and colonialism, Sunseri analyzes research methodologies and the power relations they structure and maintain. In Chapter Eighteen, Taiaiake Alfred's "Rebellion of the Truth" speaks to the need for substantive restitution rather than a mere recognition of Aboriginal rights. By rejecting Aboriginalism as a legal integrationist ideology that undermines indigeneity, Alfred argues for an anarcho-indigenist approach that is non-capitalist, non-statist, pro-feminist, and committed to a sustainable relationship with nature.

The concluding chapter by Maria Wallis, Lina Sunseri, and Grace-Edward Galabuzi reprises the key themes of the book. In the final analysis, the book seeks to explore the extent to which the persistent legacy of colonization and racialization represent a continuing threat to equal citizenship and democracy and what is being done about it at the State, institution, and community level to address that reality.

NOTES

1. S. Razack, *Race, Space, and the Law: Unmapping a White Settler Society* (Toronto: Between the Lines, 2002), 1–2.
2. B. Lawrence and E. Dua, "Decolonizing Antiracism," *Social Justice* 32, no. 4 (2005): 20–125.
3. Ibid.
4. Ibid.
5. Stephen Harper, "'We recognize that this policy of assimilation was wrong': Text of apology given to aboriginal Canadians by Prime Minister Stephen Harper in the House of Commons, June 11, 2008," *Toronto Star*, 12 June 2008, AA12.
6. Council of Canadians, September 2007 Media Release, http://www.canadians.org/media/other/2007/25-Sept-07.html.
7. www.UN.org/WCAR/durban.pdf.
8. See J. Stein et al., *Uneasy Partners: Multiculturalism and Rights in Canada* (Waterloo, On: Wilfrid Laurier University Press, 2007); Bouchard-Taylor Commission on Practical Accommodation Practices Related to Culture Differences in Quebec, 2007 (also known as The Royal Commission on Reasonable Accommodation).
9. E. Russell and M. Dufour, *Rising Profit Shares, Falling Wages Shares* (Ottawa: Centre for Policy Alternatives, June 2007).
10. Gail Fawcett and Katherine Scott, *A Lost Decade: Urban Poverty in Canada, 1990 to 2000,* Part of the Council's Urban Poverty Project (Ottawa: Canadian Council on Social Development, 2007).
11. Armine Yalnizyan, *The Rich and the Rest of Us: The Changing Face of Canada's Growing Gap* (Ottawa: Canadian Centre for Policy Alternatives, 2007). See also René Morissette, "Revisiting Wealth Inequality," *Perspectives on Labour and Income* 7, no. 12, Statistics Canada, December 13, 2006.
12. Press Release: International Reports Highlight Canada's Lack of Progress in Addressing First Nations Poverty: UNICEF Ranks Canada 19th Out of 26 Countries in Child Poverty Report. Assembly of First Nations, March 1, 2005.
13. C. Linday, *Profiles of Ethnic Communities in Canada,* Statistics Canada, Catalogue no. 89-621-XIE. Data is drawn from the 2001 Census and the 2002 Ethnic Diversity Survey.

14. United Way of Greater Toronto, *Poverty by Postal Code: The Geography of Neighbourhood Poverty 1981–2001* (Toronto: UWGT/CCSD, 2004).
15. Michael Ornstein, *Ethno-Racial Inequality in the City of Toronto: An Analysis of the 1996 Census* (Toronto: City of Toronto, Access and Equity Unit, 2000).
16. Media Release: September 28, 2007. *$ 725 Million Tax Cut Cannot Be Justified—Anti-poverty Group* (Ottawa: National Anti-Poverty Organization).
17. Steven Staples and Bill Robinson, *More than the Cold War: Canada's Military Spending 2007–08* (Ottawa: Canadian Centre for Policy Alternatives, 2007).

PART 1

Historical Analysis

We begin this book with a historical review of racism and colonialism in Canadian society so that the reader can best contextualize contemporary social relations as well as the current demands of justice made by historically marginalized and oppressed groups in Canada. It is important to recognize that contemporary events and issues of race are both a reflection of and a resistance to a long history of racial and colonial relations. This historical perspective will provide the reader with a comprehensive understanding of colonialism and racism in Canada and the ability to connect present social conditions to the past. These tools will hopefully motivate the reader to be an agent of change. As the philosopher Karl Marx said centuries ago: "Men make their own history, but they do not make it just as they please; they do not make it under circumstances chosen by themselves, but under circumstances directly found, given and transmitted from the past."[1] Similarly, Sotsisowah, a member of the Haudenosaunee, one of the Indigenous peoples of North America also known as the Iroquois Confederacy, argues that "we need to look to history primarily because the past offers us a laboratory in which we can search to find the inherent process of Western Civilization that paralyzes whole societies and makes them unable to resist the process of colonization. . . . And lastly, we need to understand that within colonization are the exact elements of social organization that are leading the world today to a crisis that promises a foreseeable future of mass starvation, deprivation, and untold hopelessness."[2] Hence, contemporary discourses and practices of "race" and racism in Canada manifested in the everyday interactions among groups are influenced by dislike, fear, distrust of the historically constructed racialized "Other" (e.g., gazing when a veiled Arab woman passes by or moving away when an African-Canadian male approaches on the sidewalk at night). Stereotypical misrepresentations in the media of some racialized groups, and consistent racial segregation in the workplace, are not exclusively new manifestations of our social reality, but they are firmly rooted in the historical make-up of the country. Moreover, while current discourses of Canada as the champion of multiculturalism, diversity, and tolerance are increasingly popular, a closer look at both current and past experiences of Indigenous peoples and other racialized "Others" might give the reader a different and fuller picture of Canada.

Part One intends to uncover the long history of race and colonialism that has existed in Canadian society to awaken it from what Henry and Tator have termed a "historical amnesia," meaning that for the most part, many Canadians "have obliterated from their collective memory

1 Karl Marx, "The Eighteenth Brumaire of Louis Bonaparte," *The Karl Marx-Engels Reader*, ed. Robert C. Tucker (New York: W. W. Norton & Company, Inc, 1978).

2 Sotsisowah, "Our Strategy for Survival," *Basic Call to Consciousness*, ed. Akwesasne Notes (Summertown, TN: Native Voices, a div. of Book Publishing Company, 2005).

the racist laws, policies, and practices that have shaped their major social, cultural, political, and economic institutions for three hundred years."[3]

We start Part One with Eva Mackey's article "Settling Differences: Managing and Representing People and Land in the Canadian National Project" that examines the nation-building project of Canada as a white-settler colony. Mackey demonstrates to us how this project entailed an imagination of Canada as a colony founded by the two nations, Britain and France. This imagination eventually meant managing the non-white groups and the Indigenous populations and excluding them from the official narratives of Canada's national identity. By the end of the article, the reader will be able to deconstruct the dominant national narrative of Canada as a benevolent, tolerant country devoid of racism and will learn of the complex and contradictory historical practices that indeed shaped the nation-building process of the Canadian nation-state.

The second article in this part, "Rewriting Histories of the Land: Colonization and Indigenous Resistance in Eastern Canada" by Bonita Lawrence, picks up on Mackey's critical analysis of the dominant narratives of the history of Canada by rewriting for us the history of Eastern Canada from an Indigenous perspective. Lawrence closely looks at the role that racism and colonialism has had on Indigenous peoples and their lands and communities. We are shown how the interaction with European settlers/colonizers has transformed Indigenous lives both within and outside their nations so that we can best contextualize later parts in the book that investigate more current social conditions of Indigenous peoples and their relationship with the State and mainstream Canadian society.

The third article within this part, "Our Mothers Grand and Great: Black Women of Nova Scotia" by Sylvia Hamilton, illustrates a concrete example of the marginalization of the contributions that Black Afro-Canadians have made in the building of this nation. By focusing on Nova Scotia, Hamilton outlines the long and rich, as well as painfully tainted by racism, history of Black/African-Canadians in this nation, even though the official national narratives have consistently made them "outsiders." The article also provides a gender analysis of the experiences of Black women that interconnects racism, colonialism, and gender while examining how racialized women have made their survival and resistance possible while continuing to be the backbone of their family and community. All three pieces within this part of the book in fact re-write the history of Canada by examining its link with racism and colonialism while demonstrating that Indigenous peoples and non-Indigenous non-white racialized peoples have always resisted and challenged their marginalization and oppression.

CRITICAL THINKING QUESTIONS

1. Mackey states that "nation-building is a dual process, entailing the management of populations and the creation of national identity." Exactly how has Canada "managed" its populations? How has the construction of "others" interplayed in this process of management?

2. What does Mackey mean by the "myth" of national tolerance? How has it been the central foundation of Canada's national identity? And how does this myth contrast

3 F. Henry and C. Tator, *The Colour of Democracy: Racism in Canadian Society*, 3rd ed. (Toronto: Nelson, 2006).

with the real historical experiences of racialized non-British groups in the construction of Canada's national identity?

3. Can you think of contemporary examples of the exclusion and/or invisibility of Indigenous peoples and/or non-British groups in the construction of Canada's national identity?

4. Hamilton's article reveals that despite the absence of Black African Canadians in the official narratives of Canada's history, Black African women have a long history of existence in Canada. How has that history been shaped by both discrimination and resistance/survival?

5. How is a history of Canada written from an Indigenous perspective crucially different from that written by a non-Indigenous one? What would be included in the first case and probably excluded in the second case? What would be some other general differences?

FURTHER RESOURCES

Anderson, Benedict. *Imagined Communities: Reflections on the Origin and Spread of Nationalism.* revised edition. London: Verso, 1991.

Backhouse, Constance. *Colour-Coded: A Legal History of Racism in Canada, 1900–1950.* Toronto: University of Toronto Press, 1999.

Dua, Enakshi and Angela Robertson, eds. *Scratching the Surface: Canadian Anti-Racist Feminist Thought.* Toronto: Women's Press, 1999.

Kymclicka, Will and Wayne Norman. *Citizenship in Diverse Societies.* Toronto: Oxford University Press, 2000.

Mackey, Eva. *The House of Difference: Cultural Politics and National Identity in Canada.* Toronto: University of Toronto Press, 2002.

Thobani, Sunera. *Exalted Subjects: Studies in the Making of Race and Nation in Canada.* Toronto: University of Toronto Press, 2007.

FILMS

A Journey to Justice. 2004. Producer: Karen King-Chigbo. National Film Board.
Description: Pays tribute to a group of Canadians who took racism to court. Focusing on the 1930s to the 1950s, this film documents the struggle of six people who refused to accept inequality.

Indecently Exposed. 2004. Producer: Jane Elliott. Blue Eyes Production.
Description: Challenges Canadian attitudes towards Native Canadians.

No Turning Back. 1996. Producer: Michael Doxtater, Carol Geddes, and Jerry Krepakevich. National Film Board.

Description: In 1991, Prime Minister Brian Mulroney announced the creation of a Royal Commission on Aboriginal Peoples. The Commission, comprised of independent men and women, travelled to more than 100 communities, and heard from more than 1,000 Aboriginal representatives. For two-and-a-half years, director Greg Coyes, with two teams of Native filmmakers, followed the Commission on its journey from coast to coast. *No Turning Back* documents this trip.

Speakers for the Dead. 2000. Producer: Jennifer Holness and David Sutherland. National Film Board.

Description: In the 1930s in rural Ontario, farmer Bill Reid buried the tombstones of a Black cemetery under a pile of broken rocks to make way for a potato patch. In the 1980s, descendants of the original settlers, Black and White, came together to restore the cemetery where there were hidden truths no one wanted to discuss. Deep racial wounds were opened. Scenes of the cemetery excavation, interviews with residents, and re-enactments—including one of a baseball game where a broken headstone is used for home plate—add to the film's emotional intensity. *Speakers for the Dead* reveals the turmoil stirred up by desecrated graves and underlines the hidden history of Blacks in Canada.

KEY CONCEPTS/TERMS

Nation-Building as a Dual Process

White-Settler Colony

Myth of National Tolerance

Icy White Nationalism

Indian Act

Cultural Genocide

Settling Differences: Managing and Representing People and Land in the Canadian National Project

1

EVA MACKEY

Nation-building is a dual process, entailing the management of populations and the creation of national identity. The mythology of *how* Canada managed its different cultural groups (the belief in and representation of itself as tolerant) was one key feature of an emerging national identity believed to differentiate Canada from the USA. This chapter examines this dual process of managing populations and representing difference. It focuses on how, from early colonial times up to the Second World War, white Anglophone settlers in Canada mobilised representations of others and managed non-British cultural groups as part of the project to create a nation and a national identity.

Nationalism often depends upon mythological narratives of a unified nation moving progressively through time—a continuum beginning with a glorious past leading to the present and then onward to an even better future. These mythical stories require that specific versions of history are highlighted, versions that re-affirm the particular characteristics ascribed to the nation. In Canada, nationalist mythmakers draw upon particular versions of national history to explain the nation's 'fairness' and 'justice' today.

A government report on Canadian identity released in 1992 points back to history to argue that certain 'inescapable conditions' set Canadian life in 'an initial pattern' the 'logic' of which has obliged the nation to 'uncover gradually . . . the requirements of justice in a Canadian setting'. The report argues that two constitutional documents served as 'cornerstones of early Canadian life'. In the Royal Proclamation of 1763 and the Quebec Act of 1774, it suggests, 'much of what was to follow in Canadian life was laid down'. It suggests that the Royal Proclamation of 1763 determined that Canadian governments would treat Aboriginal peoples 'as autonomous and self-governing peoples'. It also claims that the Quebec Act of 1774 established that legal provision would be made for 'a distinctive society in Quebec', in that it could have 'institutions, laws and culture quite different than those of the surrounding English speaking societies'. The report then links history to the present, and *compares* Canadian identity with that of the USA. One of the consequences of the nation's historical trajectory, says the report, was that

> Canada could never really embrace what later came to be known in the United States as the concept of the 'melting pot.' Canadians chose instead to support linguistic and cultural diversity . . .
>
> These initial commitments . . . led us to discover additional standards of fairness for individuals, for communities, for regions, and cultures within the Canadian family. Thus Canada's social fabric is now interwoven with programmes of income security, social insurance, pensions and old-age security, equalization, regional development, measures which have made Canada the

> envy of much of the world, and that have come to define many Canadians' own sense of identity.
>
> (Beaudoin and Dobbie 1992: 5–6)

The report traces Canada's pluralist legacy from its genesis in the late eighteenth century to today's social programmes, thereby reiterating the myth of national tolerance, the central foundational myth of Canadian nationhood and identity. Whereas in Australia the current policy of multiculturalism is represented as a clear break with an overtly racist past (Hage 1994a: 22), in Canada the cultural pluralism of the present is often represented as on a natural continuum with Canada's history, even heritage, of tolerance. A variety of historical work in Canada challenges this mythological version of history, and indeed many people, especially Aboriginal people and Franco-Canadians, legitimately see Canadian history as 'one long confrontation' (Bothwell 1995: 7). Yet myths of tolerant nationhood are ubiquitous, and have a certain kind of authority, as exemplified by their appearance in a government document on national identity.

This dual process of management and representation is a complex and contradictory process of inclusion and exclusion, of positive and negative representations of Canada's internal and external others. As Stuart Hall argues, identity is formed in the 'simultaneous vectors of similarity, continuity, and difference' (in Chabram and Fregoso 1990: 206). We can see this at times conflicted process of identity formation, as Aboriginal people and non-British cultural groups are managed, located, let in, excluded, made visible or invisible, represented positively or negatively, assimilated or appropriated, depending on the changing needs of nation-building. The 'heritage of tolerance' is actually a heritage of contradictions, ambiguity, and flexibility.

ALLIES AND OTHERS IN THE COLONIAL ENCOUNTER

From the early days of Canadian historical writing, historians liked to portray the colonisers of Canada as more generous than those of the USA. According to these histories, while the Americans violently and brutally conquered their 'Indians', the Native peoples of Canada never suffered similar horrors of conquest. English Canadian historians relished comparing the brutal treatment of Aboriginal people by the Americans with the apparently 'generous' treatment they had received from Canadians (Trigger 1986: 321). Yet, these same historians 'expressed gratitude that Providence had sent epidemics and intertribal wars to sweep away the native peoples of Southern Ontario and Quebec', thus leaving the regions for French and English settlers who, 'unlike their American neighbours, had no wrongs against native people for which to atone' (ibid.: 321). In fact, such interpretations, as Trigger suggests, depended on great self-deception and hypocrisy on the part of writers whose governments were 'treating their former allies with much the same mixture of repression and economic neglect as American governments were treating defeated enemies' (ibid.: 321).

Of interest to me in these claims is not so much their truth-value, but rather the way in which they indicate a push to construct a settler national identity perceived as innocent of racism. While it is true that to a certain extent early Canadian historical writing treated Native people 'respectfully', and gave them a prominent role (Trigger 1986: 316), this 'respect' did not result from any natural Canadian form of tolerance. It emerged because of the essential role Native people played in the early years of the colonial project.

The main goal of the early Imperial presence was resource extraction through the fur-trade, an economic activity that absolutely depended on Native people's labour and knowledge. Finley and Sprague argue that '*Without Indian goodwill* there could be *no fur rush, not even a fur trade*' (1979: 22, emphasis mine). Native people were also necessary as allies in battles between successive colonisers (between the French and the British and then between the British and the Americans in the War of 1812) in the fight to stake claims on relatively uninhabited territory (Trigger 1986). The particularities of colonial rule in Canada, therefore, required a certain amount of respect for and the building of alliances with, Native peoples. Although such 'generosity' and 'tolerance' may now be constructed as an early example of a natural characteristic of the nation and its citizens, it was more a matter of expediency, and a response to the specifics of the colonial economic project, especially as it developed in Canada.

These modes of representation also reflect broader trends in Western perceptions of cultural and racial 'others'. In the early centuries of contact between Europe and the 'new world', the 'discovery' of people in the Americas was a 'problem' that needed explaining for the Europeans: how to account for the racial and cultural differences encountered? Were Native people even part of the human race? How did they fit into Christian cosmology? Could 'Indians' be traced back to Adam and Eve? In order to be consistent, orthodox Christian thinkers had to grant Native people souls and humanness because scriptural history dictated God's creation of all humankind at one time in a specific spot (Berkhofer 1978: 35). In this monogenetic interpretation, all people were seen as descendants from Adam and Eve, and all were considered to have the same human potential (Banton 1987; Berkhofer 1978: 42; Curtin 1964: 32–3; S. Hall 1992b; Stocking 1968; Todorov 1984).

The manner in which 'others' were represented in Western discourse changed over time, and became a matter of serious contest and debate (Banton 1967, 1987; Goldberg 1993; Stocking 1968). Up to the nineteenth century, dominant interpretations of physical and cultural difference were often based on stereotypes and hierarchies in which Europeans were considered superior (S. Hall 1992b). These interpretations justified conquest and sometimes the genocide and enslavement of humans considered by Europeans to be inferior. However, they were not based on supposedly scientific notions of race as 'type' or subspecies', ideas that were to develop in the nineteenth century (discussed later in the chapter). In this sense early representations were *relatively* positive and relations more flexible, in comparison with later 'racial' ideas.

Therefore, the 'positive', 'generous' and 'tolerant' treatment and representations of Native people that historians credited Canadians with must be seen within the contexts both of the specifics of the colonial project in Canada, and also of broader global trends in European representations of others. I would argue that these historical relationships have been interpreted and re-shaped within a national tradition in order to create a mythology of white settler innocence, a mythology that exists in various forms today. Such myths do not simply hide inequalities and oppression, but contribute to a mythology of national identity.

As the context of nation-building changed, so did representations of Aboriginal people, and the policies created to manage populations. By the mid- to late 1700s, the colonial presence increasingly established itself through settlement rather than through resource extraction. Canada's European population increased with an influx of people from the USA during and after the American Revolution in the 1770s. Further, in 1763 the English colonisers conquered the French, and Native people began to rebel against the colonisers (Bumsted 1992a: 141).

At this time, first the Royal Proclamation and then the Québec Act were passed in Britain. Such colonial policies, as discussed in the beginning of this chapter, are today defined as the origins of the Canadian characteristic of tolerance. The Royal Proclamation of 1763 is credited with establishing that Canadian governments would treat Aboriginal peoples as autonomous and self-governing. It was primarily designed, however, as a form of colonial cultural policy for Québec, the French colony the British had conquered. Intended to promote cultural assimilation of the French to the English language and the British legal system (Cook 1963: 4), the policy granted certain land rights in the western areas of North America to Native peoples, with the aim of preventing Westward expansion of the American colonists. The aim was that Americans would instead move into Québec, 'in sufficient numbers to submerge the existing French speaking population' (ibid.: 4). The Proclamation also established English law and representative institutions in the colony, in part to limit the power of Roman Catholicism (McNaught 1970: 46). The British conquerors imagined a situation in which 'a French population, diminished in importance by emigration, would be quickly swallowed by a northward-moving wave of Protestantism' (ibid.: 47). The Royal Proclamation, therefore, played different populations against each other in the interests of the British colonial project, paradoxically giving Native peoples land rights in order to control and assimilate members of another European culture. My point here is that the policy, now appropriated into the idea of Canada's historical 'respect' for Native peoples, did not recognise Aboriginal people in a manner that might threaten the colonial project. Instead, the recognition of difference and the granting of land rights were part of a flexible strategy to manage the colonial project at a difficult and potentially dangerous time.

Despite attempts to manage population movements with the Royal Proclamation, by 1774 it was clear the assimilation policy would not work because the wave of immigrants had not arrived in the numbers expected (Cook 1963: 4–6). Further, there was growing discontent with Britain in the southern colonies (not yet, but soon to become, the United States). The British therefore wanted to secure the loyalty of Québec, and especially the powerful Catholic clergy, in case of trouble (Cook 1963: 4–6; McNaught 1970: 46–7).

This changed situation demanded a different approach to managing cultural difference. The Québec Act of 1774 recognised and guaranteed the position of the Roman Catholic Church, and the French language in Québec. Again, colonial policy was flexible, recognising and enabling certain populations in the interest of controlling others (in this case the French against the southern colonies) as part of the broader project.

RACE AND SETTLEMENT

Up to the Confederation of Canada in 1867, a series of rapid changes transformed the relatively unsettled territory of what is now known as Canada, and in the process altered both the lives and the colonial representations of Native peoples. This transformation can be traced, in part, to a shift to colonisation through settlement rather than primarily through resource extraction (furs), a process in which Native people had been central. Changing attitudes to Aboriginal people can also be traced to global shifts in beliefs about race, cultural difference, progress and civilisation.

Settlement increased during the late eighteenth and early nineteenth century, and as a result, by the end of the War of 1812, few Native people lived in the southernmost parts of Upper and Lower Canada (now Ontario and Québec), the most populated areas of Canada. Many who did had retreated northward to escape the encroaching settlement (Bumsted 1992a: 229), and those who remained were increasingly isolated on reserves (Trigger 1986: 318). In Nova Scotia, Native people were 'given' land and provisions, but the land was in the form of unsurveyed local reserves on which white settlers would often squat at will (Bumsted 1992a: 229).

As territorial boundaries began to take a more institutionalised form, cultural and racial boundaries began to harden, both in Canada and globally. Marriages between Native women and white men had been common during the fur trade. The frequency of intermarriage declined and the practice became less socially acceptable when, after 1820, the state institutionalised the migration of single white women as prospective wives in a push to settle the country (Van Kirk 1980). Similar processes occurred in other colonial situations such as Sumatra (Stoler 1989: 143–6) and Africa (Cairns 1965: 246), indicating in a broader way that racial ideology and control of sexuality intersected within the colonial project (McClintock 1995; Stoler 1995). As settlement increased, Canadian historical writing began to place Native peoples in a more restricted role. Negative stereotypes began to appear, with greater emphasis given to scalping, torture, massacres, and sexual promiscuity. Whereas earlier it had been assumed Native people could change and develop, they were increasingly presented as incapable of civilisation and therefore doomed to perish in the natural march of 'progress' (Trigger 1986: 318–21). The hardening and sharpening of racial boundaries were consistent with so-called 'scientific' ideas of 'race' developing in the nineteenth century (Banton 1987; Stocking 1968).

Central to theories of evolution and 'race' emerging in the late eighteenth and early nineteenth century was the idea of progress—the notion that societies move through consecutive stages to reach the ultimate pinnacle of evolution: European-style 'civilisation'. Although the idea of civilisation as the acme of progress had been present in much earlier thought, up to the late eighteenth century it was assumed that 'civilisation' was the potential and destined goal of *all* humankind. In the nineteenth century, more and more people began to see civilisation as only achievable by certain 'races' (Stocking 1968: 35–6). Polygenetic interpretations of human origins began to gain more authority than Christian based monogenetic interpretations (Berkhofer 1978: 56; Stocking 1968), as several trends came together to provide a seemingly scientific explanation for difference (ibid.: 36–7). In Canada these trends were reflected in the near universal belief amongst whites that Native people, as they had existed, were disappearing with the inevitable march of progress (Francis 1992: 53). In tandem with global trends of the late nineteenth century, this assumption meant that Aboriginal cultural artefacts were collected and classified. Aboriginal people were also photographed, painted and written about, in an attempt to document what was perceived as civilisation's picturesque yet less developed past (Dippie 1992; Francis 1992: 11–44; Street 1992).

EMERGING NATIONAL IDENTITY

Formed in 1867, the Dominion of Canada was a political unification of the three separate provinces of British North America: Canada (now Ontario and Québec), New Brunswick, and Nova Scotia (Bumsted 1992a: 371). Newfoundland and Prince Edward Island joined the Dominion later, as did the Western provinces and Northern territories.[1] Canada was the product of a political alliance, and the emergence of nationalist sentiment was not guaranteed. From the earliest days, local and provincial loyalties and interests threatened national unity. Like present-day Canadians, not all nineteenth-century Canadians shared the same vision of the nation (ibid.: 382). Despite these divisions, soon after Confederation a search began for a Canadian national identity, resulting in an increase in the production of literature and painting, and the formation of national cultural, educational, and sports institutions (ibid.: 381). At the same time Confederation was seen as fragile, because by the mid-1880s it had been shaken by the Riel Rebellion, a great deal of French resistance to confederation and a depression. The threat of 'continentalism' was also profound, because the Liberal government of the time had committed itself to the policy of freer trade with the United States, a situation that to many meant the inevitable assimilation of Canada into the United States (Berger 1970: 3–4). These challenges to the nation 'galvanise[d] the defenders of Canada and the British connection into action', and brought out into the open 'the ideas and sentiments, traditions and hopes, that constituted their sense of nationality' (ibid.: 4), and resulted in the formation of the Canada First Movement.

ICY WHITE NATIONALISM

The early nationalism of the Canada First Movement was grounded in the belief that Canada was a 'Britain of the North', a 'northern kingdom' whose unique and distinctive character derived from its northern location, its ferociously cold winters, and its heritage of 'northern races' (Berger 1966: 4). This racialised 'Canadianness' was mobilised to create links between Canada and Britain and other northern and 'civilised' nations, to differentiate Northern and Southern peoples (races), and to distinguish Canada from the USA. It also drew on specific forms of racial ideology that had developed on a global scale.

In his 1869 speech, 'We Are the Northmen of the New World', Robert Grant Haliburton, an associate of the Canada First Movement, asserted that the distinct characteristic of the new Dominion was that it was and always should be 'a Northern country inhabited by the descendants of the Northern races' (in Berger 1966: 6). A corollary to the idea of the northern race as superior was the construction of the South as other: inferior, weaker, and also essentially female. While the adjective 'northern' symbolised the masculine virtues of 'energy, strength, self-reliance, health and purity', southern was equated with 'decay and effeminacy, even libertinism and disease' (Berger 1970: 129). The northern (more masculine) races were, in this view, also naturally more oriented towards freedom and liberty, and the southern races with tyranny (Berger 1966: 15). The Canada First Movement believed that the traditions, customs, and unwritten codes which preserve civil and political liberty were 'rooted in the elemental institutions' of the Northmen and all northern nations (ibid.: 7). It was declared, therefore, that Canada was destined to become a pre-eminent power because of its superior racial characteristics.

The Canada First Movement drew upon environmental relativism, a discourse that linked environment and character—an idea proposed by Montesquieu as early as 1748)—and had prominence at the time (Banton 1987: 8, 40–5). What is most interesting, in the context of current thinking, is that the Canada First Movement constructed the *United States*, a nation now often considered the quintessential modern Western nation, as the degenerate and decaying south. Canada was constructed as different from and superior to the United States because of environment, a discourse that functioned through a complex set of ideas which helped to produce Canadian national identity.

One of the most persistent benefits articulated about Canada was that, unlike the United States, its northern climate would keep the country 'uncontaminated' by weaker southern races. Parkin suggested that the northern climate was one of Canada's 'greatest blessings', 'a fundamental political and social advantage which the Dominion enjoys over the United States', because it resulted in a 'persistent process of natural selection'. According to Parkin, Canada's climate would ensure that it would avoid the 'Negro problem', a problem 'which weighs like a troublesome nightmare upon the civilisation of the United States' (in Berger 1970: 131). Canada, he emphasised, would be a nation of the 'sturdy races' of the North (in Berger 1966: 8–9).

The Canada First movement did not consider the United States an Anglo-Saxon country, in part because of the effects of the southern climate and immigration on racial characteristics. They thought the southern climate not only made the northern races of the USA deteriorate, but also attracted ever growing 'multitudes of the weaker races from Southern Europe', and provided a home to 'the large Negro element' (Berger 1966: 14). Canada was considered naturally superior because it had not diluted its northern blood. Canadian or British liberty was also considered 'far superior to the uproarious democracy of the United States' (ibid.: 16). Parkin, when discussing US immigration, suggests that these immigrants from southern climes were 'unaccustomed to political freedom, unaccustomed to self-government' (in Berger 1966: 17).

This is an instance of the still current idea that tolerance is a Canadian national characteristic (albeit inherited from the British Empire). Today, as at the turn of the century, the notion of Canada's tolerance is mobilised in a discourse of differentiation that constructs the USA as other. The flexibility of nationalist discourse is apparent, because although cultural pluralism was central to ideas of nationhood, *how* it was constructed was not consistent. French cultural differences were sometimes erased: when Canada, for example, was defined as *homo*geneously northern, unlike the *hetero*geneous USA. At other times the French presence was made paramount, to mark difference from the USA: when it was argued, for example, that Canada allowed the retention of local traditions and *hetero*geneity, unlike the *homo*geneous USA. Yet, as Berger suggests, the Canada First Movement could be tolerant of the French partly because they were convinced of their inevitable decline as a significant proportion of the population. He argues that in fact, 'the more one was convinced of their numerical unimportance, the more one could glorify their character' (1970: 147). Some kinds of differences were valued, as long as there was no threat to the dominance of the Anglo-Saxon, especially, as in the case of the French, if it helped to construct Canada as different from the USA.

One interesting silence in the writings of the Canada First Movement is the absence of a discussion of the way in which Native peoples fit into the idea of the natural superiority of Northern races. If one followed the logic of the 'northern thinking' of the Canada First Movement, wouldn't Native people, having lived in the Northern climates of Canada for longer, be *even more* hardy, superior, and naturally inclined to freedom, liberty and hard work, than Anglo-Saxon immigrants who come from a relatively warmer climate? Perhaps the assumption that Aboriginal people were vanishing (discussed above) meant they did not have to be considered.

The United States was also used as a negative example in the heated debates at the end of the nineteenth century on whether to let in 'southern races', discussions inspired by increased immigration into Canada. One observer, drawing attention to the influx of southern Europeans, queried if there was anything to prevent the nation from becoming entirely Americanised. In 1899 the editor of *Canadian Magazine* feared Canada might become as 'rude, as uncultured, as fickle, as heterogeneous, as careless of law and order and good citizenship as the United States' (cited in Berger 1970: 148–9).

NATION-BUILDING, IMMIGRATION AND NATIONAL IDENTITY

As the discussions above illustrate, immigration policy is more than simply a process of importing the labour needed for nation-building. It is also an 'expression of a political idea of who is, or could be, eligible to receive the entitlements of residence and citizenship' (Smith 1993: 50). One characteristic of the ideal nation-state, as it was characterised in the nineteenth century, was the conception that geographic and cultural boundaries should be synonymous (Anderson 1991; Gellner 1983). Immigration policy is therefore a site that articulates a potential contradiction between the material needs of nation-building and the attempt to create an ideal 'imagined community'. Immigration was essential for nation-building, yet also perceived as potentially dangerous if it threatened the development and maintenance of a national population and a national identity.

Throughout the formative years of Canadian nation-building, immigration policies favoured Britons and northern Europeans. At the same time the need for labour, especially agricultural labour, often outstripped the supply of preferred 'racial stock'. In the first three decades of Confederation, the westward expansion of the nation was a failure. It was difficult to get British immigrants to come to Canada, and many who did eventually migrated to the United States. The Macdonald government of the 1880s tried unsuccessfully to develop immigration schemes with the British government, whose major aim, argues Whitaker (1991: 5), was to remove their 'turbulent Irish subjects at Canada's expense'. Canada simply could not compete with the United States in attracting immigrants until the development of weather-resistant strains of wheat and the exhaustion of the US agricultural frontier lands (Whitaker 1991: 5). Despite Canadian preferences for British immigrants, during the 1890s the Canadian government fulfilled the need for settlers and labour by recruiting Slavs from the Austro-Hungarian Empire. Minister of the Interior, Clifford Sifton, managed to find a place for them in the Canadian imaginary. They became the necessary immigrant of 'good quality', the 'stalwart peasant in sheepskin coat, born on the soil, whose forefathers have been farmers for ten generations, with a stout wife and a half-dozen children' (in Harney 1989: 53).

Immigration laws were therefore flexible as well as racial, exemplified in Canada's treatment of Chinese labourers. The transcontinental railway was built chiefly on the labour of 6,500 Chinese men specifically imported for this job, many of whom died in the process (Bumsted 1992b: 29). In 1883 Prime Minister Macdonald defended Chinese immigration, arguing that it would be 'all very well to exclude Chinese labour when we can replace it with white labour, but until that is done, it is better to have Chinese labour than no labour at all' (in ibid.: 29). The Chinese were not treated as regular immigrants, not being granted even potential citizenship rights (Whitaker 1991: 9). When the railway was finished in 1885, Chinese labour was no longer needed, and between 1878 and 1899 the British Columbia legislature passed over twenty-six bills aimed at preventing or restricting the settlement of Asians (Palmer 1991: 9).

Up until the Second World War, when most immigration ceased, Canada had a strict hierarchy of preferred racial groups for immigration. While Canadians often point to the White Australia Policy as an example of an overt racism which never existed in Canada, Canada had what Hawkins calls a 'White Canada Policy' (1989: 17). This policy drew upon an environmental racism similar to the Canada First Movement, excluding Blacks and Asians on the grounds that they were unsuited to the cold climate of Canada. The notion of 'climactic unsuitability' was enshrined in immigration law as a reason for barring non-whites until 1953 (Kallen cited in Smith 1993: 55). Yet, showing the flexibility of immigration law, notions of race and culture were often interchangeable. While Section 38, Clause c, of the Immigration Act of 1910 focused on Canada's right to exclude races on the basis of climatic unsuitability, amendments to the Act in 1919 expanded these criteria to include cultural elements. Immigrants could then also be excluded if they were 'deemed undesirable owing to their peculiar customs, habits, modes of life and methods of holding property, and because of their probable inability to become readily assimilated' (cited in Hawkins 1989: 17).

From 1890 to the present, immigration policy has, according to Harney, had several relatively constant goals 'appearing under different flags of convenience and employing different idioms'. Canada has needed immigrants to populate the country so as to discourage American 'expansionary tendencies' and to 'protect the Pacific Rim' from heavy Asian immigration. Immigration was also intended to populate the country enough to create 'economies of scale and a rational East-West axis' to bolster an independent polity and a viable economy. Immigration was also intended to 'combat separatism and maintain British hegemony'. More recently, immigration has been mobilised to create an image of Canada as a land of opportunity, characterised first by the 'fairness of British institutions', and now by the 'civility of state-sponsored pluralism in the form of official multiculturalism'

(Harney 1989: 53). Therefore, immigration was an essential, and yet necessarily flexible, aspect of nation-building.

SETTLING THE WEST: GENTLE MOUNTIES AND PICTURESQUE 'INDIANS'

After Confederation in 1867, the Canadian government promoted agricultural settlement in the North-west of Canada, and as a result needed to push Native people aside as quickly as possible. They negotiated a series of treaties that extinguished Native entitlement to land in exchange for reserves on marginal and unattractive land (Bumsted 1992a: 391–3). The settlement of the west in the late nineteenth century had grave consequences for the lives of Native peoples. It also shaped how they came to be represented in the Canadian national imaginary.

Royal Canadian Mounted Police (or 'The Mounties') and the transnational railway are central and revered symbols of Canada, resonating with romantic notions of national progress and expansion. The formation of the North-west Mounted Police in 1873, to act as a quasi-military agent of the government in Western Canada, is one of the most romanticised events in Canadian popular history. In histories of the RCMP, the west was painted as a lawless and frozen land where both 'Indians'[2] and whites were uncivilised. Into this environment burst the just arm of the Canadian Government, actualised in the red-coated Mounties who would calmly protect uncivilised whites and 'Indians' from each other. In his 1912 history of the Mounted Police, A.L. Haydon writes that they quickly created 'a bloodless revolution' in which over 30,000 Aboriginal people, hostile to the white invasion and at war with each other, were 'transformed into a peaceful community showing every disposition to remain contented and law-abiding' (in Francis 1992: 64). What actually happened, according to some historians, was that Westward expansion did not initially require use of force, because the capitulation of Native land rights was brought about by disease (smallpox) and loss of subsistence (buffalo) (Finley and Sprague 1979: 216).

In the mythology, however, the frontier—now calmed and civilised by the loyal red-coated representatives of British justice—was then ready for the next stage of what was commonly seen as its inevitable progression from savagery to civilisation: the arrival of thousands of hard-working farmers. This miracle of civilisation was brought about, according to the histories, by the fair and impartial administration of British Justice; a version of history which depends on constructions of Aboriginal people as child-like, trusting, and ultimately friendly to their Canadian government invaders. In both popular and historical literature of the time, a typical scene of confrontation showed Native people like children, easily cowed by a display of authority by an RCMP officer. The classic story had an unarmed officer confronting a band of armed and angry 'braves'. The officer, apparently not noticing the rifles aimed at his head, would coolly dismount and walk up to the leader of the gang. In the story the Aboriginal people would be completely disarmed by the Mountie's reckless courage and self-confidence, and would do meekly as they were told (Francis 1992: 66).

This image of the friendly and meek 'Indian', trusting of the police, was an integral part of an emerging national identity tied to ideas about the distinction between Britain and the USA. According to the legend, the most important thing the Mounties brought to the west was their impartiality, one of the basic elements of British Law. The popular conception was that police were fair and just, and that Native people recognised this impartial justice and were grateful for it. According to Francis, the image of 'the grateful Indian'

> allowed Canadians to nurture a sense of themselves as a just people, unlike the Americans south of the border who were waging a war of extermination against their Indian population. Canadians believed that they treated their natives justly. They negotiated treaties before they occupied the land. They fed the Indians when they were starving and shared with them the great principles of British justice.
> (1992: 69)

The story of the Mounted Police had, as Francis argues, a 'powerful influence' on the way Canadians felt themselves to be 'distinct from, and morally superior to', the United States (1992: 69).

However, both US and Canadian methods of expansion and the treatment of Aboriginal people reflect similar Enlightenment assumptions abut the inevitable march of progress and civilisation, and the inevitable passing away of the 'Indian' way of life. The main difference with the Canadians was that, with the implementation of British forms of justice, it was believed the inevitable and necessary process would occur peacefully, without hardship, injustice, or violence. The supposedly fairer British/Canadian method of dealing with Native people became institutionalised in the Indian Act of 1876 (Francis 1992: 200–18). The Indian Act encouraged agricultural labour and Christianity and discouraged Aboriginal cultural practices. Over the course of the late eighteenth and early twentieth centuries, for example, governments attempted to ban the making of Totem poles, potlatch, dancing and other ceremonies (ibid.: 98–103, 183–9, 200, 218). There were many contradictions in these processes. The Act, paradoxically, sought to 'civilise' the Native peoples by assimilating them into dominant life and culture, yet at the same time segregated them onto reserves. Further, just as Native people were being prevented from public displays of their cultures, whites increasingly romanticised and appropriated Aboriginal practices by dressing up as 'Indians', by collecting their cultural 'artefacts', and even pretending to be 'Indians', as 'Grey Owl' (Englishman Archie Belaney) did in the 1930s (Francis 1992).

At the same time the Canadian Pacific Railway was trying to sell images of pure and authentic 'Indians' in feathers and buckskin, the government was enacting, through the Indian Act, a form of cultural genocide on Native peoples, who, as a result, did not necessarily appear as advertised. Many of the visitors were disappointed with the Native people they saw on their tours because they did not look like the romantic and idealised images of 'Indians' they had been promised. Francis suggests that the search to see Native people in their 'uncivilised' and 'natural' state was made all the more urgent and poignant because, in the settler imagination, 'Indians' were soon to vanish with the march of progress (1992: 182).

ASSIMILATION AND APPROPRIATION

Between the world wars, assimilation policies attempted to destroy Native cultures. The advertisement for War Bonds epitomises the ideal 'Indian' of the time, because as assimilation policies progressed, the collecting and salvaging of Native artefacts and the romanticising of 'Indians' reached a peak. In combination with tourism, the collection and romanticisation of Native people's artefacts and culture were also now integrated into more official forms of national identity. Aboriginal peoples' artefacts became national treasures, providing a visible and tangible link with the country's first peoples, a part of the nation's heritage that had to be preserved (Francis 1992: 184–5). The Parliament Buildings of Canada, a symbol of Canadian nationhood, highlight Aboriginal people as part of Canada's heritage. Constructed originally between 1859 and 1866, the Buildings, except the library, were burned to the ground in 1916. Re-completed in Gothic Revival style in 1922, the buildings were purposely built with huge blank stones, to be carved later. The impression one has today is of the incredible diversity of symbols of different cultures, carved into the very walls of the buildings. The symbolic diversity is intentional: a pamphlet put out by the Public Information Office of the House of Commons states that the many images of the Parliament Buildings 'express who we are, where we come from and what we do. Like a mirror, the Parliament Buildings reflect our diversity and uniqueness yet, at the same time, they are an enduring symbol of our unity'.

As early as 1932, the Dominion Memorial Sculpture was placed in the Hall of Honour. Designed by Dr. R. Tait McKenzie, the sculpture has images of Native people representing Canada's heritage and past. The official Parliament Buildings Guide Manual describes the first panel of the sculpture.

> In the foreground are four figures. On the left, is Canada enthroned; her right hand on a shield emblazoned with the current Arms of Canada. Her left hand is outstretched to receive the offerings of her children. A youthful figure, Canada wears a headdress of a Caribou mask with antlers, a short chiton and moccasins.
> (Parliament Buildings Guide Manual 1994: 202)

The sculpture has the nation beginning its career as a child-like Native person. Aboriginal imagery here represents the early foundations of Canada: its youth, its past. As the narrative moves on, the settlers re-shape, categorise, map and civilise nature itself. The sculpture, as it moves progressively from left to right, shows engineers measuring and cartographers mapping the Canadian wilderness. Lumberjacks and fishermen build and harvest the land. Further back and to the side of one panel is a family, which, according to the Guidebook, represents Canada's pioneers. An 'Iroquois Indian' watches them (Parliament Buildings Guide Manual 1994: 201–2).

Aboriginal people here represent the early foundations of Canada: its youth, its past, and, as is characteristic of a colonising and orientalist aesthetic (S. Hall 1992a), are idealised as nature itself. Pure untouched nature can be constructed as the raw material for the civilising work of settlement; it re-affirms the settlers' sense of themselves as those who transform raw nature into developed civilisation. Indeed, in this image, the early nature/Native roots of Canada are then integrated into a narrative of settlement and progress, as the settlers categorise, map, re-shape and civilise nature itself. In this narrative of Canada's development, inscribed onto the walls of the Parliament Buildings in the 1930s, Aboriginal people are not invisible, erased, or absent. In a similar manner, a monument of one of Canada's earliest explorers, Samuel de Champlain, situated in a prominent place in Ottawa, does not erase Native people. In the sculpture Champlain stands tall and heroic on the pedestal, while below him, seated beneath the pedestal, is an unnamed 'Indian'.

As my discussion has shown, from the days of colonialism through to the early days of nation-building, Native people have played important supporting roles in defining Canada. Often images of Aboriginal people have been mobilised to help differentiate Canada from the United States because through them British Canada can construct itself as gentle, tolerant, just and impartial. Aboriginal people also represent Canada's heritage and past, providing a link between the settlers and the land and helping to negotiate the rocky terrain of creating Canada as 'Native land' so settlers. Further, although Native people are present in these images of discovery and settlement, their presence is limited: either symbolising Canada's natural beginnings, or as the passive Iroquois watching the pioneers or the unnamed Indian at the feet of Samuel de Champlain. Canada's Aboriginal peoples were not erased in Canada's nationalist narratives, but were important supporting actors in a story which reaffirms settler progress.

NORTHERN WILDERNESS AND SETTLER NATIONAL IDENTITY

In the half-century between Confederation and the end of the First World War, Canada was transformed from a frontier nation to a Western industrial nation. Simultaneous with this 'march of progress' and 'civilisation', the very phenomena which were being destroyed were now invoked to represent Canadian nationhood. Native people, as discussed, began to be integrated into official national iconography. Meanwhile the Northern wilderness would come to represent Canada, in particular through the paintings of the Group of Seven. Northern wilderness and Canada's cold climate remain constant images in public representation and myth in Canada, 'the only sure token . . . of collective identity' (Berland 1994: 99), although as Margaret Atwood suggests, the 'values ascribed to them have varied considerably' over time (1995: 9).

The landscape paintings of the Group of Seven do not sustain and construct colonial national identity by *inviting* colonising humans to penetrate nature, as the picturesque tradition does. Instead, their paintings reject the European aesthetic in favour of a construction of a 'Canadian' aesthetic based on the obliterating and overpowering sense of uncontrollable *wilderness*. A central feature that differentiates the wilderness paintings of the Group of Seven from European traditions of painting is that the paintings are unpeopled, not just of human subjects but also of human traces. Bordo suggests that the 'deliberate and systematic absenting of human presence has no consistent precedent in the European landscape tradition' (1992: 109). Further, these nationalistic northern wilderness paintings are specifically empty of signs not just of people, but of *Native people*. The absence of Aboriginal people in this wilderness art constitutes a serious rupture with the nineteenth-century wilderness ethos, in which wilderness was identified with the Native presence (Bordo 1992: 122). In the Group of Seven's paintings, the erasure of Aboriginal presence, and the production of a notion of uninhabited wilderness, were necessary to create 'wilderness landscape' as a signifier of Canada that differentiated Canada's northernness from European northernness.[3]

The obliteration of human presence and the foregrounding of nature as savage and dangerous do not mean, however, that it was not a colonising aesthetic. The aesthetic of the Group of Seven reflects the view of European *settlers*, only in a different way than the Picturesque or the Sublime. Landscape has been described as the 'dreamwork of imperialism', disclosing both 'utopian fantasies of the perfected imperial prospect' and images of 'unresolved ambivalence and unsuppressed resistance' (Mitchell 1994: 10). The paintings of the Group of Seven embody the dreamwork of settler nationalism in Canada. Their wilderness aesthetic articulates some of the contradictory themes in Western concepts of self and other in a language of emergent nationhood. Wilderness as a symbol of Canadianness is polysemic; it can be interpreted in many ways.[4]

The systematic absence of human presence in the Group of Seven paintings can be interpreted as an expression of what I call the 'civilisation versus wilderness' opposition. In the mind of one member of the Group, Canada was 'a long thin strip of civilisation on the southern fringe of a vast expanse of immensely varied, virgin land reaching into the remote north' (in Berger 1966: 21). Bordo makes a link between the absence of people in work of the Group of Seven and the criteria of a wilderness park, a place in which human presence is deliberately erased (Bordo 1992: 114–19). He argues that wilderness is created through rules and regulations that prohibit and control not just human presence, but specifically Aboriginal presence. The 'wilderness' was inhabited for centuries by complex societies of Aboriginal people, and was not a 'wilderness' in the way we think of it today. It was not distinctly and conceptually separated and distinct from urban life, rural farmland, and human 'civilisation'. It was not, as it is now, a site marked out for leisure, a space of untouched nature in which to recuperate from one's 'real' life. Aboriginal people lived and worked in that so-called emptiness.

Wilderness—unpeopled and savage—can also work as an 'other' to 'civilisation', a comparison within Canada which reinforces the settlers' sense of the difficult struggle to civilise the wilderness and the success they have had in controlling it. If the wilderness park works as a leisure site—a controlled 'precinct' which acts as an 'other space' to modern cosmopolitan lives—the north can be seen as that kind of 'other' to the majority of Canadians who live in 'the long thin strip of civilisation'.

Margaret Atwood and Northrop Frye, two influential Canadian writers, also use wilderness as a key to articulate the elements and characteristics of Canadian culture. Atwood's *Survival* (1972) and Frye's *The Bush Garden* (1971), are central to the canon of Canadian literature taught in schools and universities. In both works, as in the aesthetic of the Group of Seven, the settler viewpoint of nature—not as 'noble' but as '*ignoble* savage'—plays a key role in defining Canadianness. Historically, in colonial discourse, links have been made between Native peoples and a purer state of nature. Indeed, the construction of Native people as more pure and natural was an important contribution to the creation of a 'civilised' Western identity. Nature was at first idealised and projected upon by early visitors to North America, as were Native people. Part of this idealisation was the construction of stereotypes of Native people, and the splitting of those stereotypical images into the 'noble' and the 'ignoble' savage (S. Hall 1992a). There are striking similarities between the dualism of such images of nature in Canadian nationalist discourse and the historical construction of women as virgins or whores, or blacks as problem or victim (Gilroy 1987).

In the earlier constructions of Canadian northernness by the Canada First Movement, the landscape is noble; it is somehow tamed, mirroring the traits and values of the people who have colonised it. It then produces offspring who (by nature of their northernness) maintain those superior traits. The picturesque landscapes of much of the early painting in Canada construct the relationship between nature and settlers in a similar mode. Nature is tamed; it reflects and reproduces a colonial mindset. The vision of nature is a projection of the viewer and the occupier.

In *Survival,* Atwood examines types of landscapes common in Canadian literature and the attitudes they express. The main argument in her chapter 'Nature the Monster' is that in Canadian literature nature is perceived as a 'monster', an evil betrayer. She suggests that this sense of betrayal and distrust of nature emerged because of the disjuncture, during the late eighteenth and early nineteenth century, between expectations imported with the settlers from England about the nature of nature, and the harsh realities of Canadian settlement. In the late eighteenth century the dominant mode in nature poetry, inspired by Edmund Burke, was that of the picturesque and the sublime. This shifted in the first half of the nineteenth century to Wordsworthian Romanticism. Although each movement was distinct, both constructed nature as essentially good, gentle and kind (1972: 50). The themes she describes in literature are similar to those I have discussed in regard to European painting.

Towards the middle of the century, under the influence of Darwinism, literary images of nature began to change. Nature remained female but became 'redder in tooth and claw' (Atwood 1972: 50). However, according to Atwood, most of the English settlers were by that time already in Canada, 'their heads filled with diluted Burke and Wordsworth, encountering lots and lots of Nature'. If Wordsworth had been right, Canada ought to have been 'the Great Good Place. At first, complaining about the bogs and mosquitoes must have been like criticising the authority of the Bible' (1972: 50).

For Atwood, these writings show a tension between 'what you were officially supposed to feel and what you were actually encountering when you got here'—a sense of being betrayed somehow by the 'divine Mother'. This theme of betrayal has been worked out, according to Atwood, through two central preoccupations in Canadian literature and fiction: victims and survival. A recurring theme is 'hanging on' or 'staying alive'. Canadian stories, according to Atwood, 'are likely to be tales, not of those who made it but of those who made it back, from the awful experience—the North, the snowstorm, the sinking ship—that killed everyone else' (1972: 35). The survivor is not as a hero, but barely alive, aware of the power of nature, and the inevitability of losing at some stage. Later, a constant theme in Canadian literature is what she calls 'Death by Nature' and even 'Death by Bushing' (ibid.: 55–6). In this scenario, if Canada were David and Nature were Goliath, Goliath would win. Submission to, and victimisation by, nature as ignoble, untamed and monstrous savage define Canadianness.

The image of Canadians being lost in the wilderness is also mobilised by Northrop Frye to contrast Canada and the United States. He argues, in *The Bush Garden,* that

Canada has, for all practical purposes, 'no Atlantic seaboard'. The traveller from Europe, he suggests, 'edges into it like a tiny Jonah entering an inconceivably large whale'. The traveler slips

> past the straight of Belle Isle into the Gulf of St. Lawrence, where five Canadian provinces surround him, for the most part invisible. Then he goes up the St. Lawrence and the inhabited country comes into view, mainly a French speaking country, with its own cultural traditions. To enter the United States is a matter of crossing an ocean; to enter Canada is a matter of being *silently swallowed by an alien continent.*
> (Frye 1971: 217)

Frye's description of 'entering' and 'being swallowed'—a tiny Jonah engulfed by a huge whale—constructs Canada as a devouring, dangerous and alien female, even a *vagina dentata* (toothed vagina). Part of this femaleness is that she is everywhere, unconquerable, and somehow not definite or definable. The USA, on the other hand, is more 'male', more definite and phallic. The Canadian land as alien and female assumes and generalises a male settler's point of view. Such tropes of the colonial landmass as sexualised, as the terrifying female being penetrated by or engulfing male colonisers were also used in Africa during the late nineteenth century (Stott 1989). In Fry's description of Canada, the settlers are uncomfortable because *they* don't *penetrate* and control the (natural/female) foreign space; nature *engulfs* and swallows *them*.

In Frye's argument, the frontier—huge, alien, unconquerable and quintessentially Canadian—is different from the USA and again, imagined as essentially female.

> To feel 'Canadian' was to feel part of a no-man's-land with huge rivers, lakes, and islands that very few Canadians had ever seen . . . One wonders if any other national consciousness has had so large an amount of the unknown, the unrealised, the humanly undigested, so built into it. Rupert Brooke speaks of the 'unseizable virginity' of the Canadian landscape. What is important here, for our purposes, is the position of the frontier in the Canadian imagination. In the United States one could choose to move out to the frontier or retreat from it back to the seaboard. In the Canadas, even the Maritimes, the frontier was all around one, a part and a condition of one's whole imaginative being . . . Such a frontier was the immediate datum of his imagination, the thing that had to be dealt with first.
> (Frye 1971: 220)

Atwood's main thesis is that Canada's essential identity is that of 'the exploited victim' (1972: 35, 36). She suggests that Canada is a colony, and that a partial definition of a colony is 'a place from which a profit is made, but *not by the people who live there*' (ibid.: 36). The notion of being lost in the wilderness—in an undefined and unknown territory—is extremely paradoxical when mobilised in a discourse of victimisation to colonialism or imperialism, since it is itself a perspective of a coloniser, a settler, not one who is colonised. The Canadian literature Atwood examines is,[5] for the most part, written by and expresses a world-view of those who settled in Canada from other countries. These are people who, although they may have felt lost and victimised by the environment or the Empire, were representatives of the colonial power that victimised Native people. Would being lost in the 'unknown territory' of the wilderness be a central metaphor for Aboriginal Canadian writers? When Frye uses the term 'no-man's land' he must mean European settlers, for it is only to them that Canada could be an 'alien continent'. The versions of northern Canadianness I discuss above utilise a settler point of view (lost in the wilderness) which erases Aboriginal people. Yet, paradoxically, white settlers take up a subject position more appropriate to Native people, in order to construct Canadians as victims of colonialism and US imperialism, and to create Canadian identity.

During the colonial period and in the early decades of nation-building, the dual process of creating Canadian identity and managing diverse populations involved complex and contradictory representations of internal and external others, and of Canada's land itself. Some nationalist narratives and government policies discussed in this chapter are structured through the inclusion of or nominal respect for Native or French peoples. Others are nationalist discourses structured through the erasure of cultural difference. Yet despite their differences, all these versions of national identity mobilise internal differences and similarities (either through erasure, inclusion, or appropriation) in order to differentiate from external others such as the USA and Britain. They all reinforce British or white settler hegemony and construct a settler (usually British) nationalist identity. Paradoxically, often this settler national identity is created through constructing others as friends or allies, or using the culture or subject positions of those others.

The politics of all this flexibility are complex. The important question here is not whether Aboriginal people, French-Canadians, or immigrants are erased in Canadian mythology, or even whether they are represented positively. The central issue is to examine *who* decides *when* and *how* Aboriginal people, French people or more recent immigrants, are or aren't represented, or are or aren't managed, in the interests of the nation-building project. These cultural groups become infinitely manageable populations as well as bit players in the nationalist imaginary, always dancing to someone else's tune. They become helpmates in the project of making a Canadian identity that defines itself as victimised by outsiders and tolerant of insiders. This process became more institutionalised during and after the Second World War, and is epitomised in the emergence of 'multiculturalism' as an official national policy and identity.

NOTES

1. Apparently the British supported the idea of unification because they hoped it would mean that the colonials could then organise their own military defence and 'become sufficiently strong militarily to discourage the Americans from adventurism in the north'. This was in tandem with their desires for less colonial responsibility and expense (Bumsted 1992a: 328–9). Kilbourne suggests the British public was thoroughly bored with the idea of Canada and has been ever since. The House of Commons was only

one quarter full when it passed the British North America Act, 'though it filled up immediately after for a debate on the dog-tax bill' (Kilbourne 1989: 13).

2. A note on terminology: Following Francis (1992), I use the term 'Indians' when I am discussing *images* of Native people, and the terms Native people, First Nations people, or Aboriginal people when discussing real people.
3. The appropriation of the work of the Group of Seven into a *national* iconography was also problematic because their images of the Pre-Cambrian shield, Algonquin, and 'north of superior' were seen by some Canadians as an imposition of 'one set of foreign images by another equally removed, emanating from metropolitan Toronto, an "Uppity Canada"' (Osborne 1988: 173). The paintings are 'centralist' in that they reflect a vision of central Canada—not the East or the West—as if the centre represented the whole.
4. Landscape imagery has been the subject of debate and critique in the disciplines of Art History, English and Geography. On nature and landscape see Cosgrove and Daniels (1988); Cronon (1995); Daniels (1993); Merchant (1994, 1995); Oelschlaeger (1992); Plumwood (1993); Schama (1995); Turner (1980).
5. Recently there have been attempts to add Aboriginal people's and 'multicultural' writing to the canon of Canadian literature—see Hutcheon and Richmond (1990) for example.

BIBLIOGRAPHY

Anderson, B. (1991) *Imagined Communities: Reflections on the Origin and Spread of Nationalism*, (revised edition) London: Verso.

Atwood, Margaret (1972) *Survival: A Thematic Guide to Canadian Literature*, Toronto: Anansi.

—— (1995) *Strange Things: The Malevolent North in Canadian Literature*, Oxford: Clarendon Press.

Banton, Michael (1967) *Race Relations*, London: Tavistock.

—— (1987) *Racial Theories*, Cambridge: Cambridge University Press.

Beaudoin, Gerald and Dorothy Dobbie (1992) *Report on the Special Joint Committee on a Renewed Canada*, 28 February 1992, Ottawa: Supply and Services Canada.

Berger, Carl (1966) 'The true north strong and free', in Peter Russell (ed.) *Nationalism in Canada*, Toronto: McGraw-Hill, 3–26.

—— (1970) *The Sense of Power: Studies in the Ideas of Canadian Imperialism 1867–1914*, Toronto: University of Toronto Press.

Berkhofer, Robert F. (1978) *The White Man's Indian: Images of the American Indian from Columbus to the Present*, New York: Alfred A. Knopf.

Berland, Jody (1994) 'On reading "The Weather"', *Cultural Studies*, 8(1):99–114.

—— (1995) 'Marginal notes on cultural studies in Canada', *University of Toronto Quarterly*, 64(4): 514–25.

Bordo, Jonathan (1992) 'Jack Pine — wilderness sublime, or the erasure of the aboriginal presence from the landscape', *Journal of Canadian Studies*, 27 (4): 98–128.

Bothwell, Robert (1995) *Canada and Quebec: One Country, Two Histories*, Vancouver, UBC Press.

Bumsted, J.M. (1992a) *The Peoples of Canada: A Pre-Confederation History*, Toronto: Oxford University Press.

—— (1992b) *The Peoples of Canada: A Post-Confederation History*, Toronto: Oxford University Press.

Cairns, H. Alan C. (1965) *Prelude to Imperialism: British Reactions to Central African Society 1840–1890*, London: Routledge and Kegan Paul.

Chabram, Angie, and Rosa Linda Fregoso (1990) 'Chicana/o cultural representations: reframing alternative critical discourses', Introduction in Special Issue, *Cultural Studies*, 4(3): 203–12.

Cook, Ramsay (1963) *Canada: A Modern Study*, Toronto: Irwin Publishing.

Cosgrove, Dennis and Stephen Daniels (eds) (1988) *The Iconography of Landscape*, Cambridge: Cambridge University Press.

Cronon, William (ed.) (1995) *Uncommon Ground: Toward Reinventing Nature*, New York: W. W. Norton.

Curtin, Phillip D. (1964) *The Image of Africa: British Ideas and Action, 1780–1850*, Madison: University of Wisconsin Press.

Daniels, Stephen (1993) *Fields of Vision: Landscape Imagery and National Identity in England and the United States*, Princeton, NJ: Princeton University Press.

Dippie, Brian W. (1992) 'Representing the other: the North American Indian', in Elizabeth Edwards (ed.) *Anthropology and Photography 1860–1920*, London: Yale University Press and Royal Anthropological Institute, 132–6.

Finley, J.L. and D.N. Sprague (1979) *The Structure of Canadian History*, Scarborough: Prentice-Hall of Canada.

Francis, Daniel (1992) *The Imaginary Indian: The Image of the Indian in Canadian Culture*, Vancouver: Arsenal Pulp Press.

Frye, Northrop (1971) *The Bush Garden: Essays on the Canadian Imagination*, Toronto: Anansi.

Gellner, Ernest (1983) *Nations and Nationalism*, London: Cornell University Press.

Gilroy, Paul (1987) *There Ain't no Black in the Union Jack*, London: Routledge.

Goldberg, David Theo (1993) *Racist Culture: Philosophy and the Politics of Meaning*, Oxford and Cambridge: Blackwell.

Hage, Ghassan (1994a) Locating multiculturalism's other: a critique of practical tolerance', *New Formations*, 24: 19–34.

Hall, Stuart (1992a) 'The west and the rest: discourse and power', in Stuart Hall and Bram Gieben (eds) *Formations of Modernity*, Cambridge: Polity Press in Association with the Open University, 275–332.

—— (1992b) 'The question of cultural identity', in Stuart Hall, David Held and Tony McGrew (eds) *Modernity and its Futures*, Cambridge: Polity Press in Association with the Open University, 273–326.

Harney, Robert F. (1989) '"So Great a Heritage as Ours": immigration and the survival of the Canadian polity', in Stephen R. Graubard (ed.) *In Search of Canada*, New Jersey: Transaction Publishers, 51–98.

Hawkins, Freda (1989) *Critical Years in Immigrations: Canada and Australia Compared*, Kensington: McGill-Queens University Press.

Hutcheon, Linda and Marion Richmond (eds) (1990) *Other Solitudes: Canadian Multicultural Fictions*, Toronto: Oxford University Press.

Kilbourne, William (1989) 'The peacable kingdom still', in Stephen R. Graubard (ed.) *In Search of Canada*, New Jersey: Transaction Publishers, 1–30.

McClintock, Anne (1995) *Imperial Leather: Race, Gender and Sexuality in the Colonial Contest*, New York: Routledge.

McNaught, Kenneth (1970) *The Pelican History of Canada*, London: Penguin.

Merchant, Carolyn (ed.) (1994) *Ecology: Key Concepts in Critical Theory*, New Jersey: Humanities Press.

—— (1995) *Earthcare: Women and the Environment*, New York: Routledge.

Mitchell, W.J.T. (ed.) (1994) *Landscape and Power*, Chicago: University of Chicago Press.

Oelschlaeger, Max (ed.) (1992) *The Wilderness Condition: Essays on Environment and Civilization*, Washington, DC: Island Press.

Osborne, Brian S. (1988) 'The iconography of nationhood in Canadian art', in Dennis Cosgrove and Stephen Daniels (eds) *The Iconography of Landscape*, Cambridge: Cambridge University Press, 162–78.

Palmer, Howard (1991) *Ethnicity and Politics in Canada Since Confederation*, Saint John, NB: Canadian Historical Association.

Parliament Buildings Guide Manual (active document—in state of continual update), used by tour guides of the Parliament Buildings; obtained from the Public Information Office, House of Commons, Ottawa, April 1994.

Plumwood, Val (1993) *Feminism and the Mastery of Nature*, London: Routledge.

Schama, Simon (1995) *Landscape and Memory*, Toronto: Random House.

Smith, Susan (1993) 'Immigration and nation-building in Canada and the United Kingdom', in Peter Jackson and Jan Penrose (eds) *Constructions of Race, Place and Nation*, Minneapolis: University of Minnesota Press, 50–77.

Stocking, George W. Jr (1968) *Race, Culture, and Evolution: Essays in the History of Anthropology*, New York: The Free Press.

Stoler, Ann Laura (1989) 'Rethinking colonial categories: European communities and the boundaries of rule', *Society for the Comparative Study of Society and History*, 31: 134–61.

—— (1995) *Race and the Education of Desire*, Durham: Duke University Press.

Stott, Rebecca (1989) 'The Dark Continent: Africa as female body in Haggard's Adventure Fiction', *Feminist Review*, 32: 68–89.

Street, Brian (1992) 'British popular anthropology: exhibiting and photographing the other', in Elizabeth Edwards (ed.) *Anthropology and Photography*, London: Yale University Press in Association with The Royal Anthropological Institute, 122–31.

Todorov, Tzvetan (1984) *The Conquest of America*, New York: HarperPerennial.

Trigger, Bruce G. (1986) 'The historian's Indian: native Americans in Canadian historical writing from Charlevoix to the present', *Canadian Historical Review*, LXVII (3): 315–42.

Turner, Frederick (1980) *Beyond Geography: The Western Spirit Against the Wilderness*, New York, The Viking Press.

Van Kirk, Sylvia (1980) *'Many Tender Ties': Women and Fur-Trade Society in Western Canada, 1670–1870*, Winnipeg: Watson and Dwyer Publishers.

Whitaker, Reg (1991) *Canadian Immigration Policy Since Confederation*, Saint John, NB: Canadian Historical Society.

2

Rewriting Histories of the Land

BONITA LAWRENCE

Canadian national identity is deeply rooted in the notion of Canada as a vast northern wilderness, the possession of which makes Canadians unique and "pure" of character. Because of this, and in order for Canada to have a viable national identity, the histories of Indigenous nations,[1] in all their diversity and longevity, must be erased. Furthermore, in order to maintain Canadians' self-image as a fundamentally "decent" people innocent of any wrongdoing, the historical record of how the land was acquired—the forcible and relentless dispossession of Indigenous peoples, the theft of their territories, and the implementation of legislation and policies designed to effect their total disappearance as peoples—must also be erased. It has therefore been crucial that the survivors of this process be silenced—that Native people be deliberately denied a voice within national discourses.[2]

A crucial part of the silencing of Indigenous voices is the demand that Indigenous scholars attempting to write about their histories conform to academic discourses that have already staked a claim to expertise about our pasts—notably anthropology and history. For many Aboriginal scholars from Eastern Canada who seek information about the past, exploring the "seminal" works of contemporary non-Native "experts" is an exercise in alienation. It is impossible for Native people to see themselves in the unknown and unknowable shadowy figures portrayed on the peripheries of the white settlements of colonial Nova Scotia, New France, and Upper Canada, whose lives are deduced solely through archaeological evidence or the journals of those who sought to conquer, convert, defraud, or in any other way prosper off them. This results in the depiction of ancestors who resemble "stick figures"; noble savages, proud or wily, inevitably primitive. For the most part, Indigenous scholars engaged in academic writing about the past certainly have little interest in making the premises of such works central to their own writing—and yet the academic canon demands that they build their work on the back of these "authoritative" sources. We should be clear that contemporary white historians have often argued in defence of Aboriginal peoples, seeking to challenge the minor roles that Native people have traditionally been consigned in the (discursively created) "historical record." What is never envisioned, however, is that Indigenous communities should be seen as final arbiters of their own histories.

What is the cost for Native peoples, when these academic disciplines "own" our pasts? First of all, colonization is normalized. "Native history" becomes accounts of specific intervals of "contact," accounts which neutralize processes of genocide, which never mention racism, and which do not take as part of their purview the devastating and ongoing implications of the policies and processes that are so neutrally described. A second problem, which primarily affects Aboriginal peoples in Eastern Canada, is the longevity of colonization and the fact that some Indigenous peoples are considered by non-Native academics to be virtually extinct, to exist only in the pages of historical texts. In such a context, the living descendants of the Aboriginal peoples of Eastern Canada are all too seldom viewed as those who should play central roles in any writing about the histories of their ancestors.

Most important, however, is the power that is lost when non-Native "experts" define Indigenous peoples' pasts—the power that inheres when oppressed peoples choose the tools that they need to help them understand themselves and their histories:

> The development of theories by Indigenous scholars which attempt to explain our existence in contemporary society (as opposed to the "traditional" society constructed under modernism) has only just begun. Not all these theories claim to be derived from some "pure" sense of what it means to be Indigenous, nor do they claim to be theories which have been developed in a vacuum separated from any association with civil and human rights movements, other nationalist struggles, or other theoretical approaches. What is claimed, however, is that new ways of theorizing by Indigenous scholars are grounded in a real sense of, and sensitivity towards, what it means to be an Indigenous person . . . . Contained within this imperative is a sense of being able to determine priorities, to bring to the centre those issues of our own choosing, and to discuss them amongst ourselves.[3]

For Indigenous peoples, telling our histories involves recovering our own stories of the past and asserting the epistemological foundations that inform our stories of the past. It also involves documenting processes of colonization from the perspectives of those who experienced it. As a result, this chapter, as an attempt to decolonize the history of Eastern Canada, focuses on Indigenous communities' stories of land theft and dispossession, as well as the resistance that these communities manifested towards colonization. It relies primarily on the endeavours of Indigenous elders and scholars who are researching community histories to shape its parameters. Knowledge-carriers such as Donald Marshall Senior and Indigenous scholars who carry out research on behalf of Indigenous communities such as Daniel Paul, Sakej Henderson, and Georges Sioui are my primary sources. For broader overviews of the colonization process, I draw on the works of Aboriginal historians such as Olivia Dickason and Winona Stevenson. In some instances, I rely on non-Native scholars who have consulted Native elders, such as Peter Schmalz, or who have conducted research specifically *for* Indigenous communities involved in resisting colonization (where those communities retain control over ownership of the knowledge and how it is to be used), such as James Morrison. In instances where no other information is available, the detailed work of non-Native scholars such as Bruce Trigger and J. R. Miller is used to make connections between different events and to document regional processes. The issues at hand are whether the scholar in question is Indigenous and the extent to which the scholar documents the perspectives of Indigenous communities about their own pasts.

As history is currently written, from outside Indigenous perspectives, we cannot see colonization *as* colonization. We cannot grasp the overall picture of a focused, concerted process of invasion and land theft. Winona Stevenson has summarized how the "big picture" looks to Aboriginal peoples: "Mercantilists wanted our furs, missionaries wanted our souls, colonial governments, and later, Canada, wanted our lands."[4] And yet, this complex rendition of a global geopolitical process can obscure how these histories come together in the experiences of different Indigenous nations "on the ground." It also obscures the *processes* that enabled colonizers to acquire the land, and the *policies* that were put into place to control the peoples displaced from the land. As a decolonization history, the perspectives informing this work highlight Aboriginal communities' experiences of these colonial processes, while challenging a number of the myths that are crucial to Canadian nation-building, such as the notion that the colonization process was benign and through which Canada maintains its posture of being "innocent" of racism and genocide. Other myths about Native savagery and the benefits of European technologies are challenged by Native communities' accounts of their own histories and are explored below.

MERCANTILE COLONIALISM: TRADE AND WARFARE

The French and early British trade regimes in Canada did not feature the relentless slaughter and enslavement of Indigenous peoples that marked the Spanish conquest of much of "Latin" America. Nor did they possess the implacable determination to obtain Indigenous land for settlement by any means necessary that marked much of the British colonial period in New England. Thus the interval of mercantile colonialism in Canada has been portrayed as relatively innocuous. And yet, northeastern North America was invaded by hundreds of trade ships of different European nations engaged in a massive competition for markets; an invasion instrumental in destabilizing existing intertribal political alliances in eastern North America. It is impossible, for example, to discount the central role that competition for markets played in the large-scale intertribal warfare that appears to have developed, relatively anomalously, throughout the sixteenth, seventeenth, and eighteenth centuries in much of eastern Canada and northeastern United States. Oral history and archeological evidence demonstrate that these wars were unique in the history of these Indigenous nations.

It is important to take into consideration the extent to which the new commodities offered by the Europeans gave obvious material advantages to those nations who successfully controlled different trade routes. Inevitably, however, as communities became reliant on trade to obtain many of the necessities of life, access to trade routes became not only desirable but actually necessary for survival (particularly as diseases began to decimate populations, as the animal life was affected, and as missionaries began to make inroads on traditional practices).[5] These pressures resulted in such extreme levels of competition between Indigenous nations that an escalation into continuous warfare was almost inevitable. For example, in the seventeenth century, the Mi'kmaq people of the Gaspé began killing Iroquoians who crossed into their trade territories. They also fought a trade war with the Abenaki in 1607.[6] Meanwhile, the Innu nation in the sixteenth century became embroiled in two different trade wars—the Naskapi fought the Inuit for access to furs in Labrador while the Montagnais fought the Iroquois for control of the rich Saguenay River route to James Bay and the Great Lakes.[7] But most profoundly, trade wars (in conjunction with the diseases accompanying the traders) decimated populations in what is now southern Ontario and Quebec.[8]

The nations of the Iroquois Confederacy, which waged much of this warfare for control of trade territories, were themselves devastated by the century-long struggle between Britain and France for control of the fur trade in the Great Lakes region.

They were ultimately defeated by the Ojibway nation after a lengthy conflict that left "mounds of bones" at certain sites in southern Ontario. In 1701, the Iroquois signed a peace treaty with the Ojibway.[9] In the face of encroaching European trade, this treaty developed into a mutual non-aggression pact that, until the land was taken, was never violated.

Warfare and trade among Indigenous nations profoundly changed the ecology of the land and way of life for nations of many regions. Yet these should not be seen as evidence of Indigenous savagery or of a breakdown of Indigenous values;[10] rather, these profound changes, in part, resulted form the severe pressures caused by the intense competition of European powers during mercantile colonialism to depopulate entire regions of all fur-bearing animals.

Intertribal wars were also carried out for another reason, according to the oral traditions of many of the nations whose homelands were assaulted by these processes. According to Georges Sioui, warfare was also waged by many of the nations weakened by disease and trade warfare as a way to replenish their population base. Often, Indigenous nations adopted their captives:

> We used our alliance to the French to go and attack the English colonies to the south with the primary intent of capturing people, especially young and female, and ritually, through adoption, giving them a new life in our Nations. As it was, clan-mothers and matriarchs had the principal say in these military undertakings; they had the primary responsibility of maintaining and restoring the integrity and composition of the societies which, as woman-leaders, they headed. White, and other, captives were given over to clan-mothers who had organized war expeditions through approaching and commissioning war chiefs. The captives were then ritually and factually nationalized and, then, brought up and treated as full members of their adoptive social communities . . . some of our Aboriginal Nations survived almost only because of our traditional mother-centred thinking. Had we, at that time, had leaders formed in patriarchal colonial institutions, as is so often the case nowadays, many of our nations simply would not have survived beyond the eighteenth century.[11]

DISEASE AND CHRISTIANIZATION IN THE HURON-WENDAT NATION

Although French colonial policies focused primarily on the fur trade, under the terms of the Doctrine of Discovery, the monopolies they granted to different individuals in different regions included the mandatory presence of Christian missionaries.[12] The missionaries relied on trade wars (and the epidemics frequently preceding or accompanying them) to harvest converts from Indigenous populations physically devastated by mass death. Nowhere is this more obvious than among the Huron-Wendat people.

The Wendat, whom the Jesuits labeled "Huron," were the five confederated nations of the territory known as Wendake (now the Penetanguishene Peninsula jutting into Georgian Bay). It was made up of twenty-five towns, with a population that peaked at thirty thousand in the fifteenth century.[13] The Wendat relied both on agriculture and fishing, and until extensive contact with French traders began in 1609, they enjoyed remarkable health and an abundance of food.

Georges Sioui suggests that Wendat communities first came into contact with disease through the French, who were dealing with large groupings of Wendat living together as agricultural people. It was not until 1634, however, when the Jesuits, who had visited in 1626, returned to set up a mission that the Wendat encountered a continuous wave of epidemics, which culminated in the virulent smallpox epidemic of 1640 that cut their population in half.[14] So many elders and youths died in the epidemics that the Wendat began to experience serious problems in maintaining their traditional livelihoods and grew extremely dependent on French trade for survival. The epidemics also had a catastrophic effect on the Wendat worldview. The psychological shock of such an extreme loss of life was experienced as sorcery, as the introduction of a malevolent power into the Wendat universe.[15]

It was into this weakened population that the Jesuits managed to insinuate themselves, using their influence in France to have French traders withdrawn and replaced by Jesuit lay employees. The Jesuits sought to impress the Huron with their technological superiority and allowed their traders to sell certain goods, particularly guns, only to Christian converts.[16] As the number of Christian converts grew in response to such virtual blackmail, the Jesuits gradually obtained enough power in the communities to forbid the practising of Wendat spiritual rituals. In response, a traditionalist Wendat faction developed in an attempt to resist the Jesuits and, indeed, any dealings with the French. Instead, they proposed a trading pact with the Iroquois but were unable to obtain sufficient influence to achieve this. Despite the growing power of the Jesuits, the mission could not protect the Christian Wendat against attacks by Seneca and Mohawk war parties in the winter of 1648 and the spring of 1649. The Confederacy was shattered and the Wendat abandoned their villages, dying of starvation and exposure by the thousands.

About six hundred destitute survivors followed the Jesuits back to Quebec, forming a community at Loretteville, which still exists today. A large number were captured by the Iroquois and adopted into the Seneca, Onondaga, and Cayuga nations. Others joined traditional allies and migrated to Ohio in the United States, where they acquired land as "Wyandot" people. In the 1830s, the Wyandot were forcibly relocated from Ohio to Kansas as part of the U.S. government's "clearing" of Eastern Native peoples from the land. Many lost their tribal status in Kansas, but a small group of Wyandot acquired a reserve in northeastern Oklahoma where they continue to live today. A small number of Wendat remained in Ontario and maintained two reserves in the Windsor region. In the early nineteenth century, both reserves were ceded and sold by the Crown. A small acreage remained and was occupied by a group known as the Anderdon band. This band, consisting of the remaining forty-one Wendat families in Ontario, were enfranchised under the *Indian Act* in 1881, at which point they officially ceased to exist as "Indians." Their land base was divided up into individual allotments. Despite the loss of a collective land base and "Indian" status, the descendants of the forty-one families in Windsor still consider themselves Wendat.[17]

Georges Sioui is a member of the Loretteville Huron-Wendat band. He reflects on his people's experience of Jesuit conversion and how central missionaries were to the economic order of colonial agendas:

> Throughout history, the process of conversion has been a process of subversion and destruction. All over the world, in whatever climate, missionaries have considered disease and death as particularly effective means of undermining the pride of circular-thinking peoples . . . and impoverishing them, thereby making them submit to the socio-economic order of the invader. Impoverishment is the natural outcome of disunity, which is hastened by the poverty brought on by disease. Brebeuf [the Jesuit who established a mission among the Wendat known as Huronia] foresaw an abundant harvest of souls following the disastrous epidemic that cut down thousands of Wendats . . . . Such a harvest could only be realized among a people who walked in constant fear of death and the spectre of annihilation. By 1636, the Jesuit missionaries were already conscious of wielding almost absolute power over Wendat leaders. At the end of the Feast of the Dead that he attended, Brefeuf publicly refused a present from the hands of the great chief of the Attignawantans (and possibly of the entire Wendat confederacy). *The only acceptable gift was the abandonment of their culture by all the savages.*[18]

It is clear, when we explore what happened at Wendake, that the confluence of disease, dependency on trade, and the missionary crusade weakened the Wendat people so drastically that attacks by the Seneca and Mohawk were sufficient to shatter their Confederacy beyond repair. It is also clear that the presence of the missionaries played a significant role in narrowing the options of Confederacy members and that later legislation, such as enfranchisement policies, caused further dispersal and dissolution. All of these processes were part and parcel of the colonial strategies to assert and formalize European presence and authority on the land.

From the perspective of the Huron-Wendat people, the destabilization that mercantilism brought was in many respects almost as deadly a process for indigenous societies as the actual land acquisition project that followed it. Indeed, when land acquisition is not marked by direct military force, mercantile trade is usually the preliminary step.

Huron-Wendat history that emphasizes the destabilization of mercantilism directly challenges the notion that the large-scale warfare between Native peoples during colonization was a function of savagery or a total breakdown of Native values. Rather, it can be understood as an effort to survive in the face of the massive pressures of European trade interests and from the need to repopulate nations devastated by disease and invasion. Perhaps more important, this history challenges the ubiquitous notion that the small nations devastated by colonization have ceased to exist. As Sioui demonstrates, the Huron are more than a tragic remnant of history:

> from the Native's point of view, the persistence of essential values is more important than change, just as from the non-Native point of view, it has always been more interesting to shore up the myth of the disappearances

> of the Amerindian . . . . An examination of the Amerindian philosophical tradition will show the persistence, vivacity, and universality of the essential values proper to America . . . if history is to be sensitive to society's needs, it must also study and reveal to the dominant society what is salutary, instead of continuing to talk about "primitive" cultures that are dead and dying . . . . In short, Amerindians think that while they have changed, like everything in the world, they are still themselves.[19]

The catastrophic changes that the Huron-Wendat have undergone are perhaps less important than the fact that they have survived as a people, and that their worldview has changed but remains fundamentally Wendat. These myths of savagery and of a "loss of culture" form an essential part of contemporary settler ideology—a justification for the denial of restitution for colonization, the backlash against Aboriginal harvesting rights, and policies of repression against Native communities. Through exploring Huron-Wendat history informed by their own realities, a culture regarded as "dead" by the mainstream speaks to us about its contemporary world.

THE MI'KMAQ: DIPLOMACY AND ARMED RESISTANCE

Not all nations faced the Wendat experience of Christianization. The Mi'kmaq nation was perhaps unique in the way it used Christianity as a source of resistance to colonization in the earliest years of contact with Europeans.

Mi'kmaki, "the land of friendship," covers present-day Newfoundland, St-Pierre and Miquelon, Nova Scotia, New Brunswick, the Magdalen Islands, and the Gaspé Peninsula. It is the territory of the Mi'kmaq, which means "the allied people." The Mi'kmaq nation became centralized during a fourteenth-century war with the Iroquois Confederacy. Since then it has been led by the Sante Mawiomi, the Grand Council, and has been divided into seven regions, each with its Sakamaws or chiefs. It is part of the Wabanaki Confederacy, which includes the Mi'kmaq, the Abenakis in Quebec, the Maliseets in western New Brunswick, and the Passamaquoddies and Penobscots in New England.[20] The Wabanaki Confederacy is only one of seven Confederacies, the others being the Wampagnoag, Pennacook, Wappinger, Powhattan, Nanticoki, and Leanape. These Confederacies represent the thirteen surviving Indigenous nations along the Eastern seaboard which have asserted their sovereignty over the entire Maritime and New England regions of Canada and the United States.[21]

The Mi'kmaq people were the first Native people in North America to encounter Europeans, and were aware of the political implications of contact. The French entered their territory in earnest in the sixteenth century and had set up small maritime colonies by the early seventeenth century. Knowledge of the genocide of Indigenous peoples in the Caribbean and Mexico by the Spanish travelled along the extensive trade networks that existed across North America at the time and reached the Mi'kmaq by the mid-sixteenth century. In response to this information, and to the spread of disease that increased with greater contact, the Mi'kmaq avoided the French coastal settlements and consolidated their relationships with other Eastern nations of the Wabanaki Confederacy.[22] However,

Messamouet, a Mi'kmaw[23] scholar and prophet who had travelled to France and learned of how the Europeans conceptualized law and sovereignty, developed another option known as the "Beautiful Trail," which would involve the Mi'kmaq nation negotiating an alliance with the Holy See in Rome.

When the French began their colonial ventures into North America, the Catholic monarchs of Europe were in a power struggle over sovereignty issues with the Holy Roman Empire, which had been instrumental in consolidating independent European tribal groups into nations governed by the Holy See.[24] The discovery of the existence of the Americas by explorers such as Columbus amplified the power struggle between the Holy See and the ascendant monarchs of Europe over the question of whose authority would prevail in these "discovered" lands—that of the nations of Europe or the Church?

By building an alliance with the Holy See, the Mi'kmaq nation sought recognition as a sovereign body among the European nations. In this way, Mi'kmaki could resist the authority of the French Crown. In 1610, Grand Chief Membertou initiated an alliance with the Holy See by negotiating a Concordat that recognized Mi'kmaki as an independent Catholic Republic. As a public treaty with the Holy See, the Concordat had the force of international law, canon law, and civil law. Its primary effect was to protect the Mi'kmaq from French authority "on the ground." The process, however, initiated a gradual centralization of authority in the Mawiomi—prior to the Concordat, Mi'kmaw families were organized into regional self-governing units; under the Concordat, the Mawiomi attained political authority for the Mi'kmaq nation. The terms of the Concordat also gave individual Mi'kmaw people the freedom to choose or reject Catholicism. Priests under the authority of the Holy See, rather than those under the French Crown, baptized each family and its extended family from one hunting district to the next within Mi'kmaki territory. Within twenty years, most Mi'kmaw families had been baptized. As well, the alliance granted the Church access to Mi'kmaki where it could build churches and promised Mi'kmaki protection from the Holy See against all other European monarchies. Under the Concordat and alliance, the Mawiomi maintained a theocracy which synthesized Catholic and Mi'kmaq spirituality and maintained Mi'kmaq independence form the French Crown.[25]

In 1648, the Treaty of Westphalia ended the Holy See's rule over European monarchies. The treaty's settlement of territorial claims placed some lands under the control of nation-states and others under the control of the Holy See: Mi'kmaki "reverted" to Mi'kmaq control and all protections ceased to exist.

Unfortunately for the Mi'kmaq, the French were not the only colonial power to invade their world. What the British sought was not furs and missions but land where they could build colonies for their surplus populations. The colonies they developed along the Atlantic seaboard, therefore, practised open extermination against the Indigenous peoples around them. Once they had taught the English colonists how to survive in the new land, Indigenous peoples were deemed superfluous at best, and an impediment to settlement at worst. Despite the fact that massive outbreaks of illness had already decimated the populations they encountered (in certain regions, nineteen out of twenty Indigenous people who came into contact with the British succumbed to disease), the British initiated a number of attacks against Indigenous villages, attacks which often escalated into full-scale wars. British slavers scoured the Atlantic coast for Indigenous people who were sold in slave markets all over the world. Indeed, they began raiding Mi'kmaq territory for slaves in the mid-1600s.[26]

As the British encroached north from New England to Nova Scotia, the Mi'kmaq responded with open resistance. From the mid-1650s until the peace treaty of 1752 (which was reaffirmed in the treaty of 1761), they waged continuous warfare against the British, fighting land battles and capturing almost one hundred British ships. As the long war proceeded, and the Mi'kmaq were gradually weakened, the ascendant British developed policies to exterminate the Mi'kmaq. They used a variety of methods, including distributing poisoned food, trading blankets infected with diseases, and waging ongoing military assaults on civilian populations.[27] These acts of genocide were in addition to the diseases that the European soldiers introduced. In 1746, a typhus and smallpox epidemic spread through Nova Scotia, killing one-third of the Mi'kmaq population. (As late as the 1920s, Mi'kmaw people were still telling stories about the mounds where those who perished were buried.) It was into this decimated population that the British introduced scalping policy as another method of extermination. For two decades, the British paid bounty for Mi'kmaq scalps and even imported a group of bounty hunters known as Goreham's Rangers from Massachusetts to depopulate the surviving Mi'kmaq nation.[28]

Those who survived this genocide were destitute, left with no food and without the necessary clothing to keep warm in a cold climate. Many were reduced to begging. Thousands died of starvation and exposure until limited poor relief was implemented on a local basis. Others eked out a bare existence selling handicrafts, cutting wood for whites, or working as prostitutes (which resulted in outbreaks of venereal disease). Those who struggled to acquire individual land plots were denied title; as a result, it was not uncommon for Mi'kmaw families to engage in the backbreaking labour of clearing and planting a patch of land, only to find that when they returned from fishing, hunting, or gathering excursions that white squatters had taken the land.[29] When the British opened up the region for white settlement, they refused to set aside land for the Native peoples. In British legal thinking of the day, non-Christians who had been defeated by and were subject to the British had no rights to land under the "Norman Yoke," an aspect of the Doctrine of Discovery that the British inserted for their own benefit.[30]

Since the signing of the 1752 treaty, which brought an end to warfare, the Mi'kmaq have sought to resolve the ongoing land and resource theft, with little success. In 1973, the *Calder* case decision forced the Canadian government to recognize that it had some obligation to deal with land claims. Although comprehensive claims were held to be worthy of consideration in Western Canada, the Mi'kmaq claims were held to be "of a different nature" and Mi'kmaq participation in this process was denied. In 1977, the Mi'kmaq Grand Council made a formal application for land and compensation under the 1973 policy. The federal government, however, insisted that Mi'kmaq rights had somehow been "superceded by law."[31]

In exploring Mi'kmaq resistance efforts—negotiating a Concordat with the Holy See, waging the longest anti-colonial war in North America, surviving policies designed to exterminate them—we see a picture of Native peoples as resourceful and capable of engaging a powerful enemy in armed conflict for a significant period of time. Perhaps even more important, we see Mi'kmaq people as actors on an international stage, engaging the

European powers not only through warfare but through diplomacy, signing international treaties as a nation among nations. Mi'kmaq perspectives of their own history, then, reinstate Native peoples as global citizens and challenge the colonial perspective of Native peoples as powerless victims.

It is impossible to understand contemporary struggles for self-determination without this view of Native peoples as nations among other nations. Today, the spirit that enabled the Mi'kmaq to resist genocide is being manifested in the continuous struggles over the right to fish and in their challenge to their rights under the Concordat with the Holy See. Although the Concordat came to an end in the late seventeenth century, the memory of the alliance has come into play in recent years. When Pope John Paul II committed the Catholic Church to supporting the liberation struggles of Aboriginal peoples in Canada in the late 1990s, the Mawiomi launched its campaign to clarify the settlement terms of the Treaty of Westphalia. It is believed that Canada usurped lands accorded to the Mi'kmaq under the Concordat's international law. By re-establishing communication with the Holy See, the Mawiomi wish to recreate its partnership in ways that enhance the autonomy and spiritual uniqueness of the Mi'kmaq.[32]

GEOPOLITICAL STRUGGLES BETWEEN THE COLONIZERS AND INDIGENOUS RESISTANCE IN THE GREAT LAKES REGION

The British entered the territory now known as Canada from two fronts: the East Coast region (primarily for settlement purposes) and Hudson's Bay (under the charter of the Hudson's Bay Company for the purpose of the fur trade). As British traders spread south into the Great Lakes region, expanding their fur trade, competition with the French escalated. From 1745 onwards, the British adopted policies that regulated trade, protected "Indian" lands from encroachment, and secured military allies among Native peoples through the distribution of gifts. In 1755, the Indian Department was officially established under imperial military authority.[33]

The struggle between Britain and France over the Great Lakes region had profound effects on the Iroquois and Ojibway peoples who lived there. The trade struggle between Europeans forced first one party and then the other to lower the prices of trade goods relative to the furs that were traded for them. Ultimately, when warfare broke out, the effect was devastating, as colonial battles fought in Native homelands destroyed these regions and drew Native peoples into battles, primarily to ensure that a "balance of power" resulted (which would ensure that both European powers remained deadlocked and that one power would not emerge victorious over another).[34]

In 1763, the warfare between France and Britain ended when France surrendered its territorial claims in North America. The trading territory from Hudson's Bay west to the Rockies was claimed by the Hudson's Bay Company and was considered by the British to be its property. Meanwhile, Britain laid claim to most of the territory of eastern North America formerly held by the French, although it lacked any real ability to wrest the land from the Native nations who occupied it. Nor could it control how the nations of these regions would choose to act. Because it was important for Britain to reassert its formal adherence to the Doctrine of Discovery and to ensure that its claims to eastern North

America would be respected by other European regimes, the British government consolidated its imperial position by structuring formal, constitutional relations with the Native nations in these territories. The Royal Proclamation of 1763 recognized Aboriginal title to all unceded lands and acknowledged a nation-to-nation relationship with Indigenous peoples which the Indian Department was in charge of conducting. Department agents could not command; they could only use the diplomatic tools of cajolery, coercion (where possible), and bribery.[35] The nation-to-nation relationship was maintained until the end of the War of 1812 when the post-war relationships between Britain and the American government became more amicable and made military alliances with Native nations unnecessary.

In the meantime, Britain's ascendancy in the Great Lakes region marked a disastrous turn for Native peoples. Most of the nations were devastated by years of French and British warfare that they had been drawn into. Once the British had control of the fur trade, they began to drive prices down and violate many of the long-standing trade practices that had been maintained while the French were active competitors. It was also obvious to Indigenous people that one unchallenged European power was far more dangerous to deal with than a group of competing Europeans. During this desperate state of affairs, a number of Indigenous nations attempted to form broad-ranging alliances across many nations in an effort to eliminate the British presence from their territories, culminating in the Pontiac uprising of 1763.

Pontiac, an Odawa war chief, was inspired by the Delaware prophet Neolin. He wanted to build a broad-based multinational movement whose principles involved a return to the ways of the ancestors and a complete avoidance of Europeans and their trade goods. At least nineteen of the Indigenous nations most affected by the Europeans shared this vision. Their combined forces laid siege to Fort Detroit for five months, captured nine other British forts, and killed or captured two thousand British. Within a few months, they had taken back most of the territory in the Great Lakes region from European control.

Between 1764 and 1766, peace negotiations took place between the British and the alliance. The British had no choice in the matter; the Pontiac uprising was the most serious Native resistance they had faced in the eighteenth century.[36] As a consequence, the British were forced to adopt a far more respectful approach to Native peoples within the fur trade and to maintain far more beneficial trade terms. However, the dependency of many of the Indigenous nations on British trade goods and their different strategies in dealing with this dependency weakened the alliance and it could not be maintained over the long term.[37] This unfortunately coincided with the British plan to devise ways of removing the military threat that Native peoples clearly represented without the cost of open warfare. The primary means they chose were disease and alcohol.

There is now evidence to suggest that the smallpox pandemic—which ravaged the Ojibway and a number of the Eastern nations including the Mingo, Delaware, Shawnee, and other Ohio River nations, and which killed at least one hundred thousand people—was deliberately started by the British.[38] The earliest evidence of this deliberate policy is the written request of Sir Jeffrey Amherst to Colonel Henry Bouqet at Fort Pitt. In June 1763, Amherst instructed Bouqet to distribute blankets infected with smallpox as gifts to the

Indians. On June 24, Captain Ecuyer of the Royal Americans noted in his journal, "We gave them two blankets and a handkerchief out of the smallpox hospital. I hope it will have the desired effect."[39]

Peter Schmalz suggests, however, that germ warfare was not as widespread against the Ojibway who held the territory north of the Great Lakes as it was against the Native peoples of the Ohio valley and plains. The "chemical warfare" of alcohol, however, was waged against the Ojibway in a highly deliberate manner. Major Gladwin articulated this policy clearly: "The free sale of rum will destroy them more effectively than fire and sword." The effects of widespread alcohol distribution were immediate. Factionalism increased and the Ojibway could no longer unite for an adequate length of time to solve a common threat to their well-being. By 1780, a split developed between the Ojibway of southwestern Ontario (who had to struggle against devastating social disintegration under the effects of alcohol and their resulting dependency on the British) and the Ojibway situated closer to Lake Huron, Georgian Bay, and Lake Superior (who maintained a stronger cultural cohesion and independence from Europeans).[40]

In the Great Lakes region, chemical and germ warfare were used by the British as the primary means to acquire land and impose control. Despite this, the Pontiac uprising demonstrated the power of Indigenous nations organized in armed resistance to colonization. At the same time, we can see the divisive effect that dependency on British trade goods brought to the nations involved in the fur trade. From these perspectives, these changes to Indigenous ways of life had long-term and highly significant effects on the possibilities of maintaining sovereignty and resistance to European expansion. The centuries-long fur trade changed the course of Indigenous history in Eastern Canada, as the considerable military power of the Indigenous nations was subverted by their need for trade goods to support their changing way of life.

OJIBWAY EXPERIENCES OF COLONIZATION IMMIGRATION, DECEPTION, AND LOSS OF LAND

As the fur trade spread further west, the British government consolidated its hold over the Great Lakes area by implementing settlement policies. At the end of the American Revolution, Loyalists poured into the territory that had become known as Upper Canada, bringing new epidemics of smallpox that decimated the Ojibway around Lake Ontario. Immigration from the United Kingdom was openly encouraged as a means of cementing the British hold on the territory. In the thirty-seven-year interval from 1814 until the census of 1851, the white population of Upper Canada multiplied by a factor of ten—from 95,000 to 952,000. The Native population, formerly one-tenth of the population, declined into demographic insignificance in the same interval.[41] At the same time local government encouraged the migration of as many American Indians into the colony as possible. This achieved a two-fold goal: securing the potential services of these Native nations against the Americans in times of war, and overwhelming local Ojibway communities with thousands of incoming Ojibway and Pottawatomi from south of the border. Walpole Island, for example, with a population of three hundred, took in over eight hundred new people between 1837 and 1842.[42] This influx of refugees, while welcomed by their relations to the north, threw many Ojibway communities into disarray at the precise moment when their lands were being taken from them.

Between 1781 and 1830, the Ojibway gradually ceded to the British most of the land north of what is now southern Ontario. The British knew that the Ojibway were aware of the warfare being committed against Native peoples in the United States, where uncontrolled, violent settlement and policies of removal were being implemented. Using this knowledge to their advantage, the British presented land treaties as statements of loyalty to the Crown and as guarantees that the lands would be protected from white settlement. Through the use of gifts and outright lies, to say nothing of improperly negotiated and conflicting boundaries, most of the land of southern Ontario was surrendered over a fifty-year period. The British used the following procedures to negotiate land treaties:

1. By the Proclamation of 1763, the rights of Indigenous peoples to the land were acknowledged.
2. The Indigenous peoples of each area were called to consider a surrender of lands, negotiated by traders or administrators that they already knew and trusted.
3. Only the chiefs or male representatives were asked to sign.[43]
4. The surrender was considered a test of loyalty.
5. The area ceded was deliberately kept vague.
6. Some compensation, in the form of gifts, was given.
7. In many cases, the land was left unsettled for a few years, until disease and alcohol had weakened potential resistance. When the settlers began to come in and the Native people complained, they were shown the documents they had signed and told there was no recourse.[44]

The British knew that deceit could drive a better bargain—but as the Ojibway learned to drive harder bargains themselves, the price of land went up. For example, one hundred townships obtained in earlier surrenders might cost the same price as ten townships obtained once the Ojibway understood how negotiations worked. The Ojibway were not, as is commonly believed, fooled by the British. By the late eighteenth century, they were fully aware that Europeans treated land differently than they did. They were familiar with the highly populated centres of British, French, and American settlement and had extensive communication with Native peoples who had already been dispossessed. Earlier land cessions were easier for the British because of the trust that many Ojibway had in the traders and diplomats who had demonstrated a degree of honesty and goodwill in the past. This was combined with the Ojibways' need for trade goods, a certain degree of loyalty, and, especially in the face of the wars of extermination being waged in the U.S., a lack of alternatives. Furthermore, prior to 1825, the Ojibway did not read or write in English and the treaties were ambiguous, incomplete, and often badly worded.[45]

For example, on October 9, 1783, Captain W. R. Crawford negotiated the infamous "gunshot treaty" or "walking treaty"—this was a "blank treaty" which did not specify any boundaries to the land the treaty claimed. Colonial officials were forced to investigate the complaints of the Mississaugas whose lands were taken by this treaty. Crawford argued that they had surrendered all the land "from Toniato or Onagara to the River in the Bay of Quinte within eight leagues of the bottom of the said Bay, including all the islands, extending from the lake back as far as a man can travel in a day."[46] Although later treaties clarified the boundaries (in the interests of Europeans), clear definitions of boundaries were intentionally omitted from the text of treaties and those who signed them were encouraged to include within the ceded lands the territory of other bands.

These practices should not be seen as part of the distant past. As late as 1923, despite attempts to clarify boundaries of the "walking treaty" negotiated by Crawford, the federal government realized that almost half of the City of Toronto, as well as the towns of Whitby, Oshawa, Port Hope, Cobourg, and Trenton were on land that had not yet been ceded. At that point, the government gave $375 to the Ojibway of Alnwic, Rice Lake, Mud Lake, and Scugog for the land.[47] The government showed no more scruples in 1923 than it had in 1783—the land not yet legally ceded was illegally bought for a pittance, two centuries after the fact.

SETTLER VIOLENCE AND LOSS OF LAND

When the first two waves of land cessions were over in what is now southern Ontario, two million acres remained in the hands of Native peoples. Over the next fifty years, the British exerted continuous pressure on the Saugeen Ojibway, whose territories of the Bruce Peninsula and its watershed were still unceded. Eager to acquire their land, the British developed a new way of obeying the letter of the law while violating its spirit—they began to use the threat of settler violence to force land surrenders. The constant encroachment of armed, land-hungry settlers forced the Saugeen Ojibway to continuously retreat, negotiating small land surrenders a piece at a time. Often the treaties were negotiated with individuals who had no authority within their communities to negotiate treaties; these treaties, therefore, were illegal. At one point, the Saugeen Ojibway bypassed their angry and unco-operative Indian Agent and petitioned Lord Durham in England to protect their remaining lands. Durham refused, noting that the 1837 rebellion confirmed that the will of the white colonists, especially for cheap land, took precedence over the will of the Indigenous peoples. However, the Saugeen Ojibway continued to petition and, in 1846, obtained a Royal Deed of Declaration, which stated that they and their descendants were to possess and enjoy the Saugeen (Bruce) Peninsula in perpetuity, in addition to receiving regular monies for other surrenders.[48]

By this point, the Bruce Peninsula had assumed major importance in the eyes of the colonial government. A large influx of settlers, primarily refugees from the Irish potato famine and from English industrial slums, put pressure on the colony for even greater tracts of land. Once again, armed squatters were allowed to invade and seize lands. The government, in response to the protests of the Saugeen Ojibway, claimed that the scale of immigration made it impossible for them to control the new settlers. The non-stop violent encroachment forced the Saugeen Ojibway to surrender the Bruce Peninsula in 1854, leaving for themselves five large reserves, many of which were thriving farm communities, and all of the islands along the Bruce Peninsula, which had been their traditional fishing grounds. Three years later, white settlers used extensive force to move in on the farming communities. Within the next twenty years, almost all of the islands were sold by the federal government and all but two of the reserves that still exist today were surrendered. After the British relinquished control of the Indian Department to Canada in 1860, hundreds of acres continued to be carved off the Ojibway reserves in southern Ontario to create farmland for settlers, for the building of roads, and for the expansion of towns and cities.[49]

The above discussion demonstrates how the British fur trade interests in Upper Canada were gradually supplanted by settlement policies, which allowed the Crown to use whatever means were at hand to consolidate its hold over former "Indian" territories. These policies resulted in the endless misery of relocation and land loss for the Ojibway people of what is now southern and central Ontario and left many unresolved claims for restitution of stolen lands. These claims include the efforts of the Caldwell Ojibway to obtain a reserve[50] after being forced off their land near Lake Erie during the first wave of land grabs in the early 1800s, and the monumental struggles around fishing rights waged by contemporary Saugeen Ojibway communities.[51]

MOVING NORTH: RESOURCE PLUNDER OF OJIBWAY AND CREE TERRITORIES

The consolidation of the land and resource base of what is now northern Ontario became crucial to Canada's westward expansion. This phase of colonization, which took place within the twentieth century, is still being pursued. In many respects, colonial acquisition of the land is still disputed, and as a result, Indigenous histories of this region have yet to be written.

Once the land base in southern Ontario was secure, business interests in the colony looked to the rich resources in the north. Within a few years, the vast timber forests were being cut, and the growing presence of mineral prospectors and mining operations in northern Ontario caused a number of Ojibway leaders to travel to Toronto to register complaints and demand payment from the revenues of mining leases. When there was no response to these or other entreaties, the Ojibway took matters into their own hands and forcibly closed two mining operations in the Michipicoten area. Soon troops, which were not called in to protect the Saugeen Ojibway from violent white settlers, were on the scene to quell the "rebellion," and government investigators began to respond to the issues that leaders were bringing to them.[52] The Ojibway wanted treaties, but they demanded a new concession—that reserve territories be specified before the treaties were signed. After considerable discussion and many demands from the Ojibway leaders, the Robinson-Huron and Robinson-Superior Treaties were signed in 1850. These treaties ceded a land area twice the size of that which had already been given up in southern Ontario, set aside reserves (although much smaller than the Ojibway had hoped for), and provided the bands with a lump-sum payment plus annual annuities of $4 per year per person. Most important, hunting and fishing rights to the entire treaty area were to be maintained.

With these treaties, the colony gained access to all the land around Lake Huron and Lake Superior, south of the northern watershed. All land north of this was considered Rupert's Land, the "property" of the Hudson's Bay Company, and so, by 1867, one of the first acts of the newly created Canadian government was to pay a sum to the Hudson's

Bay Company in order to acquire "the rights" to this land. Inherent in the concept of "Canada," then, was the notion of continuous expansion, a Canadian version of "manifest destiny," no less genocidal than the United States in its ultimate goals of supplanting Indigenous peoples and claiming their territory.

Under section 91(4) of the *Constitution Act, 1867,* the Canadian federal government was given constitutional responsibility for "Indians and Lands reserved for the Indians," while section 109 gave the provinces control over lands and resources within provincial boundaries, subject to an interest "other than that of the Province in the same."[53] Through these unclear constitutional provisions, the newly created province of Ontario struggled for the right to open lands far north of its boundaries for settlement and development without a valid surrender of land from the Ojibway and Cree people living there. In 1871, Ontario expanded its borders to seize control of a huge northern territory rich in natural resources. "Empire Ontario," as then premier Oliver Mowat termed it, stretched as far north as Hudson's Bay (although it took a series of legal battles with the federal government to accomplish it), and spurred a rivalry with the neighbouring province of Quebec when it expanded north along the east side of Hudson's Bay.

In the late 1890s, the Liberal regime of Oliver Mowat, dominated by timber "barons" whose immense profits had been made through logging central Ontario and the Temagami region, was succeeded by the Conservative regime of James Whitney. Proponents of modern liberal capitalism, the Conservatives pushed aggressively ahead with northern development, focusing on railways, mining, and the pulp and paper industry.[54] Three northern railways were constructed to access timber, develop mineral resources, and access potential hydroelectric sites to power the resource industries. The railways opened up the territory to predators at an unprecedented rate. As a rule, if the presence of Cree or Ojibway people hindered development, the newly created Department of Indian Affairs relocated them away from the area.

It is important to understand the scale of the mineral wealth taken from the lands of the Ojibway and Cree in the past century, at great disruption to their lives and without any compensation. Since the early 1900s, the Cobalt silver mines brought in more than $184 million; Kirkland Lake gold mines produced $463 million; and Larder Lake produced $390 million.[55] Meanwhile, the Porcupine region, one of the greatest gold camps in the world, produced over $1 billion worth of gold and had the largest silver, lead, and zinc mines in the world.[56]

Across northeastern Ontario, hydroelectric development was sought primarily for the new mining industry. In 1911, however, timber concessions for the pulp and paper industry were granted, mainly to friends of government ministers, on condition that hydroelectric dams be built to power them out of the industry's money. In many cases, pulp cutting and dam construction proceeded well before permits were granted to do so.[57]

REASSERTING A SILENCED HISTORY

This chapter has introduced only a few examples of Indigenous writers, or non-Native historians working with Elders, who have recorded Indigenous nations' stories of their past. These stories introduce new perspectives to what is considered "Canadian" history.

For example, a detailed exploration of colonization in Mi'kmaki is frequently left out of most "Canadian" history texts. In part this is because of the Wabanaki Confederacy's links that tied the Mi'kmaq to wars with the Thirteen Colonies and the British Crown, events normally considered outside the purview of "Canadian" history. As Indigenous nations write their histories, we can expect to see more work that disregards the border, insists on the integrity of the Anishinabek nation or the Blackfoot Confederacy or Haudenosaunee territories that span both sides of the international border, and that disregards "Canadian" history.

Writing from the perspectives of the Indigenous nations enables specific communities to give a full and honest account of their struggles with colonizers intent on their removal and elimination as peoples, and to name the racism, land theft, and policies of genocide that characterize so much of Canada's relationships with the Indigenous nations. Even more important, Indigenous peoples are not cast as faceless, unreal "stick figures" lost in a ferment of European interests, but as the living subjects of their own histories.

While accounts of colonization from these nations are important, much remains missing from their perspectives. The Mi'kmaq people, for example, speak of the Jenu, the interval of the Ice Age when the Mi'kmaq retreated to what is now Central America until their lands became liveable again.[58] For Indigenous nations, colonization is only one part of a much older story. The ancient histories and cosmologies of our nations need to be removed from their current mythologized and depoliticized locations inside childrens' anthologies and from the mystical and colourful "origin myths" that inevitably precede television documentaries of "Canadian" history, and written back in as the organizing concepts of the histories of this land. The work of individuals such as Basil Johnston is a valuable beginning. Johnston links Anishinabek stories to the languages, cultures, and contemporary realities of the region that is now considered Quebec, Ontario, Manitoba, and parts of the United States, an area which has been and continues to be occupied by the Anishinabek people.[59] A detailed exploration of the traditional symbols used by many Indigenous nations and how these symbols have shaped the worldviews of the people is another history of the land that needs to be included.

It perhaps goes without saying that the histories of Indigenous nations will decentre the histories of New France and Upper Canada as organizing themes to the histories of this land. Canadian historians who are currently considered the experts could work in conjunction with Indigenous peoples wanting to tell their stories of the land. But the works of the experts alone, which provide powerful and detailed histories of the Canadian settler state, do not represent the full picture. It is the voices of Indigenous peoples, long silenced but now creating a new discourse, which will tell a fuller history.

NOTES

1. I have used a number of terms interchangeably to describe the subjects of this article. Generally, I use the term "Indigenous peoples," as it is the international term most commonly selected *by* Indigenous peoples to describe themselves. However, Indigenous peoples in Canada often use the term "Aboriginal" or "Native" to describe themselves; as a result, I have included these terms as well, particularly when focusing on the local

context. Occasionally, the term "Indian" is included when popularly used by Native people (such as the term "American Indians").

2. Emma Larocque, "Preface—of "Here Are Our Voices—Who Will Hear?" in Jeanne Perrault and Sylvia Vance, eds., *Writing the Circle: Native Women of Western Canada* (Edmonton: NeWest Publishers, 1993).
3. Linda Tuhiwai Smith, *Decolonizing Methodologies: Research and Indigenous Peoples* (London: Zed Books, 1999), p. 38.
4. Winona Stevenson, "Colonialism and First Nations Women in Canada,: in Enakshi Dua and Angela Robertson, eds., *Scratching the Surface: Canadian Antiracist Feminist Thought* (Toronto: The Women's Press, 1999), p. 49.
5. Losing access to have been devastating for many communities. In *The Ojibwa of Southern Ontario* (Toronto: University of Toronto Press 1991), Peter S. Schmalz recounts how Captain St. Pierre arrived at Madeline Island in 1718 to find an isolated community of Ojibway who had, over the past twenty-two years, lost access to the fur trade as a result of geographic isolation, war with the Iroquois, and the deadly trading competition between the French and the English, which involved continuously cutting off each others' markets. After a century of growing dependence on European technology, the community no longer had the endurance to hunt without guns or the skills to make stone, bone, and wood tools and utensils to replace the metal ones they had become dependent on using. The women had lost many of the skills of treating skins (when they were able to obtain them) for clothing. St. Pierre found a ragged and starving community, desperate to enter into a matter, after all, simply of individuals "roughing it" and re-adapting to Indigenous forms of technology, Indigenous communities had to be able to live off the land on a scale that would keep whole communities viable.
6. James (Sakey) Youngblood Henderson, *The Mi'kmaw Concordat* (Halifax: Fernwood Publishing, 1997), p. 80.
7. Olivia Patricia Dickason, *Canada's First Nations: A History of Founding Peoples from Earliest Times* (Toronto: Oxford University Press, 1992), pp. 103-7
8. Between Jacques Cartier's first visit to the Montreal region in 1534 and Samuel de Champlain's establishing of a colony in1608—a matter of seventy years—disease and warfare with the Haudenosaunee Confederacy resulted in the disappearance of the extensive St. Lawrence Iroquoian communities (each village numbering up to two thousand people), which had populated what is now the Montreal region. The highly strategic nature of this location for control over the St. Lawrence trade route cannot be discounted as the reason for such continuous warfare. However, it is also the case that many of the former residents of these communities were then incorporated into the Mohawk nation, as a means of repopulating communities weakened by disease—and in this respect, the St. Lawrence Iroquoians did not become extinct but were incorporated into the Six Nations Iroquois Confederacy. See Bruce G. Trigger, *Natives and Newcomers: Canada's "Heroic Age" Reconsidered* (Montreal: McGill-Queen's University Press, 1985), p. 147; and James V. Wright, "Before European Contact," in Edward S. Rogers and Donald B. Smith, eds., *Aboriginal Ontario: Historical Perspectives on the Firsts Nations* (Toronto: Dundurn Press, 1994), p. 35.

9. Schmaltz, *The Ojibwa of Southern Ontario*, p. 16.
10. Contemporary attacks on Aboriginal harvesting, as well as the distrust that many environmentalists apparently hold for Native communities' abilities to maintain ecological relationships with the environment, have only been accelerated by the interest on the part of some historians in "debunking" notions of the viability of Aboriginal ecological relationships in the past. Calvin Martin, for example, has advanced theories that suggest Aboriginal peoples lost their respect for animals during the fur trade because of the breakdown of their spiritual framework, which was caused by illness contracted from Europeans.
11. Georges E. Sioui, "Why We Should have Inclusivity and Why We Cannot Have it," *Ayaangwaamizin: The International Journal of Indigenous Philosophy* 1, 2 (1997), p. 56.
12. The Doctrine of Discovery was the formal code of juridical standards in international law that had been created by papel edict to control the different interests of European powers in the different lands they were acquiring. For its primary tenets, see Ward Churchill, *Struggle for the Land: Indigenous Resistance to Genocide, Ecocide and Expropriation in Contemporary North America* (Toronto: Between the Lines, 1992), p. 36.
13. Georges E. Sioui, *Huron Wendat: The Heritage of the Circle* (Vancouver: University of British Columbia Press, 1999), pp. 84-5.
14. Bruce G. Trigger, "The Original Iroquoians: Huron, Petun, and Neutral," in Rogers and Smith, eds., *Aboriginal Ontario*, p. 51.
15. Sioui, *Huron Wendat*, p. 86.
16. Trigger, "The Original Iroquoians," p. 54.
17. Ibid., pp. 55-61.
18. Sioui, *Huron Wendat*, p. 153.
19. Georges E. Sioui, *For an American Autohistory* (Montreal: McGill-Queen's University Press, 1992), p. 22
20. Donald Marshall, Sr. (Grand Chief), Grand Captain Alexander Denny, and Putus Simon Marshall of the Executive of the Grand Council of the Mi'kmaw Nation, "The Govenant Chain," in Boyce Richardson, ed., *Drumbeat: Anger and Renewall in Indian Country* (Toronto: Summerhill Press and the Assembly of First Nations, 1989), p. 78.
21. Gail Boyd, "Struggle for Aboriginal Mobility Rights Continues," *The First Perspective* (March 6, 1998), p. 6; "Wabanaki Confederacy Conference at Odenak—1998," *Micmac Maliseet Nations News*, Aug. 3, 1998, p. 3.
22. Henderson, *The Mi'kmaw Concordat*, pp. 80-1.
23. Mi'kmaq people generally wish to be referred to in the terms of their own language, rather than through the generic term "Micmac," which had been applied to them. My limited understanding of the Mi'kmaq language suggests to me that individuals and family groups are referred to as "Mi'kmaw," while the nation and its language is referred to as "Mi'kmaq." My apologies to those who are better language speakers, for whom my use of terminology may not be accurate enough.
24. Henderson, *The Mi'kmaw Concordat*, p. 37.

25. The independence enjoyed by the Mi'kmaq under the concordat did not sit well with the Jesuits who came to Acadia to minister to both Acadian colonists and Mi'kmaqs. The Mi'kmaq rejected the Jesuits' authoritarian ways, after which the Jesuits attended only to the Catholics of New France. Mi'kmaki continued a relatively anomalous independence from French missionaries and colonists for most of the period of French ascendancy in North America and indeed, for the most part considered themselves, and were considered as, allies with the French Crown in its escalating war with the British in North America. Ibid., pp. 82-4, 85-93.
26. David E. Stannard, *American Holocaust: The Conquest of the New World* (Toronto: Oxford University Press, 1992) p. 238; Ward Churchill, *Indians Are Us? Culture and Genocide in Native North America* (Toronto: Between Lines, 1994), pp. 34-5; Jack D. Forbes, *Black Africans and Native Americans: Color, Race and Caste in the Evolution of Red-Black Peoples* (Oxford: Basil Blackwell, 1988), pp. 54-8; and Dickason, *Canada's First Nations*, p. 108.
27. Dickason, *Canada's First Nations*, p. 159; Daniel N. Paul, *We Were Not the Savages: A Mi'kmaq Perspective on the Collision between European and Native American Civilizations* (Halifax: Ferwood Books, 2000), pp. 181-2.
28. Paul, *We Were Not the Savages*, p. 2007.
29. Theresa Redmond, "We Cannot Work Without Food: Nova Scotia Indian Policy and Mi'kmaq Agriculture, 1783-1867," in David T. McNab, ed., *Earth, Water, Air and Fire: Studies in Canadian Ethno-history* (Waterloo: Wilfrid Laurier University Press, 1998), p. 116-17
30. Dickason, *Canada's First Nations*, p. 110. Britain violated a number of Doctrine of Discovery tenets by including the notion of the "Norman Yoke," which stipulated that land rights would rest upon the extent to which the owners of the land demonstrated a willingness and ability to "develop" their territories in accordance with a scriptural obligation to exercise dominion over nature. By this means, Britain claimed to be abiding by existing international law, while waging warfare for control of North American colonies with the French who had no interest in "developing" the land. See Churchill, *Struggle for the Land*, p. 37.
31. Marshall, Denny, and Marshall, "The Covenant Chain," p. 7.
32. Henderson, *The Mi'kmaw Concordat*, p. 104.
33. Stevenson, "Colonialism and First Nations Women," p. 53.
34. Many of the Indigenous nations affected by this warfare appeared to have fought strategically to ensure that a balance of power between competing Europeans was maintained. It is significant that as the French and British became locked in a death struggle, the Ojibwa appear to have signed a pact of non-aggression with the Iroquois. In general, as the extent of European interference in their affairs became crucial, many of the Great Lakes nations appear to have resisted fighting each other by the mid-eighteenth century. See Schmalz, *The Ojibwa of Southern Ontario*, p. 58.
35. John S. Milloy, "The Early Indians Acts: Developmental Strategy and Constitutional Change," in I.A. Getty and A.S. Lussier, eds., *As Long as the Sun Shines and the Water Flows: A Reader in Canadian Native History* (Vancouver: University of British Columbia Press, 1983), pp. 56-63.

36. Dickason, *Canada's First Nations*, pp. 182-4.
37. Schmalz, *The Ojibwa of Southern Ontario*, has suggested that during the Pontiac uprising, the Ojibway and other nations were too divided by their dependence on European trade goods and by the inroads that alcohol was making in the communities to successfully rout the British from the Great Lakes region, as they might have been capable of doing in earlier years. Although driving the British out of the region was undoubtedly the wish of some of the Ojibway communities, there were other communities situated far away from encroaching British settlement, but equally dependent on European technology, that were less certain of the threat the British ultimately posed.
38. Churchill, *Indians Are Us?* p. 35.
39. E. Wagner and E. Stearn, *The Effects of Smallpox on the Destiny of the Amerindian* (Boston: Bruce Humphries, 1945), pp. 44-5.
40. Schmalz, *The Ojibwa of Southern Ontario*, p. 87.
41. J.R. Miller, *Skyscrapers Hide the Heavens: A History of Indian-White Relations in Canada* (Toronto: University of Toronto Press, 1989), p. 92.
42. Schmalz, *The Ojibwa of Southern Ontario*, p. 13.
43. Excluding Native women from the process was central to its success. In eastern Canada, Native women's voices were in many cases considered extremely authoritative in matters of land use. Excluding them from the signing process made land theft that much easier, by allowing those who did not control the land to sign it over. See Kim Anderson, *A Recognition of Being: Reconstructing Native Womanhood* (Toronto: Sumach Press, 2001).
44. Schmalz, *The Ojibwa of Southern Ontario*, p. 123.
45. Ibid, pp. 124-5.
46. Peter Russell, *The Correspondence of the Honorable Peter Russell*, eds. G.A. Gruikshank and A.F. Hunter (Toronto, 1935); J.L Morris, *Indians of Ontario* (Toronto, 1943), as cited in Schmalz, *The Ojibwa of Southern Ontario*, p. 126.
47. Schmalz, *The Ojibwa of Southern Ontario*, p. 126.
48. Ibid,, pp. 136, 138-9.
49. Ibid., pp. 116, 141, 146.
50. The traditional territory of the Caldwell band is Point Pelee, which is now a national park. The Caldwell band were involved in the War of 1812 as allies to the British Crown, where they were known as the Caldwell Rangers. After the war in 1815, the British Crown acknowledged their efforts and their loyal service and awarded them their traditional territory "for even more." But it wasn't classified as a reserve, and meanwhile, British soldiers who retired after the war were awarded most of the land. By the 1860s the few remaining members of the Caldwell band that were still living on their traditional territories were beaten out of the new park by the RCMP with bull-whips. By the 1970s, the Caldwell band members dispersed throughout southern Ontario began to take part in ritual occupation of the park to protest their land claim. A settlement process is currently in effect. (Anonymous Caldwell band member, interview with author, 1999).
51. After a series of struggles towards resolving historic land claims, the Chippewas of Nawash, one of two remaining Saugeen Ojibway bands, were recognized in 1992 as having a historic

right to fish in their traditional waters. This decision led to three years of racist assaults by local whites and organized fishing interests, including the sinking of their fishing boats, the destruction of thousands of dollars of nets and other equipment, assaults on local Native people selling fish, the stabbing of two Native men in Owen Sound and the beating of two others. No charges were laid by the Owen Sound Police or the OPP for any of this violence until the band called for a federal inquiry into the attacks ("Nawash Calls for Fed Inquiry into Attacks," *Anishinabek News*, June 1996, p. 14).

Meanwhile, the Ontario Ministry of Natural Resources, in open defiance of the ruling recognizing the band's rights, declared a fishing free-for-all for two consecutive years, allowing anglers licence-free access to the waters around the Bruce Peninsula for specific weekends throughout the summer ("Fishing Free-for-all Condemned by Natives," *Anishinabek News*, July 1995, p. 1).

In 1996, despite considerable opposition the band took over the fishery using an *Indian Act* regulation that severed their community from the jurisdiction of the provincial government (Roberta Avery, "Chippewas Take Over Management of Fishery," *Windspeaker*, July 1996, p. 3). The other Saugeen Ojibway band on the peninsula, the Saugeen First Nation, announced the formation of the Saugeen Fishing Authority and claimed formal jurisdiction of the waters of their traditional territory. They demanded that sports fishermen and boaters would have to buy a licence from them to use their waters. The provincial government recognized the claims of neither bands, instead demanding they limit their catch and purchase licences from the provincial government in order to be able to fish at all (Roberta Avery, "Fishery in Jeopardy, Says University Researcher," *Windspeaker*, Aug. 1996, p. 16).

By 1997, a government study into fish stocks in Lake Huron revealed that certain fish stocks were severely impaired. While the report was supposed to be for the whole Lake Huron area, it in fact zeroed in on the Bruce Peninsula area a number of times, feeding the attitudes of non-Natives about Native mismanagement of the fishery (Rob McKinley, "Fight Over Fish Continues for Nawash," *Windspeaker*, Sept. 1997, p. 14). To add to the difficulties in 1997, Atomic Energy of Canada announced their desire to bury 20,000 tonnes of nuclear waste in the Canadian Shield. This brought to the band's attention the extent to which the fishery was already affected by nuclear contamination from the Bruce Nuclear Power Development on Lake Huron, 30 km south of the reserve (Roberta Avery, "No Nuclear Waste on Indian Land," *Windspeaker*, April 1997, p.4).

52. Dickason, *Canada's First Nations*, p. 253.
53. James Morrison, "Colonization, Resource Extraction and Hydroelectric Development in the Moose River Basin: A Preliminary History of the Implications for Aboriginal People." Report prepared for the Moose River/James Bay Coalition, for presentation to the Environmental Assessment Board Hearings, Ontario Hydro Demand/Supply Plan, November 1992, p. 4.
54. Bruce W. Hodgins and Jamie Benidickson, *The Temagami Experience: Recreation, Resources and Aboriginal Rights in the Northern Ontario Wilderness* (Toronto: University of Toronto Press, 1989), pp. 88-9.
55. Roy M. Longo, *Historical Highlights in Canadian Mining* (Toronto: Pitt Publishing Co., 1973), pp. 66-107.

56. John M. Guilbert and Charles F. Park, Jr., "Porcupine-Timmins Gold Deposits," in *The Geology of Ore Depostis* (New York: W.H. Freeman and Company, 1986), p. 863.
57. Howard Ferguson, then minister of Lands and Forests, had so consistently awarded timber and pulpwood concessions without advertisement, public tenders, or even formal agreements on price to individuals like Frank Anson who founded the powerful Abitibi Power and Paper Company, that he was found guilty in 1922 of violating the *Crown Timber Act*—one of the few whites to ever be prosecuted for disobeying federal legislation concerning Indigenous land. See Morrison, "Colonization, Resource Extraction and Hydroelectric Development."
58. Henderson, *The Mi'kmaw Concordat*, p. 17.
59. Basil Johnston, *The Manitous: The Spiritual World of the Ojibway* (Toronto: Key Porter, 1995).

Our Mothers Grand and Great: Black Women of Nova Scotia

SYLVIA HAMILTON

3

Very little of what one reads about Nova Scotia would reveal the existence of an Afro-Nova Scotian[1] population that dates back three centuries. Provincial advertising, displays, and brochures reflect people of European ancestry: the Scots, the Celts, the French, and the Irish, among others. There is occasional mention of Nova Scotia's first people, the Micmac. Yet Afro-Nova Scotians live in 43 communities throughout a province which is populated by over 72 different ethnic groups. Tourists and official visitors often express great surprise when they encounter people of African origin who can trace their heritage to the 1700s and 1800s. To understand the lives of Black women in Nova Scotia, one has first to learn something about their people and their environment.

The African presence here began in 1605 when a French colony was established at Port Royal (Annapolis Royal). A Black man, Mathieu da Costa, accompanied Pierre Du Gua, Sieur De Monts, and Samuel de Champlain to the new colony. Da Costa was one of Sieur De Monts's most useful men, as he knew the language of the Micmac and therefore served as interpreter for the French. The existence of Blacks in Nova Scotia remained singular and sporadic until the late 1700s, when 3,000 Black Loyalists arrived at the close of the American War of Independence. Though the Black Loyalists were free people, other Blacks who came at the same time with white Loyalists bore the euphemistic title "servant for life." Both groups joined the small population of Black slaves already present in the province. A second major influx of Blacks would occur following the War of 1812; approximately 2,000 former slaves, the Black Refugees, arrived in Nova Scotia during the postwar period between 1813 and 1816.

African people have a long tradition of oral history; stories about their heroes and heroines have therefore gone unrecorded. When a people begins the process of creating a written record of their champions, an initial tendency is to lionize and revere all. Since they will be paraded for all to see, faults and shortcomings are minimized and criticism is not often tolerated. The making of cultural heroes and heroines is an act of unification and empowerment. This process, just beginning among Afro-Nova Scotians, is integral to the survival of a people.

> On Saturday next, at twelve o'clock, will be sold on the Beach, two hogshead of rum, three of sugar, and two well-grown negro girls, aged fourteen and twelve, to the highest bidder.

From her first arrival in Nova Scotia, the Black woman has been immersed in a struggle for survival. She has had to battle slavery, servitude, sexual and racial discrimination, and

---

1 While the term Black is most commonly used to identify people of African origin, Afro-Nova Scotian has come into contemporary use to identify people of African descent who live in Nova Scotia.

ridicule. Her tenacious spirit has been her strongest and most constant ally; she is surviving with a strong dignity and an admirable lack of self-pity and bitterness. She is surviving, but not without struggle.

During Nova Scotia's period of slavery, Black female slaves were called upon to do more than simple domestic chores for their masters. Sylvia was a servant of Colonel Creighton of Lunenburg. On July 1, 1782, the town was invaded by soldiers from the strife-ridden American Colonies. Sylvia shuttled cartridges hidden in her apron from Crieghton's house to the fort where he was doing battle. When the house came under fire, Sylvia threw herself on top of the Colonel's son to protect him. During the battle she also found time to conceal her master's valuables in a bag which she lowered into the well for safekeeping. Typically, it was not Sylvia who was recognized for her efforts, but her master and a militia private to whom the provincial House of Assembly voted payments of money from the county's land taxes.

Another tidy arrangement involved slave-holding ministers. These men of the cloth adjusted their beliefs and principles accordingly when they purchased slaves. Lunenburg's Presbyterian minister John Seccombe kept a journal in which he noted that "Dinah, my negro woman-servant made a profession and confession publickly [sic] and was baptized, July 17, 1774." Dinah had a son, Solomon, who was brought to the province as a slave and who died in 1855 at age ninety; no record was found of the date of Dinah's death. In 1788 a mother and daughter were enslaved by Truro's Presbyterian pastor, Reverend Daniel Cock. When the mother became "unruly," he sold her but kept the daughter. In the same year, a Black woman named Mary Postill was sold in Shelburne; her price was one hundred bushels of potatoes.

Many slaves could hold no hope of being set free upon the death of their owners. Annapolis merchant Joseph Totten left his wife Suzannah the use of "slaves, horses, cattle, stock etc." and "to each of three daughters a negro girl slave . . . to her executors, administrators and assigns for ever." Amen. Others who were not given their freedom seized it for themselves. Determined owners placed newspaper ads offering rewards to their return.

While Black women slaves were being sold, left in wills, traded, and otherwise used, Black Loyalist women, ostensibly free, endeavoured to provide a livelihood for themselves and their families while at the same time labouring to establish communities. In 1787 Catherine Abernathy, a Black Loyalist teacher, instructed children in Preston, near Halifax. She taught a class of 20 children in a log schoolhouse built by the people of the community. Abernathy established a tradition of Black women teachers which would be strongly upheld by her sisters in years to come. Similarly, her contemporaries Violet King and Phillis George, the wives of ministers, carved another distinct path: Black women supporting their men and at the same time providing a stable base for their families. Even though history has documented the lives of Boston King and David George, it has remained silent on the experiences of Violet and Phillis.

What must it have been like for Phillis George in Shelburne in the late 1780s? Her husband travelled extensively, setting up Baptist churches in Nova Scotia and New Brunswick. He preached to and baptized Blacks and whites alike, not a popular undertaking at that time. The Georges had three children; money and food were scarce. On one occasion, a gang of 50 former soldiers armed with a ship's tackle surrounded their household, overturning it and what contents it had. Some weeks later, on a Sunday, a mob arrived at George's church; they whipped and beat him, driving the Baptist minister into the swamps of Shelburne. Under the cover of darkness, David George made his way back to town, collected Phillis and the children, and fled to neighbouring Birchtown.

What of Phillis George and other Black Loyalist women: unnamed women who were weavers, seamstresses, servants, bakers, and hat makers? We can in some measure recreate the society they lived in; we can even speculate on what they looked like. But except for isolated cases, their memories and experiences are their own and will remain with them fixed in time.

One of those rare, isolated instances is that of Rose Fortune. A descendant of the Black Loyalists, Rose lived in Annapolis Royal in the mid-1880s. She distinguished herself by establishing a baggage service for travellers arriving by boat at Annapolis from Saint John and Boston. A modest wheelbarrow and her strong arms were her two biggest assets. Rose's noteworthy activities were not only commercial. She concerned herself with the well-being of the young and old alike. Rose Fortune declared herself policewoman of the town and as such took upon herself the responsibility of making sure young children were safely off the streets at night. Her memory is kept alive by her descendants, the Lewis family of Annapolis Royal. Daurene Lewis is an accomplished weaver whose work is well known in Nova Scotia. She also holds the distinction of being the first Black woman elected to a town council in the province.

Black Loyalists had been promised land sufficient to start new lives in Nova Scotia. However, when the land grants were allocated, the Black Loyalists received much less than their white counterparts. Dissatisfaction with this inequity coupled with an unyielding desire to build a better future for their families provided the impetus for an exodus to Sierra Leone, West Africa, where the Black Loyalists hoped life would be different. In 1792 Phillis and David George, along with 1,200 Black Loyalists, sailed from Nova Scotia to Sierra Leone.

Four years later, 500 Jamaican Maroons were sent in exile to Nova Scotia. A proud people, the Maroons were descendants of runaway slaves who for over 150 years waged war against the British colonists in Jamaica. Upon their arrival, the men were put to work on the reconstruction of Citadel Hill. Of the Maroon women very little is recorded. We do know they were used for the entertainment of some of the province's esteemed leaders: Governor John Wentworth is believed to have taken a Maroon woman as his mistress, while Alexander Ochterloney, a commissioner placed in charge of the Maroons, "took five or six of the most attractive Maroon girls to his bed, keeping what the surveyor of Maroons, Theophilus Chamberlain, called a 'seraglio for his friends.'" The Maroon interlude ended in 1800 when they too set sail for Sierra Leone.

Between 1813 and 1816 the Black Refugees made their way to the province. It is this group whose memory is strongest in Nova Scotia, for their descendants may be found in communities such as Hammonds Plains, Preston, Beechville, Conway, Cobequid Road, and Three Mile Plains. Some of the earliest sketches and photographs of the Halifax city market show Black women selling baskets overflowing with mayflowers. Basketweaving for them was not an activity used to fill idle time: it was work aimed at bringing in money vital to the survival of the family. This tradition has endured because there are women who learned the craft from their mothers, who in turn learned it from their mothers. Edith Clayton of East Preston has been weaving maple market baskets since she was eight years old; it is a tradition which reaches back to touch six generations of her family. Not only does Edith Clayton continue to make and sell baskets, she also teaches classes in basketweaving throughout Nova Scotia as a means of preserving and passing on a significant and uniquely Afro-Nova Scotian aspect of the culture and heritage of the province.

Many and varied are the roles Black women have played and continue to play within their own and the broader community. It has often been said they are the backbone of the Black community: organizers, fund-raisers, nurturers, caregivers, mourners. When an attempt was made in 1836 by the provincial government to send the people of Preston to Trinidad, it was the women who objected:

> They all appear fearful of embarking on the water—many of them are old and have large families, and if a few of the men should be willing to go, the Women would not. It is objected among them that they have never heard any report of those who were sent away a few years ago to the same place, and think that if they were doing well some report of it would have reached them. They seem to have some attachment to the soil they have cultivated, poor and barren as it is. . . .

Nowhere has their involvement been more pronounced than in the social, educational, and religious life of the Black community. In 1917 the women of the African Baptist churches in the province gathered together to establish a Ladies' Auxiliary which would take responsibility for the "stimulation of the spiritual, moral, social, educational, charitable, and financial work of all the local churches of the African Baptist Association." These women gathered outside around a well in the community of East Preston since the church had no space for them to use; this gathering became known as the "Women at the Well." Some of these same women later organized an auxiliary to provide support for the Nova Scotia Home for Coloured Children. In 1920, for the first time in Canadian history, a convention of coloured women was held in Halifax.

A woman well-respected throughout Nova Scotia's Black communities is ninety-four-year-old Muriel V. States. She was present on that day the women gathered at the well to establish the Ladies' Auxiliary. She was present as well at another historic event: the 1956 creation of the Women's Institute of the Ladies' Auxiliary she had helped to organize 39 years before.

One hundred and five delegates were registered for a meeting whose theme was "Building Better Communities." Among the issues discussed were community health and educational standards and family relations. Muriel States, who was the Auxiliary's official organizer at the time, told her sisters their activities would not go unnoticed:

> Today, we women of the African Baptist Association have taken another step which will go down in history as the first Women's Institute held this day at this church. We feel that we as women have accomplished much and are

> aiming to do great things in the future. We are already reaping the reward of untiring and united effort in all that tends to the promotion of the church and community welfare.

Since 1956 the meeting of the Women's Institute has been an annual event. In October 1981, the Institute celebrated its twenty-fifth anniversary. Its history tells of the dedicated work of many women: Gertrude Smith, Margaret Upshaw, Pearleen Oliver, Selina David, Catherine Clarke, and many others. Today the Institute undoubtedly records the largest gatherings of Black women in the province; annual conventions draw several hundred Black women.

In 1937 the Nova Scotia teacher's college in Truro had a student population of over 100 students. My mother Marie remembers being one of the college's two Black students; her companion was Ada Symonds. Teaching was my mother's second choice for a career; nursing, her first choice, was not open to Blacks. The bar remained solidly across this door until 1945, when pressure from the Nova Scotia Association for the Advancement of Coloured People and from Reverend William and Pearleen Oliver forced its removal. Two Black women were admitted as trainees in nursing.

Teaching became the selected profession for many Black women. Some chose it because they wanted to teach, others because there were no other options open. These women are remembered in the many communities where they taught. They are especially remembered for their diligent work and commitment in the face of the hardship and adversity of a society which has tried unceasingly to deny their existence. They had to put up with one-room segregated schools, few resources, and little money. They stayed late to devote extra time to those students who had to stay home to help pick blueberries and mayflowers or to help garden. When the school day was over, another day began for them: seeing to their own children, cooking supper, ironing the children's clothes for school, preparing lessons, and attending a meeting at the church.

As they laboured at teaching, nursing, housekeeping, typing, and other jobs, Black women have not led easy lives. Nova Scotian Black women, like their counterparts elsewhere, have always known a double day. Some say the Black woman invented it. Work was and continues to be an integral part of her life. She has not had the luxury of deciding to stay home; with the current state of both our provincial and national economies, it is unlikely she will be afforded that choice in the near future.

Public attention in Canada has been increasingly riveted upon incidents of racially motivated attacks on individuals and groups in some of our major urban centres. The manner in which these cases have been described would leave one to believe such occurrences are relatively new phenomena in this country. Even the most cursory examination of the experiences of Afro-Nova Scotians will clearly demonstrate that, indeed, such is not the case.

In 1946 New Glasgow theatres were segregated; Blacks sat upstairs, whites occupied the downstairs seats. While in New Glasgow, Viola Desmond of Halifax decided to go to the theatre. She bought a ticket (balcony seat) but decided to sit downstairs. Though she was ordered to move, Viola refused, offering instead to pay the difference in price. The theatre manager declined the offer and called the police. Viola Desmond was carried away by the officer and held in jail overnight. The next day she was fined $20 and costs. She was charged with having avoided paying the one-cent entertainment tax. A year later, Selma Burke, a Black woman from the United States, was refused service in Halifax. It is not difficult to see that this environment had the power to dampen spirits, damage identities, and lessen the desire for change. But there were Black women who felt equal to the challenge.

A New Glasgow publishing venture which began as an eight-by-ten broadsheet in 1945 soon blossomed into a full-fledged newspaper. This was *The Clarion*, edited and published by Dr. Carrie Best. Published twice monthly, *The Clarion* called itself the voice of "coloured Nova Scotians." Dr. Best published timely articles on civil-rights issues in Nova Scotia and elsewhere; the paper featured a women's page and carried sports and social news. In 1949 *The Clairon* gave birth to *The Negro Citizen*, which achieved nationwide circulation. But Dr. Best was not moving down a totally untravelled path; one century before, in 1853, Mary Shadd Cary launched *The Provincial Freeman* from Windsor, Ontario. In so doing, she became the first Black woman in North America to found and edit a weekly newspaper. Dr. Best has been awarded the Order of Canada; in 1977 she published her autobiography, *That Lonesome Road.*

Other Black flowers were blossoming as well in the 1940s. When Portia White was 17 she was teaching school and taking singing lessons. Winning the Silver Cup at the Nova Scotia Music Festival paved the way for her to receive a scholarship from the Halifax Ladies Musical Club to study at the Halifax Conservatory of Music. By the time she was 31 Portia White had made her musical debut in Toronto. Four years later, in 1944, she made her debut at New York's famed Town Hall and later toured the United States, South America, and the Caribbean. Of the "young Canadian contralto's" debut, one New York critic wrote:

> as soon as she stepped on to the stage and began to sing it was obvious that here was a young musician of remarkable talents. Miss White has a fine, rich voice which she uses both expressively and intelligently. . . . The artist has an excellent stage presence . . . she was greeted with enthusiastic applause at each entrance. Miss White is a singer to watch, a singer with a bright future.

In 1969, Portia White's estate donated a gift of $1,000 to the Halifax City Regional Library to assist in setting up a music library in the city. The record collection which was subsequently installed is large and varied; few members of the borrowing public, however, know how the collection they so enjoy was originally established.

Recently Black women in Nova Scotia have begun to enter areas where their absence has heretofore been conspicuous: government, law, journalism, business, and medicine. This is not to say the struggle has ended or that we have arrived. While the attitude of the Black woman toward herself has been undergoing changes, the perceptions and attitudes of others both within her own community and beyond it require continual challenges to bring about any significant changes in how she is regarded and treated by others. As Black women begin paying tribute to themselves and their own work, others will pay tribute also. This year the family of Joyce Ross, a daycare director and longstanding community worker in East Preston, held a recognition dinner in her honour. Pearleen Oliver, author, historian, and educator, was one of three women selected to receive the YWCA Recognition of Women Award initiated in 1981. She was the first woman to serve as moderator of the African United Baptist Association and is the author of *Brief History of Colored Baptists of Nova Scotia 1782–1953* (1953) and *A Root and a Name* (1977). When the Recognition of Women Award was announced for 1982, Doris Evans, an educator and community worker, was among the three women honoured. And there are still many others who have experiences that need to be examined and stories that need to be told—women such as Ada Fells of Yarmouth, Edith Cromwell of Bridgetown, Clotilda Douglas of Sydney, Elsie Elms of Guysborough, Ruth Johnson of Cobequid Road. And there are others. . . .

Writer Mary Helen Washington, in the introduction to her book *Midnight Birds*, speaks of the process whereby Black women recover and rename their past. She talks about the monuments and statues erected by White men to celebrate their achievements, "to remake history, and to cast themselves eternally in heroic form." Yet there is no trace of women's lives. "We have," she says, "been erased from history." As research and exploration into the lives of Black women in Nova Scotia continues, a fuller view, one with dimension and perspective, will emerge. We will know then where to erect our monument. Now there are only signposts pointing the way.

Originally published in CWS/cf's Winter 1982 issue, "Multiculturalism" (Volume 4, Number 2).

Author's postscript: In the years since this article was first published, language and terminology have evolved as part of a process of self-definition and empowerment. People of African origin in Nova Scotia and elsewhere in Canada, often use African Nova Scotian, or African Canadian, in addition to Black. Aboriginal people in Nova Scotia have used Mi'Kmaq to name themselves. Readers interested in learning more about African descended women are encouraged to consult "Women and the Black Diaspora," *Canadian Woman Studies/les cahiers de la femme (CWS/cf)* 23 (2) (Winter 2004). This volume presents a rich selection of essays, poetry and book reviews by international scholars and writers. My article in this volume, "A Daughter's Journey," is a reflection on my work since the original publication of "Our Mothers Grand and Great," in *CWS/cf* in 1982.

PART 2

The Canadian State

The Canadian State holds a vaunted place in the discussion of colonialism and racism in Canada. As the most powerful institution in society and the most effective instrument in the imposition of Canadian settler rule and political order, it is highly implicated in the dynamics of colonization and the attendant racialized hierarchies that emerge from them. The chapters in this section interrogate some of the key institutional arrangements that sustain relations of colonization and hierarchies of racism and the discourses that legitimate the Eurocentric character of Canadian society, the State-society relations that reproduce it, and the unequal processes and outcomes experienced by racialized citizens.

In Chapter Four, "On the Dark Side of the Nation: Politics of Multiculturalism and the State of Canada," Himani Bannerji argues that the Canadian State uses processes of legitimation to generate consent for practices that perpetuate colonization and racialization. Bannerji identifies Official Multiculturalism as one of those high moral, neutral-sounding concepts used to mask and sustain relations of colonization and racialization, thereby maintaining an otherwise "mono-cultural" Eurocentric character of the nation.

From a critical legal theory perspective, in Chapter Five Patricia Monture-Angus in "Theoretical Foundations and the Challenge of Aboriginal Rights" suggests that "Law was and remains a central tool in delivering oppression and colonialism to First Nations." She seeks to understand the ways in which the Canadian legal system has been used to perpetuate relations of colonization and racial oppression that Indigenous peoples are subjected to.

In Chapter Six, according to Frances Henry and Carol Tator in "State Responses to Racism," the Canadian state has often played a positive role in confronting racism. As their chapter shows, however, these efforts conflict with - the traditional involvement of the state in practices that intensify racialization and raise questions about the effectiveness of existing institutions and instruments in addressing racism in Canada.[1]

Finally, in Chapter Seven, Sujit Choudhry in "Protecting Equality in the Face of Terror: Ethnic and Racial Profiling and s.15 of the Charter" identifies racial profiling as a more prevalent practice since September 11, 2001, following the terrorist attacks on the United States. In the Canadian context, however, racial profiling has a long history, including the internment of the Japanese in World War II, the persistent disproportionate stops of African Canadians in Toronto streets, and the profiling of Arabs and Muslims at border points. Choudry is interested in exploring whether, in the time of terror, threats to national security trump the Charter of Rights protections and make racial profiling an acceptable and even necessary tool of law enforcement.

1 F. Henry and C. Tator, *The Colour of Democracy: Racism in Canadian Society*, 3rd ed. (Toronto: Nelson, 2006).

CRITICAL THINKING QUESTIONS

1. Himani Bannerji argues that the Canadian State uses processes of legitimation to generate consent for practices that perpetuate colonization and racialization. She identifies Official Multiculturalism of one of those high moral, neutral-sounding concepts used to mask and sustain relations of colonization and racialization, thereby maintaining an otherwise "mono-cultural" Eurocentric character of the nation. Does Official Multiculturalism help or hinder the process of inclusion in Canadian society?

2. How would you experience the Canadian State and society from what Bannerji characterizes as the "Dark side of the nation"?

3. From a critical legal theory perspective, Patricia Monture-Angus suggests that "Law was and remains a central tool in delivering oppression and colonialism to First Nations." Do you agree? In what ways have Canadian legal systems been used to perpetuate relations of colonization and racial oppression?

4. What dual principles of law arising from *sec* 31 of the Canadian Constitution does Monture-Angus identify as embodying the possibility of overcoming the relationship of colonization between Canada and Aboriginal people by challenging the very foundation of the relations of oppression the constitutional order otherwise represents?

5. According to Frances Henry and Carol Tator, the Canadian State has often played a positive role in confronting racism. These efforts, however, are in conflict with its traditional involvement in practices that intensify racialization. How do you understand this complex and contradictory role of the Canadian state? How effective have the existing institutions and instruments been at addressing racism in Canada?

6. Sujit Choudhry identifies racial profiling as a more prevalent practice after September 11, 2001, following the terrorist attacks on the United States. In the Canadian context, however, racial profiling has a long history, including the internment of the Japanese in World War II, the persistent disproportionate stops of African Canadians in Toronto streets, and the profiling of Arabs and Muslims at border points. Does the threat to national security trump the Charter of Rights protections and make racial profiling an acceptable and even necessary tool of law enforcement? Does it help or hinder the administration of justice in Canada?

FURTHER RESOURCES

Bannerji, Himani. *The Dark Side of the Nation: Essays on Multiculturalism, Nationalism and Gender.* Toronto: Canadian Scholars' Press, 2000.

Henry, Francis, and C. Tator. *The Colour of Democracy: Racism in Canadian Society.* Toronto: Thomson/Nelson, 2006.

Tobani, Sunera. *Exalted Subjects: Studies in the Making of Race and Nation in Canada.* Vancouver: UBC Press, 2007.

Razack, Sherene. *Race, Space and the Law: Unmapping a White Settler Society.* Toronto: Between the Lines, 2002.

Cairns, Alan. *Citizen Plus: Aboriginal Peoples and the Canadian State.* Vancouver: UBC Press, 2000.

Monture-Angus, Patricia. *Journeying Forward: Dreaming First Nations' Independence.* Halifax: Fernwood Publishing, 1999.

Arsh, Michael, ed. *Aboriginal and Treaty Rights in Canada: Essays on Law, Equity and Respect for Difference.* Vancouver: UBC Press, 1997.

The Royal Commission of Aboriginal Peoples. *Report of the Royal Commission on Aboriginal Peoples.* Ottawa: Supply and Services Canada, 1996.

Li, Peter. *Destination Canada: Immigration Debates and Issues.* Toronto: Oxford University Press, 2003.

Stasiulis, Daiva and Radha Jhappan, "The Fracticious Politics of a Settler Society: Canada." In Daiva Stasiulis and Nira Yuval-Davis, eds., *Unsettling Settler Societies: Articulations of Gender, Race, Ethnicity and Class.* London: Sage Publications, 1995, 95–131.

Abele, Frances and Daiva Stasiulis, "Canada as a 'White Settler Colony': What About Natives and Immigrants?" In Wallace Clements and Glen Williams, eds., *The New Canadian Political Economy.* Montreal: McGill-Queens University Press, 1989, 240–277.

Galabuzi, Grace-Edward, *Canada's Economic Apartheid: The Social Exclusion of Racialized Groups in the New Century.* Toronto: CSPI, 2006.

FILMS

Acts of Defiance. 1992. Producer: Mark Zannis (Montreal: National Film Board).

Description: On the spot reportage details the 1990 standoff between the Mohawk people of Kahnawake and the municipality of Oka, and the Quebec and Canadian governments.

Kanehsatake: 270 Years of Resistance. 1993. Producer: Alanis Obomsawin. (Toronto: National Film Board).

Description: A record of the 1990 armed standoff in Oka Québec between the Québec and Canadian governments and the Kanehsatake Mohawks from the perspective of a native filmmaker who spent 75 days with the Mohawks.

Crash. 2004. Producer: Paul Haggis. Lions Gate.

Description: For two days in Los Angeles, a racially and economically diverse group of people collide with one another in unexpected ways. Examines fear and bigotry from multiple perspectives as characters careen in and out of each other's lives.

Domino. 1994. Producer: Silva Basmajian and Shanti Thakur. (Montreal: National Film Board).

Description: Portrays six interracial people's quests to forge their own identity. Explores the similarities and differences of interracial people. Each recounts how their identity was affected by the experience of their parents' history, family politics, the hierarchies of race, gender roles, and class. Their views demonstrate how living intimately with two cultures can be a source of strength and enrichment.

KEY CONCEPTS/TERMS

Colonization

Racialization

Indigeneity

Racial Profiling

"Mono-cultural," Eurocentric nation

Racialized Hierarchies

On the Dark Side of the Nation: Politics of Multiculturalism and the State of "Canada"

4

HIMANI BANNERJI

This paper is primarily concerned with the construction of "Canada" as a social and cultural form of national identity, and various challenges and interruptions offered to this identity by literature produced by writers from non-white communities. The first part of the paper examines both literary and political-theoretical formulations of a "two-nation," "two solitudes" thesis and their implications for various cultural accommodations offered to "others," especially through the mechanism of "multiculturalism." The second part concentrates on the experiences and standpoint of people of colour, or non-white people, especially since the 1960s, and the cultural and political formulating derivable from them.

> I am from the country
> Columbus dreamt of.
> You, the country
> Columbus conquered.
> Now in your land
> My words are circling
> blue Oka sky
> they come back to us
> alight on tongue.
> Protect me with your brazen passion
> for history is my truth,
> Earth, my witness
> my home,
> this native land.
>
> "OKA NADA": *A New Remembrance*
> Kaushalya Bannerji (1993, p. 20)

THE PERSONAL AND THE POLITICAL: A CHORUS AND A PROBLEMATIC

When the women's movement came along and we were coming to our political consciousness, one of its slogans took us by surprise and thrilled and activated us: "the personal is political!" Since then years have gone by, and in the meanwhile I have found myself in Canada, swearing an oath of allegiance to the Queen of England, giving up the passport of a long-fought-for independence, and being assigned into the category of "visible minority." These years have produced their own consciousness in me, and I have learnt that also the reverse is true: the political is personal.

The way this consciousness was engendered was not ideological, but daily, practical and personal. It came from having to live within an all-pervasive presence of the state in our everyday life. It began with the Canadian High Commission's rejection of my two-year-old daughter's visa and continued with my airport appearance in Montreal, where I was interrogated at length. What shook me was not the fact that they interviewed me, but rather their tone of suspicion about my somehow having stolen my way "in."

As the years progressed, I realized that in my life, and in the lives of other non-white people around me, this pervasive presence of the state meant everything—allowing my daughter and husband to come into the country; permitting me to continue my studies or to work, to cross the border into the U.S.A. and back; allowing me the custody of my daughter, although I had a low income; "landing" me so I could put some sort of life together with some predictability. Fear, anxiety, humiliation, anger and frustration became the wiremesh that knit bits of my life into a pattern. The quality of this life may be symbolized by an incident with which my final immigration interview culminated after many queries about a missing "wife" and the "head of the family." I was facing an elderly, bald, white man, moustached and blue-eyed—who said he had been to India. I made some polite rejoinder and he asked me—"Do you speak Hindi?" I replied that I understood it very well and spoke it with mistakes. "Can you translate this sentence for me?" he asked, and proceeded to say in Hindi what in English amounts to "Do you want to fuck with me?" A wave of heat rose from my toes to my hair roots. I gripped the edge of my chair and stared at him—silently. His hand was on my passport, the pink slip of my "landing" document lay next to it. Steadying my voice I said "I don't know Hindi that well." "So you're a PhD student?" My interview continued. I sat rigid and concluded it with a schizophrenic intensity. On Bloor Street in Toronto, sitting on the steps of a church—I vomited. I was a landed immigrant.

Throughout these twenty-five years I have met many non-white and Third World legal and illegal "immigrants" and "new Canadians" who feel that the machinery of the state has us impaled against its spikes. In beds, in workplaces, in suicides committed over deportations, the state silently, steadily rules our lives with "regulations." How much more intimate could we be—this state and we? It has almost become a person—this machinery—growing with and into our lives, fattened with our miseries and needs, and the curbing of our resistance and anger.

But simultaneously with the growth of the state we grew too, both in numbers and protest, and became a substantial voting population in Canada. We demanded some genuine reforms, some changes—some among us even demanded the end of racist capitalism—and instead we got "multiculturalism." "Communities" and their leaders or representatives were created by and through the state, and they called for funding and promised "essential services" for their "communities," such as the preservation of their identities. There were advisory bodies, positions, and even arts funding created on the basis of ethnicity and community. A problem of naming arose, and hyphenated cultural and political identities proliferated. Officially constructed identities came into being and we had new names—immigrant, visible minority, new Canadian and ethnic. In the mansion of the state small back rooms were accorded to these new political players on the scene. Manoeuvring for more began. As the state came deeper into our lives—extending its political, economic and moral regulation, its police violence and surveillance—we simultaneously officialized ourselves. It is as though we asked for bread and were given stones, and could not tell the difference between the two.

IN OR OF THE NATION? THE PROBLEM OF BELONGING

Face it there's an illegal
Immigrant
Hiding in your house
Hiding in you
Trying to get out!

* * * * *

Businessmen Custom's officials
Dark Glasses Industrial Aviation
Policemen Illegal Bachelorettes
Sweatshop-Keepers Information Canada
Says
"You can't get their smell off the walls."

Domestic Bliss
Krisantha Sri Bhaggiyadatta (1981, p. 23)

The state and the "visible minorities," (the non-white people living in Canada) have a complex relationship with each other. There is a fundamental unease with how our difference is construed and constructed by the state, how our otherness in relation to Canada is projected and objectified. We cannot be successfully ingested, or assimilated, or made to vanish from where we are not wanted. We remain an ambiguous presence, our existence a question mark in the side of the nation, with the potential to disclose much about the political unconscious and consciousness of Canada as an "imagined community" (Anderson, 1991). Disclosures accumulate slowly, while we continue to live here as outside-insiders of the nation which offers a proudly multicultural profile to the international community. We have the awareness that we have arrived into somebody's state, but what kind of state; whose imagined community or community of imagination does it embody? And what are the terms and conditions of our "belonging" to this state of a nation? Answers to these questions are often indirect and not found in the news highway of Canadian media. But travelling through the side-roads of political discursivities and practices we come across markers for social terrains and political establishments that allow us to map the political geography of this nation-land where we have "landed."

We locate our explorations of Canada mainly in that part where compulsorily English-speaking visible minorities reside, a part renamed by Charles Taylor and others as "Canada outside of Quebec" (COQ).[1] But we will call it "English Canada" as in common parlance.

This reflects the binary cultural identity of the country to whose discourse, through the notions of the two solitudes, survival and bilingualism, "new comers" are subjected.[2] Conceptualizing Canada within this discourse is a bleak and grim task: since "solitude" and "survival" (with their Hobbesian and Darwinist aura) are hardly the language of communitarian joy in nation making.

What, I asked when I first heard of these solitudes, are they? And why survival, when Canada's self-advertisement is one of a wealthy industrial nation? Upon my immigrant inquiries these two solitudes turned out to be two invading European nations—the French and the English—which might have produced two colonial-nation states in this part of North America. But history did not quite work out that way. Instead of producing two settler colonial countries like Zimbabwe (Rhodesia) and South Africa, they held a relationship of conquest and domination with each other. After the battle at the Plains of Abraham one conquered nation/nationality, the French, continued in an uneasy and subjected relation to a state of "Canada," which they saw as "English," a perception ratified by this state's rootedness in the English Crown. The colonial French then came to a hyphenated identity of "franco-something," or declared themselves (at least within one province) as plain "Québécois." They have been existing ever since in an unhappy state, their promised status as a "distinct society" notwithstanding. Periodically, and at times critically, Quebec challenges "Canadian" politics of "unity" and gives this politics its own "distinct" character. These then are the two solitudes, the protagonists who, to a great extent, shape the ideological parameters of Canadian constitutional debates, and whose "survival" and relations are continually deliberated. And this preoccupation is such a "natural" of Canadian politics that all other inhabitants are only a minor part of the problematic of "national" identity. This is particularly evident in the role, or lack thereof, accorded to the First Nations of Canada in the nation-forming project. Even after Elijah Harper's intervention in the Meech Lake Accord, the deployment of the Canadian Army against the Mohawk peoples and the long stand-off that followed, constant land claims and demands for self government/self-determination, there is a remarkable and a determined political marginalization of the First Nations. And yet their presence as the absent signifiers within Canadian national politics works at all times as a bedrock of its national definitional project, giving it a very particular contour through the same absences, silences, exclusions and marginalizations. In this there is no distinction between "COQ" or English Canada and Quebec. One needs only to look at the siege at Oka to realize that as far as these "others" are concerned, Europeans continue the same solidarity of ruling and repression, blended with competitive manipulations, that they practiced from the dawn of their conquests and state formations.

The Anglo-French rivalry therefore needs to be read through the lens of colonialism. If we want to understand the relationship between visible minorities and the state of Canada/English Canada/COQ, colonialism is the context or entry point that allows us to begin exploring the social relations and cultural forms which characterize these relations. The construction of visible minorities as a social imaginary and the architecture of the "nation" built with a "multicultural mosaic" can only be read together with the engravings of conquests, wars and exclusions. It is the nationhood of this Canada, with its two solitudes and their survival anxieties and aggressions against "native others," that provides the epic painting in whose dark corners we must look for the later "others." We have to get past and through these dual monoculturalist assumptions or paradigms in order to speak about "visible minorities," a category produced by the multiculturalist policy of the state. This paper repeats, in its conceptual and deconstructive movements, the motions of the people themselves who, "appellated" as refugees, immigrants or visible minorities, have to file past immigration officers, refugee boards, sundry ministries and posters of multi-featured/coloured faces that blandly proclaim "Together we are Ontario"—lest we or they forget!

We will examine the assumptions of "Canada" from the conventional problematic and thematic of Canadian nationhood, that of "Fragmentation or Integration?" currently resounding in post-referendum times. I look for my place within this conceptual topography and find myself in a designated space for "visible minorities in the multicultural society and state of Canada." This is existence in a zone somewhere between economy and culture. It strikes me then that this discursive mode in which Canada is topicalized does not anywhere feature the concept of class. Class does not function as a potential source for the theorization of Canada, any more than does race as an expression for basic social relations of contradiction. Instead the discursivities rely on hegemonic cultural categories such as English or French Canada, or on notions such as national institutions, and conceive of differences and transcendences, fragmentation and integration, with regard to an ideological notion of unity that is perpetually in crisis. This influential problematic is displayed in a *Globe and Mail* editorial of 29 March 1994. It is typically pre-occupied with themes of unity and integration or fragmentation, and delivers a lecture on these to Lucien Bouchard of the Bloc Québécois.

> It has been an educational field trip for Lucien Bouchard. On his first venture into "English Canada" (as he insists on calling it) since becoming leader of Her Majesty's Loyal Opposition, Mr. Bouchard learned, among other things, there is such a thing as Canadian Nationalism: not just patriotism, nor yet that self-serving little prejudice that parades around as Canadian Nationalism—mix equal parts elitism, statism and Anti-Americanism—but a genuine fellow-feeling that binds Canadians to one another across this country—and includes Quebec.

Lest this statement appear to the people of Quebec as passing off "English Canada" disguised as "the nation" and locking Quebec in a vice grip of "unity" without consent or consultation, the editor repeats multiculturalist platitudes meant to mitigate the old antagonisms leading to "separatism." The demand for a French Canada is equated with "self-serving little prejudice" and "patriotism," and promptly absorbed into the notion of a culturally and socially transcendent Canada, which is supposedly not only non-French, but non-English as well. How can this non-partisan, transcendent Canada be articulated except in the discourse of multiculturalism? Multiculturalism, then, can save the day for English Canada, conferring upon it a transcendence, even though the same transcendent state is signalled through the figure of Her Majesty the Queen of England and the English language. The unassimilable "others" who, in their distance from English Canada, need to be boxed into this catch-all phrase now become the moral cudgel with which to beat Quebec's separatist aspirations. The same editorial continues:

> Canada is dedicated to the ideal that people of different languages and cultures may, without surrendering their identity, yet embrace the human values they have in common: the "two solitudes" of which the poet wrote, "that protect and touch and greet each other," were a definition of love, not division.

But this poetic interpretation of solitudes, like the moral carrot of multicultural love, is quickly followed by a stick. Should Quebec not recognize this obligation to love, but rather see it as a barrier to self-determination, Canada will not tolerate this. We are then confronted with other competing self-determinations in one breath, some of which ordinarily would not find their advocate in *Globe and Mail* editorials. What of the self-determination of the Cree, of the Anglophones, of federalists of every stripe? What of the self-determination of the Canadian nation? Should Mr. Bouchard and his kind not recognize this national interest, it is argued, then the province's uncertainties are only beginning. In the context of the editorial's discourse, these uncertainties amount to the threat of a federalist anglophone war. The "self-determination of the Cree" is no more than an opportunistic legitimation of Canada in the name of all others who are routinely left out of its construction and governance. These "different (from the French) others," through the device of a state-sponsored multiculturalism, create the basis for transcendence necessary for the creation of a universalist liberal democratic statehood. They are interpellated or bound into the ideological state apparatus through their employment of tongues which must be compulsorily, officially unilingual—namely, under the sign of English.[3]

"Canada," with its primary inscriptions of "French" or "English," its colonialist and essentialist identity markers, cannot escape a fragmentary framework. Its imagined political geography simplifies into two primary and confrontational possessions, cultural typologies and dominant ideologies. Under the circumstances, all appeal to multiculturalism on the part of "Canada Outside Quebec" becomes no more than an extra weight on the "English" side. Its "difference-studded unity," its "multicultural mosaic," becomes an ideological sleight of hand pitted against Quebec's presumably greater cultural homogeneity. The two solitudes glare at each other from the barricades in an ongoing colonial war. But what do either of these solitudes and their reigning essences have to do with those whom the state has named "visible minorities" and who are meant to provide the ideological basis for the Canadian state's liberal/universal status? How does their very "difference," inscribed with inferiority and negativity— their otherwise troublesome particularity—offer the very particularist state of "English Canada" the legitimating device of transcendence through multiculturalism? Are we not still being used in the war between the English and the French?

It may seem strange to "Canadians" that the presence of the First Nations, the "visible minorities" and the ideology of multiculturalism are being suggested as the core of the state's claim to universality or transcendence. Not only in multiplying pawns in the old Anglo-French rivalry but in other ways as well, multiculturalism may be seen less as a gift of the state of "Canada" to the "others" of this society, than as a central pillar in its own ideological state apparatus.[4] This is because the very discourse of nationhood in the context of "Canada," given its evolution as a capitalist state derived from a white settler colony with aspirations to liberal democracy,[5] needs an ideology that can mediate fissures and ruptures more deep and profound than those of the usual capitalist nation state.[6] That is why usually undesirable others, consisting of non-white peoples with their ethnic or traditional or underdeveloped cultures, are discursively inserted in the middle of a dialogue on hegemonic rivalry. The discourse of multiculturalism, as distinct from its administrative, practical relations and forms of ruling, serves as a culmination for the ideological construction of "Canada." This places us, on whose actual lives the ideology is evoked, in a peculiar situation. On the one hand, by our sheer presence we provide a central part of the distinct pluralist unity of Canadian nationhood; on the other hand, this centrality is dependent on our "difference," which denotes the power of definition that "Canadians" have over "others." In the ideology of multicultural nationhood, however, this difference is read in a power-neutral manner rather than as organized through class, gender and race. Thus at the same moment that difference is ideologically evoked it is also neutralized, as though the issue of difference were the same as that of diversity of cultures and identities, rather than that of racism and colonial ethnocentrism—as though our different cultures were on a par or could negotiate with the two dominant ones! The hollowness of such a pluralist stance is exposed in the shrill indignation of Anglophones when rendered a "minority" in Quebec, or the angry desperation of francophones in Ontario. The issue of the First Nations—their land claims, languages and cultures—provides another dimension entirely, so violent and deep that the state of Canada dare not even name it in the placid language of multiculturalism.

The importance of the discourse of multiculturalism to that of nation-making becomes clearer if we remember that "nation" needs an ideology of unification and legitimation.[7] As Benedict Anderson points out, nations need to imagine a principle of "com-unity," or community even where there is little there to postulate any.[8] A nation, ideologically, cannot posit itself on the principle of hate, according to Anderson, and must therefore speak to the sacrificing of individual, particularist interests for the sake of "the common good" (1991, ch. 2). This task of "imagining community" becomes especially difficult in Canada—not only because of class, gender and capital, which ubiquitously provide contentious grounds in the most culturally homogeneous of societies—but because its socio-political space is saturated by elements of surplus domination due to its Eurocentric/racist/colonial context. Ours is not a situation of co-existence of cultural nationalities or tribes within a given geographical space. Speaking here of culture without addressing power relations displaces and trivializes deep contradictions. It is a reductionism that hides the social relations of domination that continually create "difference" as inferior and thus signifies continuing relations or antagonism. The legacy of a white settler colonial economy and state and the current aspirations to imperialist capitalism mark Canada's struggle to become a liberal democratic state. Here a cultural pluralist interpretive discourse hides more than it reveals. It serves as a fantastic evocation of "unity," which in any case becomes a reminder of the divisions. Thus to imagine "com-unity" means to imagine a common project of valuing difference that would hold good for both Canadians and others, while also claiming that the sources of these otherizing differences are merely cultural. As that is impossible, we consequently have a situation where no escape is possible from divisive social relations. The nation state's need for an ideology that can avert a complete rupture becomes desperate, and gives rise to a multicultural ideology which both needs and creates "others" while subverting demands for anti-racism and political equality.

Let me illustrate my argument by means of Charles Taylor's thoughts on the Canadian project of nation-making. Taylor is comparable to Benedict Anderson insofar as he sees "nation" primarily as an expression of civil society, as a collective form of self-discrimination and definition. He therefore sees that culture, community, tradition and imagination are crucial for this process. His somewhat romantic organicist approach is pitted against neo-liberal projects of market ideologies misnamed as "reform."[9] Taylor draws his inspiration, among many sources, from an earlier European romantic tradition that cherishes cultural specificities, local traditions and imaginations.[10] This presents Taylor with the difficult task of "reconciling solitudes" with some form of a state while retaining traditional cultural identities in an overall ideological circle of "Canadian" nationhood. This is a difficult task at all times, but especially in the Canadian context of Anglo-French rivalry and the threat of separatism. Thus Taylor, in spite of his philosophical refinement, is like others also forced into the recourse of "multiculturalism as a discourse," characterized by its reliance on diversity. The constitution then becomes a federal Mosaic tablet for encoding and enshrining this very moral/political mandate. But Taylor is caught in a further bind, because Canada is more than a dual monocultural entity. Underneath the "two solitudes," as he knows well, Canada has "different differences," a whole range of cultural identities which cannot (and he feels should not) be given equal status with the "constituent elements" of "the nation," namely, the English and the French. At this point Taylor has to juggle with the contending claims of these dominant or "constituent" communities and their traditions, with the formal equality of citizenship in liberal democracy, and with other "others" with their contentious political claims and "different cultures." This juggling, of course, happens best in a multicultural language, qualifying the claim of the socio-economic equality of "others" with the language of culture and tolerance, converting difference into diversity in order to mitigate the power relations underlying it. Thus Taylor, in spite of his organicist, communitarian-moral view of the nation and the state, depends on a modified liberal pluralist discourse which he otherwise finds "American," abstract, empty and unpalatable.[11]

Reconciling the Solitudes and *Multiculturalism and the Politics of Recognition* are important texts for understanding the need for the construction of the category of visible minorities to manage contentions in the nationhood of Canada. Even though Taylor spends little time actually discussing either the visible minorities or the First Nations, their importance for the creation of a national ideology is brought out by his discussion of Anglo-French contestation. Their visceral anxieties about loss of culture are offset by "other" cultural presences that are minoritized with respect to both, while the commonality of Anglo-French culture emerges in contrast. Taylor discovers that the cultural essences of COQ have something in common with Quebec—their Europeanness—in spite of the surface of diversity. This surface diversity, he feels, is not insurmountable within the European-Anglo framework, whose members' political imagination holds enough ground for some sort of commonality.

> What is enshrined here is what one might call *first level diversity*. There are great differences in culture and outlook and background in a population that nevertheless shares the same idea of what it is to belong to Canada. Their patriotism and manner of belonging is uniform, whatever their differences, and this is felt to be necessary if the country is to hold together. (1993, p. 182)

Taylor must be speaking of those who are "Canadians" and not "others": the difference of visible minorities and First Nations peoples is obviously not containable in this "first level diversity" category. As far as these "others" are concerned the Anglo-European (COQ) and French elements have much in common in both "othering" and partially "tolerating" them. Time and time again, especially around the so-called Oka crisis, it became clear that liberal pluralism rapidly yields to a fascist "sons of the soil" approach as expressed by both the Quebec state and its populace, oblivious to the irony of such a claim. It is inconsistent of Taylor to use this notion of "first level diversity" while also emphasizing the irreducible cultural ontology of Quebec as signaled by the concept of a "deep diversity" (p. 183). But more importantly, this inconsistency accords an ownership of nationhood to the Anglo-French elements. He wrestles, therefore, to accommodate an Anglo-French nationality, while the "deep diversities" of "others," though nominally cited, are erased from the political map just as easily as the similarity of the "two nations" *vis-à-vis* those "others." Of course, these manipulations are essential for Taylor and others if the European (colonial) character of "Canada" is to be held *status quo*. This is a Trudeau-like stance of dual unification in which non-European "others" are made to lend support to the enterprise by their existence as a tolerated managed difference.

This multicultural take on liberal democracy, called the "politics of recognition" by Taylor, is informed by this awareness that an across-the-board use of the notion of equality would reduce the French element from the status of "nation" to that of just another minority. This of course must not be allowed to happen, since the French are, by virtue of being European co-conquerors, one of the "founding nations." At this point Taylor adopts the further qualified notion of visible minorities as integral to his two-in-one nation-state schema. For him as for other majority ideologues they constitute a minority of minorities. They are, in the scheme of things, peripheral to the essence of Canada, which is captured by "Trudeau's remarkable achievement in extending bilingualism" to reflect the "Canadian" character of "duality" (p. 164). This duality Taylor considers as currently under a threat of irrelevancy, not from anglo monoculturalism, but from the ever-growing presence of "other" cultures. "Already one hears Westerners saying . . . that their experience of Canada is of a multicultural mosaic" (p. 182). This challenge of the presence of "others" is, for Taylor, the main problem for French Canadians in retaining their equality with English Canadians. But it is also a problem for Taylor himself, who sees in this an unsettling possibility for the paradigm of "two solitudes" or "two nations" to which he ultimately concedes. In order to project and protect the irreducible claims of the two dominant and similar cultures, he refers fleetingly and analogically, though frequently, to aboriginal communities: "visible minorities" also enter his discourse, but both are terms serving to install a "national" conversation between French and English, embroidering the dialogue of the main speakers. His placement of these "other" social groups is evident when he says: "Something analogous [to the French situation] holds for aboriginal communities in this country; their way of being Canadian is not accommodated by first level diversity" (p. 182). Anyone outside of the national framework adopted by Taylor would feel puzzled by the analogical status of the First Nations brought in to negotiate power sharing between the two European nations. Taylor's approach is in keeping with texts on nationalism, culture and identity that relegate the issues of colonialism, racism and continued oppression of the Aboriginal peoples and the oppression visited upon "visible minorities" to the status of footnotes in Canadian politics.

Yet multiculturalism as an ideological device both enhances and erodes Taylor's project. Multiculturalism, he recognizes at one level, is plain realism—an effect of the realization that many (perhaps too many) "others" have been allowed in, stretching the skin of tolerance and "first level diversity" tightly across the body of the nation. Their "deep diversity" cannot be accommodated simply within the Anglo-French duality. The situation is so murky that, "more fundamentally" we face a challenge to our very conception of diversity" (p. 182). "Difference," he feels, has to be more "fundamentally" read into the "nation":

> In a way, accommodating difference is what Canada is all about. Many Canadians would concur in this. (p. 181)
>
> Many of the people who rallied around the Charter and multiculturalism to reject the distinct society are proud of their acceptance of diversity—and in some respects rightly so. (p. 182)

But this necessary situational multiculturalism acknowledged by Taylor not only creates the transcendence of a nation built on difference, it also introduces the claims of "deep diversities" on all sides. Unable to formulate a way out of this impasse Taylor proposes an ideological utopia of "difference" (devoid of the issue of power) embodied in a constitutional state, a kind of cultural federalism:

> To build a country for everyone, Canada would have to allow for second-level or "deep" diversity in which a plurality of ways of belonging would also be acknowledged and accepted. Someone of, say, Italian extraction in Toronto or Ukrainian extraction in Edmonton might indeed feel Canadian as a bearer of individual rights in a multicultural mosaic. His or her belonging would not "pass through" some other community, although the ethnic identity might be important to him or her in various ways. But this person might nevertheless accept that a Québécois or a Cree or a Dene might belong in a very different way, that these persons were Canadian through being members of their national communities. Reciprocally, the Québécois, Cree, or Dene would accept the perfect legitimacy of the "mosaic" identity. (p. 183)

This utopian state formation of Taylor founders, as do those of others, on the rocky shores of the reality of how different "differences" are produced, or are not just forms of diversity. For all of Taylor's pleas for recognizing two kinds of diversity, he does not ever probe into the social relations of power that create the different differences. It is perhaps significant from this point of view that he speaks of the "deep diversities" of Italians or Ukrainians but does not mention those of the blacks, South Asians or the Chinese. In other words, he cannot raise the spectre of real politics, of real social, cultural and economic relations of white supremacy and racism. Thus he leaves out of sight the relations and ideologies of ruling that are intrinsic to the creation of a racist civil society and a racializing colonial-liberal state. It is this foundational evasion that makes Taylor's proposal so problematic for those whose "differences" in the Canadian context are not culturally intrinsic but constructed through "race," class, gender and other relations of power. This is what makes us sceptical about Taylor's retooling of multicultural liberal democracy by introducing the concept of "deep diversity" as a differentiated citizenship into the bone marrow of the polity, while leaving the Anglo-French European "national" (colonial and racist) core intact. He disagrees with those for whom

> ...[the] model of citizenship has to be uniform, or [they think] people would have no sense of belonging to the same polity. Those who say so tend to take the United States as their paradigm, which has indeed been hostile to deep diversity and has sometimes tried to stamp it out as "un-American." (p. 183)

This, for Taylor, amounts to the creation of a truly Canadian polity that needs a "united federal Canada" and is able to deliver "law and order, collective provision, regional equality and mutual self help . . ." (p. 183). None of these categories—for example, that of "law and order"—is characteristically problematized by Taylor. His model "Canada" is not to be built on the idea of a melting pot or of a uniform citizenship based on a rationalist and functional view of polity. That would, according to him, "straight-jacket" deep diversity. Instead,

> The world needs other models to be legitimated in order to allow for more humane and less constraining modes of political cohabitation. Instead of pushing ourselves to the point of break up in the name of a uniform model, we would do our own and some other peoples a favour by exploring the space of deep diversity. (p. 184)

What would this differentiated citizenship look like in concrete example, we ask? Taylor throws in a few lines about Basques, Catalans and Bretons. But those few lines are not answer enough for us. Though this seems to be an open invitation to join the project of state and nation-making, the realities of a colonial capitalist history—indentures, reserves, First Nations without a state, immigrants and citizens, illegals, refugees and "Canadians"—make it impossible. They throw us against the inscription of power-based "differences" that construct the self-definition of the Canadian state and its citizenship. We realize that class, "race," gender, sexual orientation, colonialism and capital cannot be made to vanish by the magic of Taylor's multiculturalism, managed and graduated around a core of dualism. His inability to address current and historical organizations of power, his inability to see that this sort of abstract and empty invitation to "difference" has always enhanced the existing "difference" unless real social equality and historical redress can be possible—these erasures make his proposal a touch frightening for us. This is why I shudder to "take the deep road of diversity together" with Charles Taylor (p. 184). Concentration and labour camps, Japanese internment, the Indian Act and reserves, apartheid and ethnic "homelands" extend their long shadows over the project of my triumphal march into the federal utopia of a multiculturally differentiated citizenship. But what becomes clear from Taylor's writings is the importance of a discourse of difference and multiculturalism for the creation of a legitimate nation space for Canada. Multiculturalism becomes a mandate of moral regulation as an antidote to any, and especially Quebec's, separatism.

ON THE DARK SIDE OF THE NATION: CONSIDERING "ENGLISH CANADA"

If one stands on the dark side of the nation in Canada everything looks different. The transcendent, universal and unifying claims of its multiculturally legitimated ideological state apparatus become susceptible to questions. The particularized and partisan nature of this nation-state becomes visible through the same ideological and working apparatus that simultaneously produces its national "Canadian" essence and the "other"—its non-white population (minus the First Nations) as "visible minorities." It is obvious that both Canada and its adjectivized correlates English or French Canada are themselves certain forms of constructions. What do these constructions represent or encode? With regard to whom or what are we otherized and categorized as visible minorities? What lies on the dark side of this state project, its national ethos?

Official multiculturalism, mainstream political thought and the news media in Canada all rely comfortably on the notion of a nation and its state both called Canada, with legitimate subjects called Canadians, in order to construct us as categorical forms of difference. There is an assumption that this Canada is a singular entity, a moral, cultural and political essence, neutral of power, both in terms of antecedents and consequences. The assumption is that we can recognize this beast, if and when we see it. So we can then speak of a "Pan-Canadian nationalism," of a Canada which will not tolerate more Third World immigrants or separatism, or of what Canada needs or allows us to do. And yet, when we scrutinize this Canada, what is it that we see? The answer to this question depends on which side of the nation we inhabit. For those who see it as a homogeneous cultural/political entity, resting on a legitimately possessed territory, with an exclusive right to legislation over diverse groups of peoples, Canada is unproblematic. For others, who are on the receiving end of the power of Canada and its multiculturalism, who have been dispossessed in one sense or another, the answer is quite different. For them the issues of legitimacy or territorial possession, or the right to create regulations and the very axis of domination on which its status as a nation-state rests, are all too central to be pushed aside. To them the same Canada appears as a post-conquest capitalist state, economically dependent on an imperialist United States and politically implicated in English and U.S. imperialist enterprises, with some designs of its own. From this perspective "Pan-Canadianism" loses its transcendent inclusivity and emerges instead as a device and a legitimation for a highly particularized ideological form of domination. Canada then becomes mainly an English Canada, historicized into particularities of its actual conquerors and their social and state formations. Colonialism remains as a vital formational and definitional issue. Canada, after all, could not be English or French in the same sense in which England and France are English and French.

Seen thus, the essence of Canada is destabilized. It becomes a politico-military ideological construction and constitution, elevating aggressive acts of acquisition and instituting them into a formal stabilization. But this stability is tenuous, always threatening to fall apart. The adjective "English" stamped into "Canada" bares this reality, both past and present. It shows us who stands on the other side of the "Pan-Canadian" project. Quebeckers know it well, and so their colonial rivalry continues. And we, the "visible minorities"—multiculturalism notwithstanding—know our equidistance from both of these conquering essences. The issue at stake, in the end, is felt by all sides to be much more than cultural. It is felt to be about the power to define what is Canada or Canadian culture. This power can only come through the actual possession of a geographical territory and the economy of a nation-state. It is this which confers the legal imprimatur to define what is Canadian or French Canadian, or what are "sub"- or "multi"-cultures. Bilingualism, multiculturalism, tolerance of diversity and difference and slogans of unity cannot solve this problem of unequal power and exchange—except to entrench even further the social relations of power and their ideological and legal forms, which emanate from an unproblematized Canadian state and essence. What discursive magic can vanish a continuously proliferating process of domination and thus of marginalization and oppression? What can make it a truly multicultural state when all the power relations and the signifiers of Anglo-French white supremacy are barely concealed behind a straining liberal democratic façade?

The expression "white supremacist,"[12] harsh and shocking as it may sound to many, encodes the painful underpinnings of the category visible minorities. The ideological imperatives of other categories—such as immigrants, aliens, foreigners, ethnic communities or New Canadians—constellate around the same binary code. There is a direct connection between this and the ideological spin-off to Englishness or Frenchness. After all, if nations are "imagined communities," can the content of this national imagination called Canada be free of its history and current social relations of power? Does not the context inflect the content here and now?

The case of Canada and its nationalism, when considered in this light, is not very different from the "official nationalism" of South Africa, erstwhile Rhodesia, or of Australia. These are cases of colonial "community" in which nation and state formations were created through the conquering imagination of white supremacy.[15] An anxiety about "them"—the aboriginals, pre-existing people—provides the core of a fantasy which inverts the colonized into aggressors, resolving the problem through extermination, suppression and containment.[16] Dominant cultural language in every one of these countries resounds with an "us" and "them" as expressed through discursivities of "minority/sub/multi-culture." A thinly veiled, older colonial discourse of civilization and savagery peeps out from the modern versions. Here difference is not a simple marker of cultural diversity, but rather, measured or constructed in terms of distance from civilizing European cultures. Difference here is branded always with inferiority or negativity. This is displayed most interestingly in the reading of the non-white or dark body which is labelled as a visible and minority body.[17] The colour of the skin, facial and bodily features—all become signifiers of inferiority, composed of an inversion and a projection of what is considered evil by the colonizing society. Implied in these cultural constructions is a literal denigration, extending into a valorized expression of European racist-patriarchy coded as white.

This inscription of whiteness underwrites whatever may be called Englishness, Frenchness, and finally Europeanness. These national characteristics become moral ones and they spin off or spill over into each other. Thus whiteness extends into moral qualities of masculinity, possessive individualism and an ideology of capital and market.[18] They are treated as indicators of civilization, freedom and modernity. The inherent aggressiveness and asociality of this moral category "whiteness" derives its main communitarian aspect from an animosity towards "others," signaling the militaristic, elite and otherizing bond shared by conquerors. The notion of Englishness serves as a metaphor for whiteness, as do all other European national essences. Whiteness, as many have noted, thus works as an ideology of a nation-state. It can work most efficiently with an other/enemy in its midst, constantly inventing new signifiers of "us" and "them." In the case of Canada the others, the First Nations, have been there from the very inception, modulating the very formation of its state and official culture, constantly presenting them with doubts about their legitimacy. Subsequently, indentured works, immigrants, refugees and other "others" have only deepened this legitimation crisis, though they also helped to forge the course of the state and the "nation."[19] "English," as an official language, has served to create a hegemonic front, but it is not a powerful enough antidote as an ideological device to undermine antagonisms that are continually created through processes of ruling; it is the ideology of whiteness/Europeanness" that serves as the key bonding element. Even though the shame of being an Italian, that is, non-English, in Canada outweighs the glory of the Italian renaissance, "Italian" can still form a part of the community of "whiteness" as distinct from non-white "others." It is not surprising, therefore, to see that one key element of white supremacy in Canada was an "Orange" mentality connecting Englishness with whiteness and both with racial purity. Books such as *Shades of Right*, for example, speak precisely to this, as does the present day right-wing nationalism of "English"-based groups. Quebec's "French" nationalism has precisely the same agenda, with a smaller territorial outreach. In fact, racialization and ethnicization are the commonest forms of cultural or identity parlance in Canada. This is not only the case with "whites" or "the English" but also with "others" after they spend some time in the country. A language of colour, even self-appellations such as "women of colour" (remember "coloured women"?), echo right through the cultural/political world. An unofficial apartheid, of culture and identity, organizes the social space of "Canada," first between whites and non-whites, and then within the non-whites themselves.

A ROSE BY ANY OTHER NAME: NAMING THE "OTHERS"

The transcendence or legitimation value of the official/state discourse of multiculturalism—which cherishes difference while erasing real antagonisms—breaks down, therefore, at different levels of competing ideologies and ruling practices. A threat of rupture or crisis is felt to be always already there, a fact expressed by the ubiquity of the integration-fragmentation paradigm in texts on Canada. Instead of a discourse of homogeneity or universality, the paradigm of multiculturalism stands more for the pressure of conflicts of interests and dynamics of power relations at work. This language is useful for Canada since imagining a nation is a difficult task even when the society is more homogeneously based on historic and cultural sharing or hegemony. Issues of class, industry and capital constantly destabilize the national project even in its non-colonial context. Gramsci, for example, in "Notes on Italian History," discusses the problem of unification inherent in the formation of a nation-state in the European bourgeois context (1971). Unificatory ideologies and institutions, emanating from the elite, posturing as a class-transcendent polity and implanted on top of a class society reveal as much as they hide. These attempts at unification forge an identifiable ideological core, a national identity, around which other cultural elements may be arranged hierarchically. It transpires that the ability and the right to interpret and name the nation's others forms a major task of national intellectuals, who are organic to the nation-state project.[20]

If this difficulty dogs European bourgeois nationalism, then it is a much more complicated task for Canada to imagine a *unificatory* national ideology, as recognized by members of the "white" ideological bloc espousing non-liberal perspectives. Ultra-conservatives in general have foresworn any pretence to the use of "multi-cultural" ideology. They view multiculturalism as an added burden to a society already divided, and accord no political or cultural importance to groups other than the French. The political grammar of "national" life and culture, as far as the near and far right are concerned, is common-sensically acknowledged as "English." According importance to multiculturalism has the possibility of calling into question the "English" presence in this space, by creating an atmosphere of cultural relativism signalling some sort of usurpation. This signal, it is felt, is altogether best removed. English-Europeanness, that is, whiteness, emerges as the hegemonic Canadian identity. This white, Canadian and English equation becomes hegemonic enough to be shared even by progressive Canadians or the left.[21] This ideological Englishness/whiteness is central to the programme of multiculturalism. It provides the content of Canadian culture, the point of departure for "multiculture." This same gesture creates "others" with power-organized "differences," and the material basis of this power lies both below and along the linguistic-semiotic level. Multiculturalism as the "other" of assimilation brings out the irreducible core of what is called the real Canadian culture.

So the meaning of Canada really depends on who is doing the imagining—whether it is Margaret Atwood or Charles Taylor or Northrop Frye or the "visible minorities" who organize conferences such as "Writing Thru 'Race.'" Depending on one's social location, the same snow and Canadian landscape, like Nellie McClung and other foremothers of Canadian feminism, can seem near or far, disturbing, threatening or benign. A search through the literature of the "visible minorities" reveals a terror of incarceration in the Canadian landscape.[22] In their Canada there is always winter and an equally cold and deathly cultural topography, filled with the RCMP, the Western Guard, the Heritage Front and the *Toronto Sun*, slain Native peoples and Sitting Bull in a circus tent, white-faced church fathers, trigger-happy impassive police, the flight and plight of illegals, and many other images of fear and active oppression. To integrate with this Canada would mean a futile attempt at integrating with a humiliation and an impossibility. Names of our otherness proliferate endlessly, weaving margins around "Canada/English/French Canada." To speak of pan-Canadian nationalism and show a faith in "our" national institutions is only possible for those who can imagine it and already are "Canada." For "others," Canada can mean the actuality of skinhead attacks, the mediated fascism of the Reform Party, and the hard-fist of Rahowa.[23]

It is time to reflect on the nomenclature extended by multiculturalism to the "others" of "Canada." Its discourse is concocted through ruling relations and the practical administration of a supposed reconciliation of "difference." The term visible minorities is a great example: one is instantly struck by its reductive character, in which peoples from many histories, languages, cultures and politics are reduced to a distilled abstraction. Other appellations follow suit—immigrants, ethnics, new Canadians and so on. Functional, invested with a legal social status, these terms capture the "difference" from "Canada/ English/French Canada" and often signify a newness of arrival into "Canada." Unlike a rose, which by any other name would smell as sweet, these names are not names in the sense of classification. They are in their inception and coding official categories. They are identifying devices, like a badge, and they identify those who hold no legitimate or possessive relationship to "Canada." Though these are often identity categories produced by the state, the role played by the state in identity politics remains unnoticed, just as the whiteness in the "self" of "Canada's" state and nationhood remains unnamed. This transparency or invisibility can only be achieved through a constellation of power relations that advances a particular groups' identity as universal, as a measuring rod for others, making them "visible" and "minorities."

An expression such as visible minorities strikes the uninitiated as both absurd and abstract. "Minority," we know from J.S. Mill onwards, is a symptom of liberal democracy, but "visible?" We realize upon reflection that the adjective visible attached to minority makes the scope of identity and power even more restricted. We also know that it is mainly the Canadian state and politics which are instrumental in this categorizing process and confer this "visibility" upon us. I have remarked on its meaning and use elsewhere:

> Some people, it implies, are more visible than others; if this were not the case, then its triviality would make it useless as a descriptive category. There must be something "peculiar" about some people which draws attention to them. This something is the point to which the Canadian state wishes to draw our attention. Such a project of the state needed a point of departure which has to function as a norm, as the social average of appearance. The well-blended, "average," "normal" way of looking becomes the base line, or "us" (which is the vantage point of the state), to which those others marked as "different" must be referred . . . and in relation to which "peculiarity" [and thus, visibility] is constructed. The "invisibility" . . . depends on the state's view of [some] as normal, and therefore, their institution as dominant types. They are true Canadians, and others, no matter what citizenship they hold [and how many generations have they lived here?] are to be considered as deviations. . . . (1993, p. 148)[24]

Such "visibility" indicates not only "difference" and inferiority, but is also a preamble to "special treatment." The yellow Star of David, the red star, the pink triangle, have all done their fair share in creating visibility along the same lines—if we care to remember. Everything that can be used is used as fodder for visibility, pinning cultural and political symbols to bodies and reading them in particular ways. Thus for non-whites in Canada,

> their own bodies are used to construct for them some sort of social zone or prison, since they cannot crawl out of their skins, and this signals what life has to offer them in Canada. This special type of visibility is a social construction as well as a political statement. (p. 149)

Expressions such as "ethnics" and "immigrants" and "new Canadians" are no less problematic. They also encode the "us" and "them" with regard to political and social claims, signifying uprootedness and the pressure of assimilation or core cultural-apprenticeship. The irony compounds when one discovers that all white people, no matter when they immigrate to Canada or as carriers of which European ethnicity, become invisible and hold a dual membership in Canada, while others remain immigrants generations later.

The issue of ethnicity, again, poses a further complexity. It becomes apparent that currently it is mainly applied to the non-white population living in Canada. Once, however, it stringently marked out white "others" to the Anglo-French language and ethos; while today the great "white" construction has assimilated them. In the presence of contrasting "others," whiteness as an ideological-political category has superseded and subsumed different cultural ethos among Europeans. If the Ukrainians now seek to be ethnics it is because the price to be paid is no longer there. Now, in general, they are white *vis-à-vis* "others," as is denoted by the vigorous participation of East Europeans in white supremacist politics. They have been ingested by a "white-Anglo" ethos, which has left behind only the debris of self-consciously resurrected folklores as special effects in "ethnic" shows. The ethnicities of the English, the Scottish, the Irish, etc. are not visible or highlighted, but rather displaced by a general Englishness, which means less a particular culture than an official ideology and a standardized official language signifying the right to rule. "Ethnicity" is, therefore, what is classifiable as a non-dominant, sub or marginal culture. English language and Canadian culture then cannot fall within the ministry of multiculturalism's purview, but rather within that of the ministry of education, while racism makes sure that the possession of this language as a mother tongue does not make a non-white person non-ethnic. Marginalizing the ethnicity of black people from the Caribbean or Britain is evident not only in the Caribana Festival but in their being forced to take English as a second language. They speak dialects, it is said—but it might be pointed out that the white Irish, the white Scots, or the white people from Yorkshire, or white Cockney speakers are not classified as ESL/ESD clients. The lack of fuss with which "Canadians" live with the current influx of Eastern European immigrants strikes a profound note of contrast to their approach to the Somalis, for example, and other "others."

The intimate relation between the Canadian state and racism also becomes apparent if one complements a discussion on multiculturalism with one on political economy. One could perhaps give a finer name than racism to the way the state organizes labour importation and segmentation of the labour market in Canada, but the basic point would remain the same. Capitalist development in Canada, its class formation and its struggles, predominantly have been organized by the Canadian state. From the days of indenture to the present, when the Ministry of "Manpower" has been transformed into that of "Human Resources," decisions about who should come into Canada to do what work, definitions of skill and accreditation, licensing and certification, have been influenced by "race" and ethnicity.[25] This type of racism cannot be grasped in its real character solely as a cultural/attitudinal problem or an issue of prejudice. It needs to be understood in systemic terms of political economy and the Gramscian concepts of hegemony and common sense that encompass all aspects of life—from the everyday and cultural ones to those of national institutions. This is apparent if one studies the state's role in the importation of domestic workers into Canada from the Philippines or the Caribbean. Makeda Silvera, in *Silenced*, her oral history of Caribbean domestic workers, shows the bonds of servitude imposed on these women by the state through the inherently racist laws pertaining to hiring of domestic workers.[26] The middle-man/procurer role played by the state on behalf of the "Canadian" bourgeoisie is glaringly evident. Joyce Fraser's *Cry of the Illegal Immigrant* is another testimonial to this (1980). The issue of refugees is another, where we can see the colonial/racist as well as anti-Communist nature of the Canadian state. Refugees fleeing ex-Soviet bloc countries, for example, received a no-questions acceptance, while the Vietnamese boat people, though fleeing communism, spent many years proving their claim of persecution. The racism of the state was so profound that even cold-war politics or general anti-communism did not make Vietnamese refugees into a "favoured" community. The story of racism is further exposed by the onerous and lengthy torture-proving rituals imposed on Latin Americans and others fleeing fascist dictatorships in the Third World. In spite of Canada's self-proclaimed commitment to human rights, numerous NGOs, both local and international, for years have needed to persuade the Canadian state and intervene as advocates of Third World refugees. Thus the state of "Canada," when viewed through the lens of racism/difference, presents us with a hegemony compounded of a racialized common sense and institutional structures. The situation is one where racism in all its cultural and institutional variants has become so naturalized, so pervasive that it has become invisible or transparent to those who are not adversely impacted by them. This is why terms such as visible minority can generate so spontaneously within the bureaucracy, and are not considered disturbing by most people acculturated to "Canada."

Erol Lawrence in his Gramscian critique "Plain Common Sense: The 'Roots' of Racism," (1986) uses the notion of common-sense racism to explain the relationship between the British blacks and the state. He displays how common sense of "race" marks every move of the state, including official nomenclatures and their implementation in social and political culture. Lawrence remarks on how hegemony works through common sense or expresses itself as such:

> The term common sense is generally used to denote a down-to-earth "good sense." It is thought to represent the distilled truths of centuries of practical experience; so much so that to say of an idea or practice that it is only common sense, is to appeal over the logic and argumentation of intellectuals

> to what all reasonable people know in their "heart of hearts" to be right and proper. Such an appeal can all at once and at the same time (serve) to foreclose any discussion about certain ideas and practices and to legitimate them. (1986, p. 48)

The point of this statement becomes clearer when we see how the Canadian state, the media and political parties are using "visible minorities," "immigrants," "refuges" and "illegals" as scapegoats for various economic and political problems entirely unrelated to them. For this they rely on common sense racism: they offer pseudo-explanations to justify crises of capitalism and erosion of public spending and social welfare in terms of the presence of "others." Unemployment, endemic to capital's "structural adjustment," is squarely blamed on "these people." This explanation/legitimation easily sticks because it replicates cultural-political values and practices that pre-exist on the ground. These labelling categories with racialized underpinnings spin-off into notions such as unskilled, illiterate and traditional, thus making the presence of Third World peoples undesirable and unworthy of real citizenship. Englishness and whiteness are the hidden positive poles of these degrading categories. They contain the imperative of exclusion and restriction that neatly fits the white supremacist demand to "keep Canada white." The multiculturalist stance may support a degree of tolerance, but beyond a certain point, on the far edge of equality, it asserts "Canadianness" and warns off "others" from making claims on "Canada." Through the same scale of values East European immigrants are seen as desirable because they can be included in the ideology of whiteness.

"Difference" read through "race," then, produces a threat of racist violence. The creation of a "minority" rather than of full-fledged adult citizens—the existence of levels of citizenship—adds a structural/legal dimension to this violence. Inequality within the social fabric of Canada historically has been strengthened by the creation of reserves, the Department of Indian Affairs, the exclusion of Jews, and the ongoing political inequalities meted out to the Chinese, the Japanese and South Asians. These and more add up to the tenuousness of the right and means to existence, jobs and politics of the "visible minorities." Being designated a minority signals tutelage. It creates at best a patron-client relationship between the state and "others" who are to be rewarded as children on the basis of "good conduct." Social behaviour historically created through class, "race" and gender oppression is blamed on the very people who have been the victims. Their problems are seen as self-constructed. The problem of crime in Toronto, for example, is mainly blamed on the black communities. Black young males are automatically labelled as criminals and frequently shot by the police. It is also characteristic that an individual act of violence performed by any black person is seen as a representative act for the whole black community, thus labelling them as criminal, while crime statistics among the white population remain non-representative of whiteness.

Visible minorities, because they are lesser or inauthentic political subjects, can enter politics mainly on the ground of multiculturalism. They can redress any social injustice only limitedly, if at all. No significant political effectiveness on a national scale is expected from them. This is why Elijah Harper's astute use of the tools of liberal democracy during the Meech Lake Accord was both unexpected and shocking for "Canadians." Other than administering "difference" differentially, among the "minority communities" multiculturalism bares the political processes of cooptation or interpellation. The "naming" of a political subject in an ideological context amounts to the creation of a political agent interpellating or extending an ideological net around her/him, which confers agency only within a certain discursive-political framework. At once minimizing the importance and administering the problem of racism at a symptomatic level, the notion of visible minority does not allow room for political manoeuvre among those for whose supposed benefit it is instituted. This is unavoidably accompanied by the ethnicization and communalization of politics, shifting the focus from unemployment due to high profit margins, or flight of capital, to "problems" presented by the immigrant's own culture and tradition. Violence against women among the "ethnics" is thought to be the result of their indigenous "traditions" rather than of patriarchy and its exacerbation, caused by the absolute power entrusted by the Canadian state into the hands of the male "head of the family." The sponsorship system through which women and children enter into the country seems calculated to create violence. Food, clothes and so-called family values are continually centre-staged, while erasing the fundamental political and economic demands and aspirations of the communities through multicultural gestures of reconciling "difference." The agent of multiculturalism must learn to disarticulate from his or her real-life needs and struggles, and thus from creating or joining organizations for anti-racism, feminism and class struggle. The agencies (wo)manned by the "ethnic" elements—within terms and conditions of the state—become managers on behalf of the state. In fact, organizing multiculturalism among and by the non-white communities amounts to extending the state into their everyday life, and making basic social contradictions to disappear or be deflected. Considering the state's multicultural move therefore allows a look into the state's interpellative functions and how it works as an ideological apparatus. These administrative and ideological categories create *objects* out of the people they impact upon and produce mainstream agencies in their name. In this way a little niche is created within the state for those who are otherwise undesirable, unassimilable and deeply different. Whole communities have begun to be renamed on the basis of these conferred cultural administrative identities that objectify and divide them. Unrelated to each other, they become clients and creatures of the multicultural state. Entire areas of problems connected to "race," class, gender and sexual orientation are brought under the state's management, definition and control, and possibilities for the construction of political struggles are displaced and erased in the name of "ethnic culture." The politics of identity among "ethnic communities," that so distresses the "whites" and is seen as an excessive permissiveness on the part of the state, is in no small measure the creation of this very culturalist managerial/legitimation drive of the state.

What, then, is to be done? Are we to join forces with the Reform Party or the small "c" conservative "Canadians" and advocate that the agenda of multiculturalism be dropped summarily? Should we be hoping for a deeper legitimation crisis through unemployment and rampant cultural racism, which may bring down the state? In theory that is an option, except that in the current political situation it also would strengthen the ultra-right. But strategically speaking, at this stage of Canadian politics, with the withdrawal and disarray of the left and an extremely vulnerable labour force, the answer cannot be so categorical.

The political potential of the civil society even when (mis)named as ethnic communities and reshaped by multiculturalism is not a negligible force. This view is validated by the fact that all shades of the right are uneasy with multiculturalism even though it is a co-opted form of popular, non-white political and cultural participation. The official, limited and co-optative nature of this discourse could be re-interpreted in more materialist historical and political terms. It could then be re-articulated to the social relations of power governing our lives, thus minimizing, or even ending, our derivative, peripheral object agent status. The basic nature of our "difference," as constructed in the Canadian context, must be rethought and the notion of culture once more embedded into society, into everyday life. Nor need it be forgotten that what multiculturalism (as with social welfare) gives us was not "given" voluntarily but "taken" by our continual demands and struggles. We must remember that it is our own socio-cultural and economic resources which are thus minimally publicly redistributed, creating in the process a major legitimation gesture for the state. Multiculturalism as a form of bounty or state patronage is a managed version of our antiracist politics.

We must then bite the hand that feeds us, because what it feeds us is neither enough nor for our good. But we must wage a contestation on this terrain with the state and the needs of a racist/imperialist capital. At this point of the new world order, short of risking an out-and-out fascism, the twisted ideological evolution of multiculturalism has to be forced into a minimum scope of social politics. Until we have developed a wider political space, and perhaps with it keeping a balance of "difference," using the avenues of liberal democracy may be necessary. Informed with critique, multiculturalism is a small opening for making the state minimally accountable to those on whose lives and labour it erects itself. We must also remember that liberalism, no matter who practises it, does not answer our real needs. Real social relations of power – of "race," class, gender and sexuality – provide the content for our "difference" and oppression. Our problem is not the value or the validity of the cultures in which we or our parents originated – these "home" cultures will, as living cultures do in history, undergo a sea-change when subjected to migration. Our problem is class oppression, and that of objectifying sexist-racism. Thinking in terms of culture alone, in terms of a single community, a single issue, or a single oppression will not do. If we do so our ideological servitude to the state and its patronage and funding traps will never end. Instead we need to put together a strategy of articulation that reverses the direction of our political understanding and affiliation – against the interpellating strategies of the ideological state apparatus. We need not forget that the very same social relations that disempower or minoritize us are present not only for us but in the very bones of class formation and oppression in Canada. They are not only devices for cultural discrimination and attitudinal distortion of the white population, or only a mode of co-optation for "visible minorities." They show themselves inscribed into the very formation of the nation and the state of "Canada." Thus the politics of class struggle, of struggle against poverty or heterosexism or violence against women, are politically more relevant for us than being elected into the labyrinth of the state. The "visible minorities" of Canada cannot attain political adulthood and full stature of citizenship without struggling, both conceptually and organizationally, against the icons and regulations of an overall subordination and exploitation.

In conclusion, then, to answer the questions "How are we to relate to multiculturalism?" and "Are we for it or against it?" we have to come to an Aesopian response of "ye, ye" and "nay, nay." After all, multiculturalism, as Marx said of capital, is not a "thing." It is not a cultural object, all inert, waiting on the shelf to be bought or not. It is a mode of the workings of the state, an expression of an interaction of social relations in dynamic tension with each other, losing and gaining its political form with fluidity. It is thus a site for struggle, as is "Canada" for contestation, for a kind of tug-of-war of social forces. The problem is that no matter who we are—black or white—our liberal acculturation and single-issue-oriented politics, our hegemonic "subsumption" into a racist common sense, combined with capital's crisis, continually draw us into the belly of the beast. This can only be prevented by creating counter-hegemonic interpretive and organizational frame-works that reach down into the real histories and relations of our social life, rather than extending tendrils of upward mobility on the concrete walls of the state. Our politics must sidestep the paradigm of "unity" based on "fragmentation or integration" and instead engage in struggles based on the genuine contradictions of our society.

NOTES

1. This division of Canada into Quebec and Canada outside of Quebec (COQ) is used as more than a territorial expression by Taylor (1993).
2. For an exposition of the notions of "solitude" and "survival," see Atwood (1972).
3. For an elaboration of these concepts, see Althusser (1977).
4. On multiculturalism, its definition and history, see Fleras and Elliot (1992).
5. On the emergence of a liberal state from the basis of a white settler colony, see Bolaria and Li (1988). Also see Kulchyski (1994) and Tester and Kulchyski (1994). For a "race"/gender inscription into a semi-colonial Canadian state, see Monture-Angus (1995).
6. For an in-depth discussion of mediatory and unificatory ideologies needed by a liberal democratic, i.e. capitalist state, see Miliband (1984) chs. 7 & 8.
7. For a clarification of my use of this concept, see Habermas (1975). This use of "legitimacy" is different from Charles Taylor's Weberian use of it in *Reconciling the Solitudes* (1993).
8. See Anderson (1991), Introduction and ch. 2. Anderson says, "I . . . propose the following definition of the nation: it is an imagined political community – and imagined as both inherently limited and sovereign. It is *imagined* because the members of even the smallest nation will never know most of their fellow members, meet them, or even hear of them, yet in the minds of each lives the image of their communion" (p. 6).
9. In *Reconciling* (ch. 4) on "Alternative Futures for Canada," Taylor fleshes out his desirable and undesirable options for Canada. This is also found in his *Multiculturalism and "the Politics of Recognition"* (1992).
10. Taylor is quite direct about his German romantic intellectual heritage. In *Reconciling,* in an essay entitled "Institutions in National Life," he states, "In Herder I found inspiration, ideas that were very fruitful for me, precisely because I was from here, I was able to understand him from the situation I had experienced outside school, outside university, and I was able to engage with his thought, internalize it, and (I hope) make something interesting out of it" (p. 136).

11. For an exposition of this idea, and Taylor's rejection of an "American" solution for "Canadian" identity, see "Shared and Divergent" in *Reconciling.*
12. On the development of active white supremacist groups in Canada, and their "Englishness," see Robb (1992). Also Ward (1978).
13. But see also the chapter on "Official Nationalism and Imperialism" (Anderson, 1991).
14. On the construction of "whiteness" as an ideological, political and socio-historical category see Allen (1994), and Roediger (1993); also Frankenberg (1993).
15. On the use of "whiteness"/Europeanness as an ideology for ruling, including its formative impact on sexuality of the ruling, colonial nations, see Stoller (1995).
16. On this theme see Joseph Conrad's *Heart of Darkness,* E.M. Forster's *A Passage to India* and Said (1993).
17. On the reading of the black, dark or "visible minority" body, see the collection of essays in Gates (1985), especially Gillman, "Black Bodies, White Bodies."
18. See Stoller (1995), but also Sinha (1995).
19. The history of immigration and refugee laws in Canada, and of the immigrants, indentured workers and refugees themselves, must be read to comprehend fully what I am attempting to say. See The Law Union of Ontario (1981). Also Canada (1974), and Canada (1986).
20. On organic intellectuals as intellectuals who are integral to any ideological and class project, see Gramsci, "The Intellectuals," in Gramsci (1971).
21. This becomes evident when we follow the controversies which are generated by writers' conferences, such as "Writing thru Race," or the black communities' response and resistance to the Royal Ontario Museum's exhibition on African art and culture, "Out of the Heart of Africa."
22. See, for example, Brand (1983), Sri Bhaggiyadatta (1993), H. Bannerji (1986), and collections such as McGifford and Kearn (1990).
23. The acronym for Racial Holy War, a neo-nazi rock band.
24. On this theme of social construction of a racialized "minority" subject and its inherent patriarchy, see Carty and Brand (1993) and Ng (1993).
25. See Avery (1995). Much work still needs to be done in this area, in which class formation is considered in terms of both "race" and gender, but a beginning is made in Brand (1991), and Brand and Sri Bhaggiyadatta (1985).
26. This is powerfully brought forth through the issue of the importation of domestic workers to Toronto from the Caribbean in Silvera (1989).

REFERENCES

Allen, Theodor. *The Invention of the White Race: Racial Oppression and Social Control.* London: Verso, 1994.

Althusser, Louis. "Ideology and Ideological State Apparatuses (Notes towards an Investigation)." In *Lenin and Philosophy and Other Essays.* London: New Left Books, 1977.

Anderson, Benedict. *Imagined Communities.* London: Verso, 1991.

Atwood, Margaret. *Survival: A Thematic Guide to Canadian Literature.* Toronto: House of Anansi Press, 1972.

Avery, Donald. *Reluctant Host: Canada's Response to Immigrant Workers, 1896–1994.* Toronto: McCelland & Stewart, 1995.

Bannerji, Himani. *Doing Time.* Toronto: Sister Vision Press, 1986.

___. "Images of South Asian Women." In *Returning the Gaze: Essays on Racism, Feminism and Politics.* Ed. Himani Bannerji. Toronto: Sister Vision Press, 1993.

___, ed. *Returning the Gaze: Essays on Racism, Feminism and Politics.* Toronto: Sister Vision Press, 1993.

Bannerji, Kaushalya. *A New Remembrance.* Toronto: TSAR Publications, 1993.

Bolaria. B. Singh, and Peter Li, eds. *Racial Oppression in Canada.* Toronto: Garamond Press, 1988.

Brand, Dionne. *Winter Epigrams.* Toronto: Williams-Wallace, 1983.

___. *No Burden to Carry: Narratives of Black Working Women in Ontario, 1920s to 1950s.* Toronto: Women's Press, 1991.

Brand, Dionne, and Krisantha Sri Bhaggiyadatta, eds. *Rivers Have Sources, Trees Have Roots: Speaking of Racism.* Toronto: Cross Cultural Communications Centre, 1985. Canada. Department of Manpower and Immigration and Information. *A Report of the Canadian Immigration and Population Study: Immigration Policy Perspective.* Ottawa: Queen's Printer, 1974. Canada. House of Commons. *Equality Now: Report of the Special Committee on Visible Minorities.* Ottawa: Queen's Printer, 1986.

Carty, Linda, and Dionne Brand. "Visible Minority Women: A Creation of the Colonial State." In *Returning the Gaze: Essays on Racism, Feminism and Politics.* Ed. Himani Bannerji. Toronto: Sister Vision Press, 1993.

Fleras, Angie, and Jean Leonard Elliot, eds. *Multiculturalism in Canada: The Challenge of Diversity.* Scarborough: Nelson, 1992.

Frankenberg, Ruth. *White Women, Race Matters: The Social Construction of Whiteness.* Minneapolis: University of Minnesota Press, 1993.

Fraser, Joyce. *Cry of the Illegal Immigrant.* Toronto: Williams-Wallace Productions International, 1980.

Gates Jr., Henry Louis, ed. *"Race," Writing and Difference.* Chicago: The University of Chicago Press, 1985.

Gillman, Sander. "Black Bodies, White Bodies." In *"Race" Writing and Difference.* Ed. Henry Louis Gates Jr. Chicago: The University of Chicago Press, 1985.

Gramsci, Antonio. *Selections from the Prison Notebooks*, edited and translated by Quentin Hoare and Geoffrey Smith. New York: International Publishers, 1971.

Habermas, Jurgen, *Legitimation Crisis.* Boston: Beacon Press, 1975.

Kulchyski, Peter, ed. *Unjust Relations: Aboriginal Rights in Canadian Courts.* Toronto: Oxford University Press, 1994.

Law of Union of Ontario, The. *The Immigrant's Handbook.* Montreal: Black Rose Books, 1981.

Lawrence, Erol. "Just Plain Common Sense: The 'Roots' of Racism," in *The Empire Strikes Back: Race and Racism in 70s Britain.* London: Centre for Contemporary Cultural Studies, Hutchinson, in association with the Centre for Cultural Studies, University of Birmingham, 1986.

McGifford, Diane, and Judith Kearn, eds. *Shakti's Words.* Toronto: TSAR, 1990.

Miliband, Ralph. *The State in Capitalist Society*. London: Quartet Books, 1984.

Monture-Angus, Patricia. *Thunder in My Soul: A Mohawk Woman Speaks*. Halifax: Fernwood, 1995.

Ng, Roxana. "Sexism, Racism, Canadian Nationalism." In *Returning the Gaze: Essays on Racism, Feminism and Politics*. Ed. Himani Bannerji. Toronto: Sister Vision Press, 1993.

Robb, Martin. *Shades of Right: Nativist and Fascist Politics in Canada, 1920–1940*. Toronto: University of Toronto Press, 1992.

Roediger, David. *The Wages of Whiteness: Race and the Making of the American Working Class*. London: Verso, 1993.

Said, Edward. *Culture and Imperialism*. New York: Vintage Books, 1993.

Silvera, Makeda. *Silenced: Talks With Working Class Caribbean Women about Their Lives and Struggles as Domestic Workers in Canada*, 2nd edition. Toronto: Sister Vision Press, 1989.

Sinha, Mrinalini. *Colonial Masculinity*. Manchester: Manchester University Press, 1995.

Sri Bhaggiyadatta, Krisantha. *Domestic Bliss*. Toronto: Five Press, 1981.

___. *The 52nd State of Amnesia*. Toronto: TSAR, 1993.

Stoller, Ann Laura. *Race and the Education of Desire: Foulcault's History of Sexuality and the Colonial Order of Things*. Durham: Duke University Press, 1995.

Taylor, Charles. *Multiculturalism and "the Politics of Recognition."* Princeton: Princeton University Press, 1992.

___. *Reconciling the Solitudes: Essays on Canadian Federalism and Nationalism*. Ed. Guy Laforest. Montreal and Kingston: McGill-Queen's University Press, 1993.

Tester, Frank, and Peter Kulchyski, *Tammarniit (Mistakes): Relocation in the Eastern Arctic*. Vancouver: University of British Columbia Press, 1994.

Ward, William Peter. *White Canada Forever*. Montreal: McGill-Queen's University Press, 1978.

5

Theoretical Foundations and the Challenge of Aboriginal Rights

PATRICIA A. MONTURE-ANGUS

Despite living the last decade of my life outside of my territory, being Kanien'kehaka remains at the center of my identity. Choosing to study Canadian law did not diminish my need to live as Kanien'kehaka. Any information or knowledge that I have gained from my involvement in Canadian universities and Canadian legal institutions over the last two decades is always brought to this core of my Kanien'kehaka identity where I compare and contrast new information to the values and teachings of my culture. This is my duty as I understand it. As a young woman, the Elders taught me that white things were not of necessity wrong but to take care and ensure that I was picking up only the "good stuff."

In order to understand the way I have processed the recognition and affirmation of existing Aboriginal (and treaty) rights in Canada's constitution, the basics of Haudenosaunee political thought must be briefly established. As I understand it, in this knowledge tradition it is difficult to separate intellectual, spiritual, political and legal realms. This is unlike the manner in which Canadian structures of state, church, law and academia are premised on separation as a fundamental and necessary value in a civilized society. John Mohawk notes of Iroquoian political thought:

> For this plan to work the Peacemaker was required to convince a very skeptical audience that all human beings really did possess the potential for rational thought, that when encouraged to use rational thought they would inevitably seek peace, and the belief in the principles would lead to the organized enactment of the vision.
>
> The test of this thinking is found in the converse of the argument. If you do not believe in the rational nature of the human being, you cannot believe that you can negotiate with him. If you do not believe that rational people ultimately desire peace, you cannot negotiate confidently with him toward goals you and he share. If you cannot negotiate with him, you are powerless to create peace. If you cannot organize around those beliefs, the principles cannot move from the minds of men into the actions of society. (1989:221)

The central object of social arrangements for my people is significantly about living in the way of peace. This peace is defined much more broadly than living without violence.

Living in peace is about living a good life where respect for our relationships with people and all creation is primary. The "Great Law of Peace," which at least partially parallels the role of law in Canadian social organization, is a principle which translates to "the way to live most nicely together."[1] Kanien'kehaka scholar Gerald Alfred expands on the leadership principles entrenched in the Kaienerekowa (Great Law) and notes:

> In the Rotinoshshonni tradition, the women of each family raise a man to leadership and hold him accountable to these principles. If he does not uphold and defend the Kaienerekowa, or if the women determine that his character or behaviour does not conform to the leadership principles, he is removed from the position. As in other traditional cultures, the moral definition of leadership focuses on a person's adherence to the values of *patience, courage, fairness, and generosity.* (1999:90, emphasis added)

If these are the qualities—patience, courage, fairness and generosity—of individuals who are (political) leaders, then these qualities are also the ones on which Kanien'kehaka politics are built.

The beliefs of the Haudenosaunee clearly influenced the Europeans who came to the shores of what we knew as Turtle Island. John Mohawk notes in the same essay quoted above:

> When the *White Roots of Peace*[2] was first published it became immediately apparent that the author had accomplished a pioneer work of sorts. Wallace exhibited astonishing insight when he alleged that prehistoric Iroquois had constructed a political philosophy based on rational thought. Not many writers on anthropology or oral history have found rational thought a prevalent theme among their subjects. Many professionals in this field operate on an expectation that rational thought is found only in the west.
>
> Such cultural blindness is unfortunate because it automatically denies most of the academic and literary world access to the best thinking of many of the world's cultures. This unspoken doctrine helps to promote the tradition in the West that non-Western people are non-rational people. Wallace's work was an honest and commendable effort to go beyond that. He saw good rational thinking in a place where such thinking was not expected to exist and he promoted his discovery, almost breathlessly, to a disbelieving world. (1989:218)

The Haudenosaunee influence on what have come to be known as western ideas about democracy has been made invisible by the operation of rules of academic discourse and the bias of those involved in that discourse.[3] Because I do not intellectually disenfranchise Aboriginal thought and intellectual traditions,[4] the continuity of Haudenosaunee influence remains an important thread in this work, which crosses conventional interdisciplinary boundaries.

Obviously then, when the categories of discourse (law, politics or academia) are inappropriately applied without consideration to the different structural bases of the worldviews of Aboriginal people, diminishing of the political sophistication of Aboriginal thought and intellectual traditions (the rational mind) occurs. When the world is looked at in a holistic way, everyone's opinion carries with it a similar weight. The way voice is legitimated in

Aboriginal society is vastly different from that of the societies that settled here. A number of consequences flow from this observation. For example, an Aboriginal academic does not have credibility in their home community based on their academic qualifications but rather on what they have earned in the Aboriginal way in their community. In fact, sometimes their academic credentials operate as a liability in relationships with other Aboriginal people. The artificial dichotomies between community member and activist, academic, politician or technician must be questioned. Any absolute dichotomy must be suspicious as no dichotomies exist in the natural world. The creation of dichotomy as a condition of existence is a colonial manifestation.

Historically, Aboriginal Peoples have been controlled by a variety of means, including our exclusion from the systems which have determined the meaning of concepts such as justice, sovereignty and rights. Howard Becker explains:

> control based on the manipulation of definitions and labels works more smoothly and costs less . . . . The attack on hierarchy begins with an attack on definitions, labels, and conventional conceptions of who's who and what's what. (1973:178)

Although the study of deviance is not analogous to the topic of this book, Becker's conclusion persuades me of the significance of his comment to the study of rights and governance. The pathway to a new relationship is paved with the long-term commitment to share the definitional power that creates the legitimacy whereby words and phrases gain their accepted meaning. It requires the free giving up of control over Aboriginal lives, with careful attention to the way language (that is English) presupposes a framework of meaning that is at least hierarchical and gendered. I also experience it as colonial. No definition can be taken for granted as inclusive. The re-examination of the way language sanctions particular worldviews and understandings is central to this process of change. This is particularly true if the framework of study is Canadian law, as law is the study of words.

As someone trained in the law, I turn to the written text of the Canadian constitution as one possible framework to consider when determining if my aspirations as an Aboriginal person and the way I understand my people's political position can be situated within that existing legal framework.[5] This may or may not be the appropriate starting point. Perhaps, it would be more logical to start with developing a true understanding of both history and the provisions of the treaties. At this time, I will leave the work of history to the historians, both tribal and/or those academically trained. I must note though that it is essential to understand one's own history or there is no future.

In 1982, when Canada repatriated its constitution, a provision which protects the rights of Aboriginal Peoples was included in that package.[6] This provision did not receive the unanimous support of Aboriginal people nor our governments, and in particular many Haudenosaunee people were opposed to it. However, given the fact that the provision does exist in Canadian law, it seems logical to use it as a starting place for discussion. The reality is that Canada's existing constitutional provisions will have a dramatic impact on the ability of and degree to which Aboriginal people and Canadians will be able to craft a new relationship based on mutual respect and co-existence. Further, it is naive to think that the entrenchment of Aboriginal and treaty rights in Canada's constitution will not have an effect on Aboriginal nations, communities and individuals. In some cases it may enhance dreaming. Or it may limit our vision making and cause fear.

There are two principles that must govern the intellectual examination of the meaning of section 35 from an Aboriginal place. This work must be completed with a respect of all peoples and in a way that brings colonialism, past and present, to account. If colonial relations cannot be challenged and changed, then there is no hope for any kind of renewed relationship. The degree to which I will eventually embrace section 35(1) depends on the degree to which it creates space for Aboriginal intellectual and political traditions. Does it foster peace? Does it create the space for the recovery of Aboriginal nations from colonial impositions and manifestations?

An examination of the historical relations between Aboriginal Peoples and Canadians (especially their government representatives), clearly indicates that the philosophical grounding of the relationship is based on a misplaced notion of Euro-Canadian superiority. More fundamentally distressing, allowing Euro-Canadian superiority to remain ingrained in the fabric of Canadian society, including Canadian legal relations, ignores the trust-like responsibility of the Canadian government to Aboriginal Peoples. A "trust" relationship does not necessarily have to be built on a notion of one party's superiority or the other's vulnerability. This is another colonial myth.

During the 1950s the courts, in fact, began to articulate the notion that Canada has a special responsibility to Indians:

> The language of the (*Indian Act*) embodies the accepted view that these aborigines are, in effect, wards of the state, whose care and welfare are a political trust of the highest obligation. (*St. Ann's Island Shooting and Fishing Club v. R.*, 1952:232)

Minimally, we would now recognize the idea of wardship as overly paternalistic (or grounded in the notion of supposed European superiority). The idea of wardship has developed into what is now legally recognized as a fiduciary responsibility (see the discussion in *Guerin v. R.* 1984:501) and more recently as a trust responsibility by the Supreme Court of Canada in *R. v. Sparrow* (1990: 180–181).

In the *Sparrow* decision, Justices Dickson and LaForest opined:

> the Government has the responsibility to act in a fiduciary capacity with respect to aboriginal peoples. The relationship between the Government and aboriginals is trust-like, rather than adversarial, and contemporary recognition and affirmation of aboriginal rights must be defined in light of this historic relationship. (1990:180)

And later in the decision the Justices declared:

> we find that the words "recognition and affirmation" incorporate the fiduciary relationship referred to earlier and so import some restraint on the exercise of sovereign power. (1990:181)

No matter how offensive the idea of "wardship" or the evolved and modern notion of "dependency" is to me, the fact of the matter is that there is a relationship of dependency between First Nations and the Crown. This *is* the reality that is the result of colonialism. As the ability to move away from these dependent and colonially inspired relations will not happen overnight, the fiduciary duty remains an interesting legal concept that might prove beneficial to First Nations' efforts to move toward truly independent and self-sustaining relationships with Canada.

In addition, the court in *Sparrow* noted:

> We cannot recount with much pride the treatment accorded to the native people of this country. (1990:177 affirming the decision of MacDonald J. in *Pasco v. Canadian National Railway Co.* 1986:37)

The assumption of superiority must be fully stripped away from all the current legal and constitutional interpretations with respect to Aboriginal Peoples and this requires more than dismissing history as unfortunate and lacking in pride. This is particularly true for section 35 as this section "recognizes and affirms" "existing Aboriginal and treaty rights." The constitutional provision requires that a standard of equality[7] be incorporated in all legal analyses.

The concept of "equality" in law has a long history in both Canadian law and in political theory.

> As used in 'liberal democracy,' the word 'liberal' connotes 'equal,' the fundamental equality of all human beings. A corollary of this ideology is the idea that each individual is free to achieve her or his own free and independent development in a free market. Liberal treatment also implicitly meant 'non-violent,' once again in reference (and in deliberate contrast) to the United States and its violent Indian policy. Chancellor Boyd, also in his *St. Catherines Milling* opinion, referred to a legal policy that promoted the immigration of Europeans in such a way so that 'their contact in the interior might not become collision.' Non-violence was implicit in the ideology of liberalism: *people treated equally, according to the law, did not need to resort to violence.* A just society was also a non-violent society. Canadians were committed to a frontier without the kind of warfare that they saw just below their border. This commitment had underlying reasons of economy as well as morality, for as some nineteenth-century observers noted, the cost of Indian wars to the United States in the decade of the 1870s exceeded the entire Canadian budget. But it was also a moral policy. (Harring 1998:12, emphasis added)

The problem with this kind of thinking, that less violence is better (and frankly I do not know how to understand the colonial process, past or present, as anything but violent) is how it acts to conceal the truth about certain events of this country's history. Carol Aylward has noted that "[w]hile the doctrine of 'colour-blindness' has promoted the myth that racism is no longer a factor in American society, the same doctrine has promoted a prevailing myth in Canada that racism was never a factor in Canadian society" (1999:76). In the United States, that "the west was won" (from the Indians) is the popular myth. In Canada, the myth is that the removal of the Indian nations from their territories was somehow peaceful (that is non-violent). Asserting a non-violent (or less violent) pattern of conquest seems to equate to suggesting a more advanced and civilized society. But, the truth of the matter is Canada has not always treated people with respect and dignity. For me, this exposes the degree to which Canada's foundations are *not* built on a principle of equality.

I understand too painfully well that asserting that Canadian notions of equality, particularly historically and legally, are more fiction than fact is a dangerous proposition. What passes for equality talk in Canadian jurisprudence is generally a conversation about power (and why one group has the power to keep it). Sherene Raczack notes:

> Rights in law are fundamentally about seeing and not seeing, about the cold game of equality staring. Talking about women's lives in the language of rights is a cold game indeed, a game played with words and philosophical concepts which bear little relationship to real life. In spite of these doubts, the game is always enticing, perhaps because it seems to hold out the promise that something about the daily realities of oppression will eventually emerge from under the ice. Equality staring, however, as Patricia Williams poetically describes, feels like a non-win situation. The daily realities of oppressed groups can only be acknowledged at the cost of the dominant group's belief in its own natural entitlement. If oppression exists, then there must also be oppressors, and oppressors do not have a moral basis for their rights claims. If, however, we are all equally human, with some of us simply not as advanced or developed as others, then no one need take responsibility for inequality. Moreover, advanced, more civilized people can reconfirm their own superiority through helping those who are less advanced. (1998:23)

From my own experiences, many times over, I know her words reflect the reality that I have lived. It is not this form of equality that I aspire to but rather the simple fact that all four human races have a fundamental right to respect and self-respect. It is through this cultural standard that I approach any kind of legal analysis (including an analysis of equality).

Section 35(1) of the constitution provides that the "existing aboriginal[8] and treaty rights" of Aboriginal Peoples are "recognized and affirmed." From the outset it is important to recognize that section 35 is not part of the *Charter of Rights and Freedoms*, nor is it a guarantee given to Aboriginal Peoples from the Government of Canada. It must be emphasized that the consent of Aboriginal nations to our position in the Canadian federation, including the application of Canadian law, is still an outstanding issue. In my mind, section 35(1) merely advances the possibility that such a conversation can now take place. Unfortunately, this is not what I see or understand to be happening in Canadian politics and Canadian courts.

A guarantee, such as the guarantees Canada provides in the *Charter*, is something qualitatively and quantitatively different than a recognition or an affirmation as found in section 35(1) of the constitution. When a recognition or an affirmation is made, the thing being recognized or affirmed already exists. In this specific case of legal-political relations, this relationship was *just* being seen by the drafters of the Canadian constitution for the first time. The rights, however, pre-exist the Canadian recognition of them. If section 35 were a grant of rights, this would ensure the ability of Canadian law to continue to embed the Euro-Canadian superiority myth. A grant is a gift of rights to the people from the Crown. As noted, section 35(1), recognizes and affirms certain rights of the Aboriginal Peoples. The words, recognize and affirm, are not intended to be the equivalent of granting rights. "Recognize" means to acknowledge something that already exists. "Affirm" means to embrace the rights which are now being recognized. To not accept that the words "recognized and affirmed," at a minimum, move us beyond thinking that western or European is superior renders the constitutional words meaningless. As a legal standard, the recognition and affirmation sets a serious goal. It requires complete rejection of the belief in Euro-Canadian superiority, the belief that denies Canadians a pride in their history.

Aboriginal Peoples have always understood that our rights are inherent. All that Canada can do is to begin to take responsibility for their historical and ongoing failure to respect the authority and legitimacy of Aboriginal governments. Now that this simple task of recognition has been completed by the entrenchment of section 35(1), the more pressing and onerous question of how to implement and respect this responsibility is without full answer in either political or legal realms. I would also point out the parallel to the situation of treaties. It is not that the agreements are "bad." It is that their implementation has been ignored.

The Canadian government has not gone (and could not go) further (such as a *Charter* style grant) than the recognition and affirmation of Aboriginal (and treaty) rights. It is because section 35(1) respects the Aboriginal view that our rights are inherent and pre-date the concept of Canada that a new relationship can be established based on (and preconditioned by) the 1982 constitutional entrenchment. This is the reason for which I suggest that section 35(1) holds promise. The way has been cleared to do Canada differently, to do Canada in a way that also includes Aboriginal people. This is the ultimate irony of section 35(1). Section 35(1), in my opinion, changes the political relationships between Aboriginal people and the state. However, section 35(1) has seen much more activity as a judicial mechanism than it has as a revolutionary political device.

In my mind the failure to secure Aboriginal consent to this constitutional provision is not fatal to the goal of establishing a renewed relationship. If all section 35(1) accomplishes is to recognize the independent relationship of Aboriginal rights to the Canadian governments, then logically no Aboriginal nation needs to consent. Consent only becomes a primary issue after the passing of section 35(1) because Canada first had to acknowledge the truth about its relationship with Aboriginal Peoples. Section 35(1) does not change, touch or interfere with in any way the Aboriginal view of the world, our values, beliefs, laws and systems of government. It is a long overdue promise by Canada, a "solemn commitment" to Aboriginal Peoples, as the court in *Sparrow* noted (1990:180).

The issue of outstanding consent to participate in Canada remains one of the principal keys to opening the door to a new and revitalized (truly a "nation to nation") relationship with Aboriginal Peoples. The entrenchment of section 35(1) creates the necessary pre-condition which now allows us to proceed to the more important question of how we will choose to relate to each other. In my view, section 35(1) is a mere invitation.

Unfortunately, as I noted before, I would not characterize the almost two decades of judicial activities since the entrenchment of section 35(1) as revolutionary. Clearly the problem with creating revolution through judicial activity is the fact that the judiciary is intended to be a stabilizing force, not a revolutionary one. Because it is adversarial by nature, litigation has not yet allowed Aboriginal Peoples and Canada to step outside "us versus them" relationships. There have been a number of attempts by the Supreme Court of Canada to specify the meaning of section 35(1). In my opinion, and this book will detail, these attempts have not yet provided a clear definition of the idea or theory of Aboriginal rights. More importantly, the judicial decisions have not lived up to the potential that exists in the words contained in section 35(1). My concern is that the fundamental principles (in the case of the Kanien'kehaka, peace, friendship and respect, as articulated by the Two-Row Wampum) required to create a "nation to nation" relationship must remain in our focus.

In *Sparrow*, the Supreme Court of Canada stated: "When the purposes of the affirmation of aboriginal rights are considered, it is clear that a generous, liberal interpretation of the words in the constitutional provision is demanded" (1990:179). In addition the law requires that the interpretation of documents (such as treaties) regarding Aboriginal Peoples must be construed liberally and any "doubtful expression be resolved in favour of the Indians" (*Nowegijick* 1983:94 and affirmed in *Sparrow* 1990:179). First, this strengthens the assertion that constitutionalizing Aboriginal rights moves us beyond assumptions of superiority. Second, it emphasizes that it is impossible to identify and deal with ambiguity without understanding the historical context in which the documents were drafted. If judicial interpretation of the section 35(1) commitment is to live up to its potential, the interpretation must also rely on Aboriginal understandings of our relationship with this land and with the state. Recognized and affirmed are relationship words more so than they are the standard jargon that lawyers associate with rights discourse.

The Aboriginal understanding of this historical context must become as important as the western understanding. I am not convinced that the Supreme Court's decision in *Delgamuukw*, where "oral history" is pronounced equal to written history in weight, is a sufficient commitment to accomplish this task. Embedded in this pronouncement is the idea that western written history is the same as oral history. It is not. Oral history must not be diminished against the standard accorded to written history in Canadian law. Oral history cannot be interpreted in the same way. Treating these forms of history as the same will likely lead to further imposition on the Aboriginal forms of history.

As a Kanien'kehaka woman and a member of two First Nations communities, I use my lived experience to help me accomplish this task of recognizing mono-culturalism in Canadian law. Obviously then, this recognition points to another problem with the reliance on Canadian law as the mechanism to resolve Aboriginal claims. Canadian law is not objective but rather grounded in Euro-Canadian cultural assumptions that are neither more nor less valid than the cultural assumptions of First Peoples which support our systems of law. I am not a "cultural" authority but rather a lawyer using my skills of legal reasoning to determine the degree to which the system I have been trained in is capable of understanding and respecting diverse cultures.

The discussion that follows is not built solely on the methods I learned during my years of legal study. Legal method was insufficient to meet my needs as an Aboriginal person, and my understanding of the proper way to look at the world. It requires that the scholar analyze the words of the judicial decision and not the cultural and social value systems embedded in the judicial pronouncements and their legal reasoning. I am concerned with both the ability of the courts to apply specific cultural and social values to Aboriginal Peoples as well as with the imposition of foreign government structures and laws on Aboriginal nations. Therefore, this realization must be accompanied by an attempt to adopt the legal method as I learned it as a student at the same time as I balance this skill against my understandings as a Kanien'kehaka woman. This skill is not necessarily unique to Aboriginal legal analysis. Dara Culhane writes:

> What I consider *not* readily accessible to common sense, and not a reflection of good sense, and therefore in need of explanation and criticism, are the Crown's positions and the evidence and theories relied upon to support them. This book is therefore a project in the anthropology of European colonialism: a study of power and of the powerful. I turn my anthropologist's spyglass on the law, an institution that quintessentially embodies and reproduces Western power. (1998:21)

This problem of the dominant cultural monopoly is pervasive throughout Canadian law and is a repeated theme in this discussion. One poignant example is offered here as demonstration of this concern. In the Gitksan and Wet'suwet'en trial (*Delgamuukw*), Chief Justice Allan McEachern, in his closing, made somewhat personal comments (and these kinds of comments are unusual for judges to make):

> The parties have concentrated for too long on legal and constitutional questions such as ownership, sovereignty, and "rights," which are fascinating legal concepts. Important as these questions are, answers to legal questions will not solve the underlying social and economic problems which have disadvantaged Indian peoples from the earliest times.
>
> ... This cacophonous dialogue about legal rights and social wrongs has created a positional attitude with many exaggerated allegations and arguments, and a serious lack of reality. Surely it must be obvious that there have been failings on both sides....
>
> It is my conclusion, reached upon a consideration of the evidence which is not conveniently available to many, that the difficulties facing the Indian populations of the territory, and probably throughout Canada, will not be solved in the context of legal rights. Legal proceedings have been useful in raising awareness levels about a serious national problem. New initiatives, which may extend for years or generations, and directed at reducing and eliminating the social and economic disadvantages of Indians are now required. It must always be remembered, however, that it is for elected officials, not judges, to establish priorities for the amelioration of disadvantaged members of society. (*Delgamuukw v. The Queen* 1991:276)

The first difficulty I have with McEachern's analysis is the manner in which he characterizes the "national problem" which Aboriginal people have become. It is a problem of "social and economic disadvantage." He, in addition, benevolently agrees "that there have been failings on both sides" accepting for the "state" a certain degree of responsibility. Nonetheless, this is a fundamental mischaracterization of the issue which was litigated in *Delgamuukw*. The Gitksan and Wet'suwet'en people were litigating (at least at the trial level) the issue of title to land and the question of jurisdiction (or which government has authority over the territory—Gitksan and Wet'suwet'en, provincial or federal).

Simply put, this is the offence in McEachern's judgment. He uses the robes of his judicial authority to manifest power, in my opinion colonial power, to trivialize the nature of the problem and the case before him. This judgment impairs the rights of Aboriginal people by turning them into "social and economic problems," when they were brought by the people as, and should be understood as, fundamental constitutional issues. These issues rightly deserve to be heard within the sphere of the "supreme law of the land." I am not suggesting that there are not pressing social and economic difficulties which Aboriginal nations must address, but rather simply that these were *not* the issues that the Gitksan and Wet'suwet'en chose to place before the courts. Not only does McEachern attempt to toss away from the judiciary any responsibility for the valid legal issues placed before the court by mischaracterizing those issues as social and economic problems, he then correctly points to the fact that programs for the amelioration of disadvantage of "members of [Canadian] society" (which I do not believe is an accurate description of how the Gitksan and Wet'suwet'en peoples see themselves) are the responsibility of "elected officials." This is tantamount to a whine suggesting that the judiciary should not have been bothered with these issues in the first place!

There are other examples in McEachern's short comments here quoted that demonstrate his unquestioned absorption of the colonial mentality. McEachern locates the economic and social disadvantage of Indian people in "earliest times." This must be a reference to the early period of European contact, as it is with contact that the conditions of disadvantage arose. It is historically incorrect, if not simply ridiculous, to refer to this as "earliest times." The people and the land were here long before European "discovery." Understanding time only through what Europeans knew about the world is one of the many ways that presumed European superiority is still being manifest.

There is a further significant issue that concerns me. I believe that Justice McEachern's quoted comments are intended to shield him from criticism of a decision that he must have known would be actively criticized. Otherwise, I am at a loss to explain the presence of those comments. He attempts to accomplish this goal in several ways. McEachern privileges his access to the evidence by stating: "it is not conveniently available to many." This may be true of the average Canadian but it is certainly not true of the access and expertise of the people themselves. McEachern, however, chooses to acknowledge his superior expertise as the source of differing opinions in an attempt to legitimize his decision and suggest that criticisms are the result of lack of information and knowledge. This strategy is reinforced by his comments in the second quoted paragraph. The logical implication of comments such as "positional attitudes" and "exaggerated allegations and arguments" is that they are also attempts to diminish the validity of critics' comments. According to McEachern, those critics operate from a position of "a serious lack of reality." If this is the degree to which Aboriginal people can rely on the judiciary to justly assess Aboriginal claims, then a serious problem of judicial credibility will continue to exist. It does not matter to me if McEachern intended these implications and consequences. Intention has little to do with racist and colonial attitude and talk.

It is precisely this form of thought that precipitates the dominant ideological monopoly, which results in the oppression of Aboriginal Peoples.[9] Justice McEachern attempts to distance law from the colonialism and oppression that First Nations have faced in Canada. Unfortunately, declaring this so does not make it true. Despite the fact that McEachern and I agree that law alone cannot fix it, law was and remains a central tool in delivering oppression and colonialism to First Nations. The judiciary, of whom McEachern is but one example, despite its convenient claim to objectivity and neutrality, must come to understand that the Canadian system of law is complicit in the oppression of Aboriginal Peoples. Further, and perhaps more important, judicial neutrality must be seen for what it is, a legal principle of great convenience. It is a principle that assists in the writing of colonial judgments such as McEachern's which are based on political judgments and moral reasoning.

Courts consider cases to be individual and isolated phenomena, and are rarely able to assess them within their historical and contemporary contexts. Yet this task is necessary to the just resolution of Aboriginal claims.

Aboriginal people turn to the courts for a number of reasons, including the recognition of their relationship with oppression and/or colonialism and the law. After all, law has always been the tool by which oppression/colonialism have been delivered to Aboriginal people in this country (Monture-Angus 1995:174). The day where Canadian law is experienced by Aboriginal people as freedom and justice has yet to arrive. It is not because Aboriginal people have faith in the Courts that legal action is commenced. It is generally because the people have no other perceived way of protecting their rights.

In addition, the artificial separation of political and legal spheres in the delineation of responsibility for ensuring the protection of "existing aboriginal and treaty rights" serves only to ensure that the government and courts are able to continue to side-step, in a complementary fashion, their concurrent responsibilities. For many Aboriginal people, the separation between courts and governments is unclear, arbitrary and of little sense. And in the result, both these branches of Canadian sovereignty can be seen and often are seen as equally oppressive.

This pattern of complimentary side-stepping of responsibility is overly familiar to Aboriginal people. During the 1970s when Aboriginal people desired to see reforms in the area of child welfare (generally a provincial responsibility), the provinces denied that they had any responsibility to "Indians." The federal government, which had the constitutional authority to legislate regarding "Indians," denied responsibility for child welfare. Children received no services beyond apprehension. From the Indian view, the federal government is our partner in the "nation to nation" relationship. Because of the failure of the federal government (and to a lesser degree provincial governments) to see Aboriginal nations as governments, we are left in the middle with no power or ability to resolve the problems that our communities face. This has been an effective state strategy for the avoidance of just resolution of the Aboriginal experience in this land since the advent of Canada.

The avoiding of personal and collective responsibility, such as evidenced by Chief Justice McEachern in the *Delgamuukw* trial, is doubly problematic when it is the supreme law of the land that is the focus of the judicial interpretation of Aboriginal (and treaty) rights. Judicial interpretation requires an awareness of how Aboriginal people have experienced our relationship with Canada, including the many ways colonialism is manifest (such as side-stepping, language, history and presumed European superiority). Accountability and responsibility for the protection of Aboriginal (and treaty) rights must be seen as personal, judicial and political. This seems like such a simple realization (and it is necessary to the solution). It is also elusive.

The solution requires that legal processes must come to embrace Aboriginal people's experience and the meaning that Aboriginal people attach to that experience. Chief Justice McEachern excused himself from any such personal responsibility in the early pages of his judgment:

> cases must be decided on admissible evidence, according to law. The plaintiffs carry the burden of proving by balance of probabilities not what they believe, although that is sometimes a relevant consideration, but rather facts which permit the application of the legal principles which they assert. The court is not free to do whatever it wishes. Judges, like everyone else, must follow the law *as they understand it.* (*Delgamuukw* 1991:6, emphasis added)

At the same time that McEachern dismisses Aboriginal beliefs, he substantiates his own belief in Canadian law as both legitimate and unquestionable. Unfortunately, because Aboriginal people see Canadian law as oppressive (and colonial) we do not necessarily share these same beliefs. Courts, law and justices have continued to fail to recognize that the Aboriginal experience of Canadian legal relations is *not* the same as their own. McEachern's understanding is not subject to the same level of scrutiny that he seems to be requiring of the Gitksan and Wet'suwet'en peoples. The Gitksan and Wet'suwet'en have only belief, whereas McEachern's cultural beliefs are *the* law and legal principles. This is an unacceptable double standard which amounts to thinly disguised racism because it exercises colonial power silently. In *Delgamuukw*, this judicial power has been exercised against Aboriginal interests and is readily apparent for those who are willing to look.

Canadian law will continue to be seen as an unjust system if this form of double standard continues to be masqueraded as legal thought and reasoning. As it stands now, all too often legal pronouncements embrace only the Euro-Canadian worldview and cultural heritage, despite the efforts of individual members of the judiciary and legal profession. This reality, of course, is not surprising in that it is substantiated and prefaced upon a misplaced notion of Aboriginal inferiority. Also of note is the fact that within the present system there are no formal or informal mechanisms, beyond the appeal process, for judicial accountability when cultural (mis)interpretation underpins the resolution of legal issues. These inadequacies in Canadian law are serious impediments to Aboriginal Peoples' attempts to utilize legal institutions to secure just resolution of our claims. The inadequacies are exposed as Aboriginal people attempt to use a system that was never intended to be used to address the kinds of issues that are endemic to Aboriginal claims.

Given the utmost importance of the task of legally interpreting these three small words, "existing aboriginal rights," it is important to focus on their meaning before we try to apply them in any kind of specific setting, such as a case. My choice to focus on "existing aboriginal rights" to the exclusion of "existing treaty rights" is not an indication that treaty rights are not as significant or as important as Aboriginal rights. Treaty rights have a history which is similar but also unique from Aboriginal rights. One does not necessarily exclude the other, that is, a right could be both Aboriginal or treaty at the same time. I start with the concept of Aboriginal rights because it is a reflection of how my own thinking on these issues developed.

When I first came to the issues presented by section 35(1) and began considering the meaning of the phrase, "existing aboriginal rights," I found it essential to consider the effect of applying western constructions of "rights" to First Peoples, whose varied cultures do not necessarily embrace such an ideological construct. What I now understand is that rights discourse is not necessarily or automatically relevant to Aboriginal cultures.[10] A system of responsibility makes more sense to the Aboriginal being. Until the parties involved can come to some form of consensus on this question of the meaning of rights then I believe the possibility exists for the constitutional affirmation to be misconstrued by conventional legal interpretation systems. There is a dangerous potential for the judicial determination of rights to occur in such a way that the difference in meaning attached to this word by Aboriginal Peoples will be invisible and therefore excluded.

Noel Lyon suggests that the shape of what is has changed with the entrenchment of Aboriginal rights in the constitution:

> Section 35 is a solemn commitment to honour the just land claims of aboriginal peoples, fulfill treaty obligations, and respect those rights of aboriginal peoples which the *Charter* ... recognizes as their fundamental rights and freedoms. What else could it be? *Constitutional reform is not done to continue the status quo.* (1988:101, emphasis added)

This passage was quoted by the Supreme Court of Canada in the *Sparrow* decision (at page 178). Constitutional scholars and lawyers (Aboriginal and Canadian) in conjunction with Aboriginal Peoples must articulate their understanding of what the status quo is before anything new can be constructed. Until we clearly understand what has been forced on Aboriginal Peoples (or what we need reject) we cannot understand what exactly requires renewing. This means that we have a lot of work to do before we can hope to interpret to its full potential the meaning of section 35 in specific cases. As the conventional legal system is absolutely not geared to this kind of analysis or process, I am fearful that the analysis which is essential to any future progress on Aboriginal rights will get lost in the rush to both prepare for litigation and keep court dockets moving. The opportunity that exists in section 35(1) for us, both as Aboriginal Peoples as well as to Canadians, may not again soon be presented.

In contemplating the meaning of section 35, I began to read about the history of rights (predominantly the liberal view), but was never satisfied. What I was reading did not fit the way that I had been brought up and the way that the Elders had taught me.

I could not situate my Aboriginal self within the discourse about rights (Monture-Angus 1995:152–168). The result was a tension that I could not initially resolve (and it continues to perplex me). It is this tension and how I have come to understand it, that I want to speak about.

What I want to avoid is constructing a competing theory of rights. I do not want to displace the western or liberal theory of rights with an Aboriginal theory of rights (particularly not a single theory, as Aboriginal Peoples are very diverse). But, at the same time I do not want a liberal or western theory forced upon me or my people. Let my people choose to pick it up if we decide that it is able to work for us.

Not belonging to the western culture, which spawned the existing theory of rights, I am not the appropriate actor to redefine the parameters of liberal rights theory. All the guidance that I can appropriately provide is to point to the fact that the theory is excluding my voice. My voice represents a separate culture and tradition; it is one of many voices within that culture, which in turn is one of the many cultures and traditions of Aboriginal Peoples. My dissatisfaction with the liberal theory of rights is probably also filtered by the fact that I am a woman.[11] The liberal theory of rights must recognize and affirm that it is possible for another theory of "rights" (in this case, more accurately, responsibilities) to exist within Aboriginal people's cultures, and that this other theory is legitimate. Equally important is the realization that these theories of rights can co-exist, as opposed to being competitive or combative.

Theoretically, we have to understand and accept that a right, to Aboriginal people, means something fundamentally different than it does within the sphere of Canadian legal relations. Recognition consists of different acts and understandings. Aboriginal people, as a consequence of centuries of being forcefully told we are inferior, must continue to remind ourselves that our experience and understanding are equally legitimate and encompass complete authority. Others have the difficult task of opening their eyes and ears for the first time. My objective here is to begin constructing, in a language that can be understood by people who have not been educated to our ways, what a theory of Aboriginal rights looks like from one Aboriginal location. Unfortunately, this is a formidable task, and more frequently I find myself articulating what the rights do not look like.

Examining the history of rights accorded to Aboriginal Peoples in Canadian jurisprudence, what I first came to understand was that it is a history bounded solely by Canadian law. The term Aboriginal rights is a term with a specific legal meaning and one that only expresses claims that have currently been accepted within Canadian law. Canadian legal process is therefore the sole gatekeeper of this process. I am concerned with the rights that have been excluded and I also refer to these as Aboriginal rights. The source of the tension is that Aboriginal rights in Canadian law do not embrace the much broader notion of Aboriginal rights that exist within my Aboriginal understanding. In effect the same words mean very different things (and the pattern of exclusion in how legal concepts are defined has already been identified as problematic).

The first recognition that must be made about Aboriginal rights in Canadian law is that they developed almost exclusively around the right to property.[12] I would assert that this right to property is more appropriately described as a struggle for the ownership and control of the land. It is not the land in and of itself that is important, but the ownership of the land. Beginning with *St. Catherines Milling* (1888) and continuing through important cases such as *Guerin* (1984) and *Calder* (1970) this pattern of "ownership of land" can be seen to be at the heart of these disputes.

A property claim in Canadian law does not have the capacity to include the Aboriginal holistic view of the land, because ownership in Canadian law is based on the linear notion of the domination of land. A holistic view of landholding enshrines not only the linear concept of ownership but also, at a minimum, spirituality and responsibility:

> With respect to the lands they lived on, many Indians felt a strong religious duty to protect their territory. Future generations would need the lands to live on, many previous generations had migrated long distances to arrive finally at the place where the people were intended to live. One could sell neither the future nor the past, and land cessions represented the loss of both future and past to most Indians.
>
> ... although Indians surrendered the physical occupation and ownership of their ancestral lands, they did not abandon the spiritual possession that had been a part of them. Even today most Indians regard their homeland as the area where the tribe originally lived. (Deloria and Lytle 1984:10–11)

The Aboriginal view of land rights encompasses both a notion of time as occupation (past, present and future) and a notion of spiritual occupation or connection. Both of these notions of Aboriginal occupation challenge the individualization of the common law system of property ownership. The relationship to land is seen not solely as a right but equally as a responsibility. In other words, the Aboriginal understanding of the relationship to land incorporates four separate (but integrated) ideas about land: individual rights and responsibilities, and collective rights and responsibilities.[13]

There have also been a number of important hunting and fishing cases heard by Canadian courts. The right to hunt or fish is one component, but not a nearly full articulation, of the right to use the land and in the case of Aboriginal people this right is one that was to exist forever—"as long as the grass grows and the rivers flow." It is often characterized in Canadian court decisions as a mere usufructuary right (a right of use, which is lesser than a right of ownership in Canadian law).[14] In Canadian law then, the history of Aboriginal rights is the history of the use and/or ownership of land rather than a relationship to be maintained to the law. Hunting and fishing rights are nothing larger, to date, than the further specification of the existing encapsulating structure of property relations as Canadian law understands them. As already indicated, this is a very narrow structure, which offends the holistic view of personal and collective relationships with the land held by Aboriginal Peoples.

It is precisely this relationship between Aboriginal rights and land that does not satisfy me. I do not think that the legally-based notion of property is a complete way of defining what Aboriginal right(s) are. The focus on property rights (or the squabble over ownership) has become misappropriately central in Canadian court decisions because it is the linear notion of ownership that is at the heart of the struggle between First Nations and Canada in the political realm. However, the consequence of dragging Canada's political inability to resolve the issue of the definition of land ownership, as well as the actual ownership rights, to the courtroom has been that Canadian law on Aboriginal rights has almost solely focused on the competing definitions of the relationship with land. This has happened with little attention paid to and little understanding of the Aboriginal intellectual tradition, which includes an understanding of land relationships as a foundational principle.

We must recognize that the competition of worldviews results in the failure to legally construct a compelling and complete theory of Aboriginal rights. Imagine the outcome if a series of contract, criminal or tort cases proceeded to adjudication without the benefit of a sound theoretical framework in which to resolve the issues. For a lawyer, this is unthinkable. The theoretical framework is in fact equally the role and *duty* of the court and not just of the legislatures.

Encapsulated in this recognition is a smaller theoretical conundrum. The larger concept of land rights is just a small portion of what must be talked about. In the common law history and the case law that we have to rely on, property is the almost exclusive place where attempts to characterize Aboriginal rights coalesce. I am not suggesting that a focus on property/land is or was wrong or even unnecessary. It was essential, as both the environment and Aboriginal Peoples' economic futures were fundamentally threatened by the loss of land. My point is simply that to locate a theory of Aboriginal rights on a borrowed construction of land rights is overly narrow, limiting, unfairly constraining and lacks vision.

When the legal analysis focuses on the framework in which individual decisions are embedded, further areas of contradiction with Aboriginal worldviews are exposed. (Legal academics are as equally responsible as the courts for this portion of the theoretical vacuum that has been created.) One such example is the way in which Canadian law is bounded by the important distinctions between matters public and private.[15] Public matters are generally matters in which one of the various arms of government is involved. Private matters on the other hand tend to involve individuals. Matters of family law between individuals are private, but matters of family law that involve the state (such as child protection hearings) are public. Aboriginal Peoples have had some success in having their traditional or customary laws regarding private-law matters, such as adoption and marriages, recognized in Canadian law.[16] Much has been made of this alleged accomplishment. However, the fact that Canadian law has recognized Aboriginal family relations only in the private-law sphere must be viewed as problematic.[17] Recent Canadian decisions where state removal of children has been the issue have not been as successful as the private-law cases. This location in the private sphere is a marginalization of Aboriginal rights and therefore is another way in which the legal structure embraces the notion of European superiority. From an Aboriginal point of view, family relations would not be seen as private-law rights. In fact, the public/private law distinction as an organizational principle of social order makes little sense to the Aboriginal mind.

Feminist writing and research, on the other hand, has pointed to the fact that so-called women's concerns tend to vest almost solely within the private sphere, and this has propagated discrimination against women in law. Care must be taken when constructing theories of Aboriginal rights to ensure that the same pattern of trivialization of women's concerns are *not* followed for Aboriginal systems of rights.

It is naive to think that Aboriginal Peoples will stop bringing their claims to court whilst this issue of the lack of a viable framework is addressed. Therefore, concurrently with continued adjudication, courts and legal academics must begin to significantly address the theoretical shortcomings and the structural obstacles in Canadian legal process and practice, and the myths (such as the myth of European superiority) which Canadian law continues to propagate.

I am more concerned about the exclusions than the inclusions (which is why a review of the case law is not central to the discussion I first wish to provoke). Interestingly enough, the case law method of Canadian law is a central problem for the kind of theoretical development that I am proposing. By wedding ourselves to the decisions of the past, we continue to entrench in present-day form, the oppressive relations of Canadian, British and French history. I understand that I am asking lawyers, judges and to a lesser degree lawmakers to do a remarkable thing, something they have never had to do in the past (as evidenced in Justice McEachern's words cited earlier in this chapter). I would hope that the legal profession looks upon it as a marvelous challenge and not a threat to the dominant legal system. After all, what we are talking about is defining and interpreting the supreme law of this land. It ought to challenge our collective and varied cultural legal imaginations.

It is very important that both the parallels and contradictions which exist between the Canadian system and the Aboriginal process of law are discussed before we go any further in interpreting section 35(1). First, such an analysis would dispel the myth that incorporating Aboriginal Peoples within Canadian legal structures (perhaps through separate systems) will somehow ghettoize the Canadian system as opposed to improving it (Monture-Okanee 1994:222–232). Second, it would alleviate the necessity of devoting (as now) the majority of our energy to telling you how different we are (which is really a discussion about why removing the boot of oppression from our necks is a good thing to do).

The recognition that the Canadian legal rights of Aboriginal Peoples are thus far merely rights about property enables one to recognize what has been excluded. As much of my work has been focused around the rights of Aboriginal offenders and the criminal justice system as it is experienced by Aboriginal people, and particularly Aboriginal women, it was easy for me to recognize what was glaringly absent from the case law on Aboriginal rights. We have not even begun to discuss what "human rights"[18] Aboriginal Peoples may have under section 35(1).

There are several reasons why this omission was fairly easy for me (and other Aboriginal people) to recognize. It is very much part of my culture to look at the history of any phenomenon to understand what it has come to mean today. The history of the development of human rights shows that it was not until 1985 and the *Bhinder* (1985) and *O'Malley* (1985) decisions, that the courts in Canada came to articulate that intent does not have anything to do with discrimination (Monture 1990:351–36). In Great Britain and to a lesser degree in the United States, intent was removed from the purview of the courts some fifteen years before the recognition was made in Canada (Vizkelety 1987:14–36). It is no wonder that sometimes it feels like it is taking us a very long time to get anywhere.

I want to tie the notion of human rights to the broader Aboriginal notion of land rights. As already discussed, the relationship that First Nations have to land is much broader than rights associated with the ownership of land. The relationship to land is spiritual and sacred.

The relationship to land reaches beyond and behind individual ownership, recognizing both past generations (the bones of my ancestors in the land) and future generations (which we refer to as "the faces in the sand"). Earth is Mother. Land is seen as part of the "human" family. It is part of all that is natural. Human rights in an Aboriginal frame of reference seem to me to include the relationship with the land. What I think human rights means in this context is the right to be self-governing or self-determining or sovereign. This right is essentially the right to be responsible. This is the most fundamental of all human rights (or responsibilities).

In attempting to understand the difficulties that arise from the failure to articulate a cohesive and complete theory of Aboriginal rights in Canadian law, I have identified three "eras" of Aboriginal rights litigation. During each of these eras, the case law developed differently. The first era is best characterized as a period where a string of small, random victories were embedded against a backdrop of political non-will. The lack of a cohesive and complete theory is most apparent in this period. This is the pre-1990 period and is fully discussed in chapter three. The second period covers the years 1990–1995. This period can best be characterized as the early years of constitutional negotiation. The Supreme Court in this period, particularly in the *Sparrow* case, took some brave, bold and new steps forward, the analysis of which is presented in chapter four. Finally, the Supreme Court in 1996 released several decisions which pulled back from the broad and solemn language of the *Sparrow* decision. The discussion in chapter five considers the *Delgamuukw* decision with a view to determining how much progress is being made. Chapter five also considers the degree to which the Supreme Court's articulation of Aboriginal rights can embrace Aboriginal demands for self-determination.

NOTES

1. Again, my thanks to Tom Porter for sharing this teaching.
2. Paul A. Wallace first published the *White Roots of Peace* in January of 1946 (see Wallace 1994).
3. The contributions of the Haudenosaunee to ideas about democracy can be found in: Jose Barreiro 1992, Laurence M. Hauptman 1986, Bruce E. Johansen 1982 and Oren Lyons et al. 1992.
4. This is Robert Warrior's phrase.
5. I also discussed this in my first book, *Thunder in my Soul,* in chapters 7 and 8 (Monture-Angus 1995).
6. Under the heading of the "Rights of the Aboriginal Peoples of Canada" a single section appears which addresses these rights:

    35. (1) The existing aboriginal and treaty rights of the aboriginal peoples of Canada are hereby recognized and affirmed.

    (2) In this Act, "Aboriginal peoples of Canada" includes the Indian Inuit and Métis peoples of Canada.

    (3) For greater certainty, in subsection (1) "treaty rights" includes rights that now exist by way of land claims agreements or may be so acquired.

    (4) Notwithstanding any other provision of this Act, the aboriginal and treaty rights referred to in subsection (1) are guaranteed equally to male and female persons.

7. Equality is a tricky word and it has many meanings. I use the word to simply mean respect for all peoples and traditions.
8. Section 35(2) provides that the Aboriginal Peoples includes the "Indian, Inuit and Métis."
   It is curious to note that the word "aboriginal" is not capitalized in the text of the constitution. This may be picking at small points but in my mind it is exemplary of the failure to respect Aboriginal Peoples that has permeated Canadian history. For this reason, I have chosen to capitalize the words in the text whenever I am not directly quoting another source.
9. I am not suggesting that this is the only problem that can be found within the reasoning of the trial court in the *Delgamuukw* case (also referred to in this discussion as the Gitksan and Wet'suwet'en case).
10. I am grateful to many Elders for helping me understand my confusion. In particular, I am grateful to Chief Jacob E. Thomas, Cayuga, Six Nations Territory.
    See also the discussion in Mary Ellen Turpel, "Aboriginal Peoples and the Canadian *Charter*: Interpretive Monopolies, Cultural Differences" (1989–90), 6 *Canadian Human Rights Yearbook,* 3–45.
11. In the view of John Locke, man's freedom is a measure of man's natural right to have equal access to the fruits of his labour. The subjection of women within the family, on the other hand, is seen by Locke as a necessary and legitimate means to ensure that private property relations are maintained. Women do not have a natural right to the fruits of their labour. Not only is this Lockean definition of equality problematic for women but it demonstrates the relationship of freedom and equality rights to relations of property ownership (or man's domination of the land). See John Locke 1980:39–42.
12. There is a relationship here between the evolution of Aboriginal rights and treaty rights. The development of specific treaty rights has also and unnecessarily been tied to the question of ownership of the land. Prior to the case of *R. v. Sioui* [1990], the federal government asserted that a treaty must involve a surrender of land. Cases such as *Sioui* and *Simon* [1986] demonstrate that fundamental change within Canadian law is possible.
13. A similar conclusion is reached by Leroy Little Bear 1976:30–34.
14. Aboriginal rights were first characterized as usufructuary rights of a personal nature (that is they are alienable only to the Crown) in the *St. Catherines Milling Case.* This view was rejected in *Guerin* where Justice Dickson stated:

    > The nature of the Indians' interest is therefore best characterized by its general inalienability, coupled with the fact that the Crown is under an obligation to deal with the land on the Indians' behalf when the interest is surrendered. *Any description of Indian title which goes beyond these two features is both unnecessary and potentially misleading.* (136, emphasis added)

    In an earlier case (*Calder*), Justice Judson opined:

    > . . . it does not help one in the solution of this problem to call it [Indian title] a "personal and usufructuary right." (156)

Scholars and judges who continue to rely on the usufruct as descriptive of Aboriginal rights rely on legal principles which Canadian courts have rejected in favour of a *sui generis* view of Aboriginal (land) rights.

15. The most famous case here is *Retail, Wholesale and Department Store Union, Local 580 v. Dolphin Deliver Ltd.* (1986).
16. See for example the following custom adoption cases: *Re Beaulieu* (1969), *Re Katie's Adoption Petition* (1961), *Re Tagornak* (1984) and the discussion in Norman Zlotkin (1984). For marriage cases please see *Connolly v. Woolrich* (1867) and the discussion in Constance Backhouse 1991:9–28; *Re Wah-Shee* (1975).
17. Although a lengthy discussion of several of the cases that respect Aboriginal customs in relation to adoption and marriage are discussed in *Partners in Confederation: Aboriginal People, Self-Government and the Constitution* by the Royal Commission on Aboriginal Peoples, 1993, pp. 5–8, this particular concern is oddly never mentioned. Of further concern is the fact that the brilliant feminist discussion pursued by Professor Constance Backhouse (1991) in her text, *Petticoats and Prejudice*, is also overlooked by the Royal Commission on Aboriginal Peoples. Neither of these oversights are acceptable as the exclusion is gender based.
18. I do not intend to limit the discussion of human rights protected under section 35(1) by defining human rights in the manner that Canada is presently accustomed to in its statutory framework. It is also ironic to note that the *Canadian Human Rights Code* section 67 provides an exception for the federal *Indian Act* regime and any regulations made thereunder. The result is that the *Indian Act* is not subject to the federal Human Rights Code.

6

State Responses to Racism in Canada

FRANCES HENRY ET AL.

> The law has been used through direct action, interpretation, silence and complicity. The law has been wielded as an instrument to create a common-sense justification of racial differences, to reinforce common-sense notions already deeply embedded within a cultural system of values . . . and to form new social constructions.
> —Kobayashi (1990:40)

This chapter explores the conflicting role of the Canadian state in both promoting and controlling racism. Public policies intended to ameliorate inequality each play a role in maintaining this conflict. On the one hand, the democratic state has a special responsibility to assert leadership and guard against the tyranny of the majority. Legislative action is the state's primary tool to promote and achieve equality and justice for all, regardless of race and colour.

On the other hand, legislation and the subordinate activities of the state can neither eliminate nor effectively control racism because the legacy of racism is so interwoven in the national culture, in its commonsense ideology and its public discourse. In this chapter, five major state responses are analyzed: multiculturalism legislation policy, the Canadian Charter of Rights and Freedoms, employment equity and human-rights codes and commissions,[1] the Employment Equity Act, and the Anti-Terrorism Act. The discussion considers the extent to which each of these political responses has delivered on its promise to diminish the legitimacy and impact of racial bias and discrimination in Canadian society.

The chapter proposes new ways of understanding the complex relationships between the state, the dominant culture, and ethno-racial minorities, particularly with respect to this book's central concern: the response to racism in a liberal democratic society. The principal thesis of this chapter is that despite the development of state responses that specifically acknowledge Canada's culturally and racially diverse population and recognize the existence of bias and discrimination, we have largely failed to achieve the goal of eliminating or even controlling racial bias and discrimination. We have failed precisely because the goal has been framed within a liberal framework and tradition.

INTRODUCTION

The state has many functions and responsibilities. One of its main roles is to proscribe behaviour. It also influences public opinion through its public-policy and legislative functions and helps thereby to define national ideology. Among its many responsibilities is the responsibility to support the social, cultural, and economic development of communities that suffer racial discrimination, by helping them to achieve full participation, access, and equity.

THE STATE'S ROLE AS PUBLIC–POLICY-MAKER AND DECISION-MAKER

The influence of state policies and practices at various levels (federal, provincial, and municipal) is critical to the eradication of racism and the promotion of racial equity. As such, the state has a special responsibility to assert leadership.

The fundamental rights and freedoms to which Canada adheres include the right of all residents to full and equal participation in the cultural, social, economic, and political life of the country. This right is based on the principle of the fundamental equality of individuals. The rights of equality of access, equality of opportunity, and equality of outcomes for all communities are therefore implicit. They are entrenched in a number of state policies and statutes, as well as in the international covenants to which Canada is a signatory.

The ideal of racial equity is a relatively new and still fragile tradition in Canada, because racism has only recently been acknowledged as a serious social concern. Both federal and provincial governments have, however, enacted legislation that in principle reflects their rejection of racism as a form of behaviour antithetical to a democratic state. The legislation includes the Canadian Charter of Rights and Freedoms, the Canadian Multiculturalism Act, the Employment Equity Act, provincial human-rights and labour codes, and the Anti-Terrorism Act, which in principle was enacted to ensure the nation's safety and security.

Many scholars in democratic liberal countries, including Canada, have pointed to the state and state apparatuses as primary sites through which racism is constructed, maintained, and preserved (Li, 199; Bannerji, 2000; Visano, 2002; Mackey, 1999). Racism occurs at several levels of governance that include the enactment and administration of laws, policies, institutions, and agencies of social control. At the same time, the state employs mechanisms of ideological control and power. Visano (2002) observes that the law functions as a set of institutional practices and discourses within the ideology of Whiteness and promotes a neutral response to racial injustices, "thereby escaping its complicities" (210). Bannerji suggests "the state of 'Canada,' when viewed through the lens of racism/difference, presents us with a hegemony compounded of a racialized common sense and institutional structures" (114).

International Declarations of Human Rights

The Canadian government has participated in several international declarations concerning human rights. The Universal Declaration on Human Rights was the first international covenant protecting human rights to be ratified by Canada. Since then, the United Nations has adopted a number of international covenants on human rights, including the International Convention on the Elimination of All Forms of Racial Discrimination, which was ratified in 1970. The convention is based on the conviction that any doctrine of superiority based on racial differentiation is scientifically false, morally condemnable, socially unjust, and dangerous.

Signing these international conventions creates the impression that Canada is committed to the development of an equitable society based on fairness and non-discrimination. International human-rights covenants provide Canada with global standards to which all federal legislation is expected to conform, but they do not bind the provinces. Moreover, most international instruments respecting human rights do not constitute a legally binding set of rules, and many contain no enforcement mechanisms. Case Study 6.1 illustrates this point.

CASE STUDY **6.1**

THE CANADIAN CHARTER OF RIGHTS AND FREEDOMS

BACKGROUND

The Canadian Bill of Rights was introduced by Prime Minister John G. Diefenbaker in 1960. Although it prohibited racial discrimination, it neither had constitutional status nor applied to provincial jurisdictions. Thus, prior to the enactment of the Constitution Act of 1982, the courts gave the Canadian Bill of Rights a very narrow interpretation. Constitutional questions about racial equality were resolved according to the "implied bill of rights" flowing from the constitutional division of powers. It is instructive that the "Fathers of Confederation" did not find it necessary to address the issue of racial equality in the provisions of the British North America Act. At the time, racial inequality was considered to be normative, and therefore the notion of providing constitutional guarantees to achieve racial equality was contrary to the collective ideology of the times.

In 1982, after a lengthy and controversial consulting process, the Canadian Charter of Rights and Freedoms was enshrined in Canada's constitution. For the first time in Canada's constitutional history, racial discrimination became unconstitutional. Enshrining the Charter in the Constitution was hailed as a triumph that was expected to put an end to many forms of overt racial discrimination in society.[2]

Section 15(1) of the Charter of Rights (the equity rights clause) came into effect in 1985 and is perhaps the most significant equality provision in the Charter.[3] It reads:

> Every individual is equal before and under the law and has the right to the equal protection and equal benefit of the law without discrimination and, in particular, without discrimination and, in particular, without discrimination based on race, national or ethnic origin, colour, religion, sex, age or mental or physical ability.

In prohibiting discrimination, section 15(1) provides five separate equality rights, namely: a right to equality before the law; a right to equality under the law; a right to equal benefit of the law; a right to equal protection of the law; and a right not to be discriminated against. It also protects affirmative action programs from constitutional litigation. Arguably, section 15(2) recognizes societal inequalities and permits affirmative action measures as a mechanism to assure equity for all Canadians.

DISCUSSION

While the Charter outlaws discrimination on the basis of race, it is seriously flawed in a number of important ways.[4] Section 1 establishes protection for all Canadians of certain basic rights and freedoms essential to a liberal democratic society; these include the protection of fundamental freedoms, democratic rights, legal rights, equality rights, Canada's multicultural heritage, Native rights, and the official languages of Canada. At the same time, however, the rights guaranteed in the Charter are subject to certain limitations: they should be reasonable, prescribed by law, demonstrably justified, and in keeping with the standards of a free and democratic society. Three of these four criteria included in section 1 are subjective and open to differing interpretations. The views of ethno-racial minorities and other disadvantaged groups on what is "reasonable" and "demonstrably justified" and on what meets the standards of a "free and democratic society" may be very different from those of the state.

The Charter does not define discrimination, racism, or race. Such interpretations have been left to the courts. Judges, justices, and lawyers have neither the expertise nor the training in social science to make determinations about the invisible network of racist discourses, beliefs, values, and norms that operate in a liberal democratic society. Many judges deny the existence of systemic racism. The courts are thus ill prepared to define the meanings and conditions of racial discrimination.

When a charge of racial discrimination has been made against an institution, a "cause of action" must initially be defined for the case to be heard in the courts. Essentially, a litigant must frame a legal claim within the legal parameter of a "cause of action." When the cause of a legal action involves racism and discrimination, it is virtually impossible to present an argument without a legal definition of such terms. Constitutional litigation involves a process of language, interpretation, and meaning.

Matas (1990) made a strong argument that the Charter is an inadequate and imperfect instrument for effectively addressing the problem of racism. He suggested that, while the Charter prohibits racial discrimination in law (section 15(1)), it does not require governments or legislatures to promote racial equality. The Charter is a passive instrument. It does not require governments or legislatures to do anything; it merely prevents them from doing certain things. The Charter prohibits racial discrimination in law, but it requires neither Parliament nor legislatures to design policies and programs to eliminate racial equality. There is no constitutional mandate to eradicate or even control racism.

For example, the Charter does not prevent one group of citizens from discriminating against another. As long as governments are not actively promoting inequality, they can legally wash their hands of what goes on in society at large (Matas, 1990). The real threat to equality does not come from legislative action but from the actions of private persons working within systems and organizations. Thus, the equality provisions in the Charter fail to address some of the real arenas of inequality in society.

Many of the same arguments can be made in discussing the weaknesses in the constitutional equality-rights process. For example, the constitutional equality-rights process. For example, the constitutional equality-rights system, through its procedures, makes the same flawed assertion as does the human-rights system, that is, that equality exists and that only the lapses from it need to be addressed. However, Kallen (1982) contended that there is a covert status hierarchy among the enumerated minorities who are eligible to receive specified protection for their human rights in section 15. Ethnic and multicultural minorities, Aboriginal peoples, and women have specified human-rights protection under other Charter provisions (sections 25, 27, and 28, respectively), whereas other enumerated minorities (such as racial minorities) do not.

Lack of a support structure for victims of inequality and the absence of a public agency with a capacity to challenge inequalities on behalf of a disadvantaged group is another major deficiency in the Charter. Perhaps one of the most important limitations of the Charter is the lack of guaranteed funding necessary to pursue a challenge. To raise an important Charter issue, a litigant must be willing and able to fight its case in the Supreme Court of Canada. To attempt a challenge of federal legislation on Charter grounds can cost $100 000 or more. Few citizens have the necessary funds. The absence of sufficient resources for minority groups exacerbates social inequalities.[5] This diminishes the section 15 provision and ensures that court actions are largely brought by wealthy litigants. It can be argued that a significant reason for the lack of Charter challenges by people of colour is that justice is economically inaccessible.

A further limitation of the Charter in relation to the protection of minority rights is the passivity of the courts. Gibson argued that the Charter "stands against a backdrop of the courts' passive tradition of self-restraint" (1985:39). Historically, the courts have deferred to democratically elected representatives. This weakness is clearly demonstrated in the repeal of the Employment Equity Act by the government of Ontario.

In sum, section 15 provides an inadequate guarantee against discrimination. Regardless of Supreme Court decisions and strong judicial statements condemning discrimination, the courts are an ineffective arena for enforcing or ensuring equality in Canada. Racial inequality cannot be eradicated by one act of legislation in one mainstream institution in society. The alleged importance of the Charter maintains the ideological fiction that the legal system can control the broader systemic biases found in Canadian society. Thus, the Charter can be said to be an instrument of the ideology of democratic racism in providing a liberal solution consistent with a liberal democratic society's view of the world. However, it provides only a template, without the authority required to implement racial equality.

The next major state policy is not aimed at addressing the problem of racism, but rather aims to protect Canadians from the impending threat of terrorism. However, as is argued in this section, the policy itself presents a significant danger to all Canadians and more specifically the rights and freedoms of Canadians of Arab, Muslim, or other Middle-Eastern backgrounds.

NEW STATE POLICY ON TERRORISM

Is the Anti-Terrorism Act an Expression of Racial Profiling by the State?

The events of 9/11 raised deep concerns about Canada's capacity to meet the threats of terrorism. Within three months, Bill C-36, the Anti-Terrorism bill came into law in December 2001. The legislation defines acts of terrorism as those actions that threaten Canadian lives or property, instill fear in society, or damage the economy, or that are targeted against political institutions and the general welfare of the country. It is an attempt by the federal government to control and contain world terrorism by identifying, prosecuting, convicting, and punishing terrorist activity. It is an omnibus bill, which amended 19 pieces of legislation. The Government of Canada Anti-Terrorism Plan has four objectives:

- stop terrorists from getting into Canada and protect Canadians from terrorist acts;
- bring forward tools to identify, prosecute, convict and punish terrorists;
- prevent the Canada-US border from being held hostage by terrorists and impacting on the Canadian economy; and,
- work with the international community to bring terrorists to justice and address the root causes of such hatred. (Government's Anti-Terrorism Act, 2001)

The new measures enshrined in the Act: allow the arrest of individuals without warrant; permit imposition of long, consecutive sentences for terrorist crimes and permit the imposition of long sentences for crimes committed during terrorist acts; make it easier for police to obtain warrants for wiretaps, which may be operated for three years; define what constitutes terrorist activity; allow the government to keep much more information secret; and increase the power of Canadian intelligence agencies to intercept electronic communication in peace and security. The Act gives the state new investigative and prosecutorial powers that include: "preventive detention," that is, the right to imprison people on the suspicion that they might commit a crime; allowing the police to compel testimony from anyone they believe has information about terrorism; closed trials of alleged terrorists; and, with a judge's permission, allowing the prosecution to deny an accused person and his or her lawyer all the knowledge of the evidence held against him.

Moreover, the definition of terrorism is extremely broad, and critics maintain that it could even be used to prosecute trade unionists involved in illegal strike activity or other harmless people involved in civil disobedience. Regular tactics such as protests, blockades, and other peaceful tactics used by many dissenting groups in society would be subject to the Act's strictures. The increased police powers and the government's ability to suppress information about its own activities are also sharply criticized. The Act also has the ability to modify 22 existing laws including the Criminal Code, the Canadian Human Rights Act, the Access to Information Act, the National Defence Act, and several others. Critics also maintain that it might lead to violations of the Charter of Rights and Freedoms. During the parliamentary debate on the Act, most of the time was spent on deciding whether to add a three-or-five-year sunset clause to some of its powers.

Since the Anti-Terrorism Bill was enacted, it has been a deeply contested issue. Middle-Eastern communities, and Canadian Muslims in particular, have felt under siege.

Organizations such as the Canadian Islamic Congress and the Canadian Arab Federation have argued that the bill is an assault on the basic civil liberties of every Canadian. Since the bill was ratified, Canadian security forces such as the RCMP and CSIS have questioned hundreds if not thousands of Muslim Canadians about travel patterns, prayer habits, associations, and other seemingly innocuous matters.

The president of the Canadian Civil Liberties Association, Alan Borovoy, suggests, "The legitimate war on terrorism doesn't require measures as broad as that." Toronto civil rights lawyer Clayton Ruby said the law "will wind up being used against peaceful protests and demonstrations. It will wind up being used to suppress opposition" (Bourrie, 2002). Human-rights groups have demonstrated grave concern over the overarching provisions of the bill. The International Civil Liberties Monitoring Group, on May 14, 2003, observed that the Anti-Terrorism Act grants police expanded investigative detention, and undermines the principle of due process. "All of these changes occur on the basis of a vague, imprecise and overly expansive definition of terrorist activity" (quoted in a speech by Senator Andreychuk, 2003). Canada's top spymaster and former CSIS chief from 1987–1991, Reid Morden, argues that "The Canadian government, in its race to catch up, went beyond the British and American legislation defining terrorist activities to include legal, political, religious and ideological protests that intentionally disrupt essential services. . . . The definition of 'terrorism' is so wide that I could easily include behaviour that doesn't remotely resemble terrorism" (quoted in a *Toronto Star* editorial, December 1, 2003).

It can be also argued, as Charles Smith does, that the Anti-Terrorism Act is yet another form of racial profiling (2004). The following are a few cases in which men of Middle-Eastern background living in Canada became victims of racial profiling by the state. In the war on terrorism they became the new enemy aliens and their democratic rights were seriously curtailed.

On the second anniversary of 9/11, two Canadian citizens, Imams Kutty and Abdool Hameed, were en route to Florida to lead an Islamic prayer service when they were detained by U.S. immigration agents, interrogated for 16 hours, and then released. Mohammed Elmasry, president of the Canadian Islamic Congress, observes that the Canadian news media reported on the incident with a degree of mild protest, "not because the men's detention was unfair and unreasonable, not because they were prominent members of a Canadian religious community, nor even because they were Canadian citizens, but because they were moderates" (*Globe and Mail*, September 30, 2003). In August 2003, the Canadian RCMP arrested 21 visa students accused of having ties to the international al-Qaeda terrorist group. All the detainees were Muslim students from Pakistan and India. They were later cleared of having any link to terrorist groups. Hassan Almrei, a 29-year-old refugee from Syria, is being held in a Canadian prison—and has been for two years—in solitary confinement without charge or bail, on a security certificate that does not allow for any of the alleged evidence against him to be shared with either Almrei or his lawyer.

Perhaps the most dramatic case of anti–Muslim bias and discrimination is that of Maher Arar, a Canadian Muslim. See Case Study 6.2.

CASE STUDY **6.2**

The Role of Canadian Officials in the Arrest and Imprisonment of Maher Arar

BACKGROUND

Maher Arar was born in Syria in 1970. In 1987 he came to Canada and later took out dual citizenship, retaining his Syrian citizenship and also becoming a Canadian Citizen. After receiving a master's degree in computer engineering, Arar worked in Ottawa as a telecommunications engineer.

In September 2002, the U.S. authorities arrested him while he was on an airline stopover in New York on his way home to Montreal. He was carrying his Canadian passport. Agents from U.S. Immigration and Naturalization alleged that Arar had links to al-Qaeda. He was first deported to Jordan and then Syria, where he was imprisoned and tortured for one year, before being released in October 2003.

All during the time of Arar's incarceration, his wife, Monia Mazigh, lobbied every level of government for his release. She appeared before the Foreign Affairs Committee. In November 2003 the RCMP began a formal investigation of its own role in the matter. The Commission for Public Complaints Against the RCMP asked "the force to respond to allegations that encouraged US officials to deport Mr. Arar . . ." (*Globe and Mail*, November 6, 2003). However, serious questions arose about the inadequate scope of the investigation, as the RCMP were unable to investigate the CSIS, the Foreign Affairs Department, or any person or institution in the United States. Also, it was pointed out that the RCMP operate without independent oversight or public representation, and thus in reality were investigating themselves.

In January 2004, Arar launched a lawsuit against the American government, alleging that U.S. officials deported Arar knowing that Syria practised torture. Later in the same month, Public Safety Minister Anne McLellan called for a public inquiry into the Arar case, to "assess the actions of Canadian officials in dealing with the deportation and detention of Maher Arar" (http://www.cbc.ca/news/background/arar). It is important to note that no charges were ever laid against Arar in Syria, the United States, or Canada. At the time of publication of this book, a public inquiry is under way.

ANALYSIS

The issues raised by this case involve questions about the rights, freedoms, and civil liberties of Canadian citizens who have been born in other countries, and are identified in some way as a threat to Canada or the United States. It can be argued that the tragic events of 9/11 have also had devastating consequences for law-abiding Canadian Muslims and other Canadians of Middle-Eastern origins.

Audrey Macklin (2002) suggests that "Members of diasporic communities, whatever their immigration status, will experience more than ever how boundaries demarcated by ethnicity, culture, religion and politicization emerge in sharp relief when viewed through the lens of the state's surveillance camera" (338). There is much anecdotal evidence to suggest that there has already been a significant impact on particular racialized groups across Canada, including Arabs and Muslims or those perceived to belong to these communities (see Smith, 2004). The Canadian Muslim Civil Liberties Association has recorded 110 hate attacks and the Canadian Islamic Congress indicates that such acts have increased by 1600 percent since 9/11. Moreover, the Anti-Terrorism Act itself, as discussed above, may put these groups at even greater risk of victimization.

The racialized discourses embedded in the anti-terrorism legislation are reflected in many other laws, particularly those related to immigration and refugees (see Chapter 3). In Canada's history, the ideology underpinning these policies has always reflected the binary polarizations separating "us" and "them," "insiders" and "outsiders." The language of the Act intimates that the nation is in a state of profound social and political crisis. Employing a discourse of moral panic, the implicit assumption is that not only is our national security at risk, but also our sense of national identity has become more fragmented and fragile.

THE STATE AS EMPLOYER: EMPLOYMENT EQUITY

Background to the Enactment of Employment Equity Legislation

Concern over employment discrimination against people of colour (visible minorities), women, persons with disabilities, and Aboriginal peoples led the federal government to establish a royal commission on equality in employment (Abella, 1984). Its task was to inquire into the employment practices of 11 designated Crown and government-owned corporations and to explore the most effective means of promoting equality for the four groups cited above. Its findings echoed earlier studies and public inquiries, that bias and discrimination were a pervasive reality in the employment system. The commissioner, Judge Rosalie Abella, observed that "strong measures were needed to remedy the impact of discriminatory attitudes and behaviours." The purpose of the Employment Equity Act was to "achieve equality in the workplace, and to correct the conditions of disadvantage in employment experienced by **designated groups**—women, Aboriginal peoples, persons with disabilities, and members of visible minorities in Canada" (Employment and Immigration Canada, 1988; Agocs, Burr, and Somerset, 1992). The remedy Abella recommended was employment equity legislation (ibid.).

Employment equity legislation provided a necessary framework to support a diverse workforce. Employment equity was intended to change the workplace by identifying systemic barriers in policies and practices that may appear neutral, but which, while not necessarily discriminatory in intent, are discriminatory in effect or result. The goal of employment equity was fair treatment and equitable representation throughout the workplace.[6]

Equality in employment means that no one is denied opportunities for reasons that have nothing to do with inherent ability (Abella, 1984). The Act applied to employers and Crown corporations with 100 employees or more under federal jurisdiction. It required all federally regulated employers to file an annual report with the Canadian Employment and Immigration Commission. The report was to provide information for a full year on the representation of all employees and members of designated groups (visible minorities, women, persons with disabilities, and Aboriginal peoples) by occupational group and salary range and on those hired, promoted, or terminated. In addition to filing the annual report, employers were required to prepare an annual employment equity plan with goals and timetables, and to retain such a plan for a period of at least three years. The Employment Equity Act was revised and adopted by Parliament in 1995, strengthening the legislation and bringing the public service, the RCMP, and the military under the purview of the Act.

There has been huge resistance to efforts to implement employment equity at every level of institutional life. For example, efforts to make the Canadian public service more representative of the Canadian public have failed. The 1996 annual report of the Canadian Human Rights Commission documents in stark numbers the huge gap between the government's commitment to a public service that mirrors the diversity of the Canadian population and its dismal record in promoting minorities.[7] In examining some of the reasons for this failure, Senator Noel Kinsella, formerly a senior bureaucrat with Heritage Canada, is quoted (in Samuel and Karam, 1996) as saying "Institutions act as a collective memory carrying forward values, principles and traditions. . . ."

A review of the data contained in annual reports of federal regulating bodies by the president of the Treasury Board in 1995 revealed that small progress had been made with respect to the hiring of members of employment equity targeted groups. For example, the percentage of women in the public service increased from 42.9 percent in 1988 to 47.4 percent in 1995. The percentage of Aboriginal representation increased from 1.7 to 2.2 percent, while visible-minority representation increased from 2.9 to 4.1 percent. In all of these categories, the available labour pool is much higher.

In 2001, the Canadian Federal Human Rights Commission found that representation of visible minorities in the private sector increased from 4.9 percent in 1987 to 11.7 percent as of December 31, 2001, and the percentage of hires increased from 12 percent in 2000 to 12.6 percent in 2001, which surpassed the 1996 census benchmark of 11.6 percent of visible minorities in the population. In the public sector, visible minorities held only 2.7 percent of all positions in the federal public service. By March 31, 2002, this had increased to 6.8 percent. Despite this improvement, the Human Rights Commission is concerned that "at the current rate of progress, the government will fail to meet its target," which, as contained in the new Embracing Change initiative, was set at a 20 percent hiring goal to be attained by March 2003. In the executive category visible minorities held only 3.8 percent of positions and of 73 new hires in this category in 2001–2002 only three went to visible minorities.

In a report released in September 2002 and prepared by the Public Service Commission, it was found that federal hiring rules are widely ignored by managers and favouritism is still very common. Managers continue to hire spouses, siblings, cousins, and people they know rather than conforming to the rules designed to ensure that public service jobs are accessible to everyone. Commonly, hiring transactions appear to be developed with the aim of appointing an already known person to the position. These conclusions are based on more than 1000 hirings that took place in eight government departments (cited in *Ottawa Citizen*, January 2, 2004). Such practices have very serious implications, because they substantially limit visible-minority access to public service positions.

In its most recent report of 2003, the Commission again found that hiring and promotion had not reached the targets set earlier by the Treasury Board. Visible-minority representation in the public service reached 7.4 percent, still less than the target, and their share of new hires, 9.5 percent, was less than half the target. In the executive category, only 7 out of 82 persons were visible minorities (Annual Report, 2003).

The inequity exists within many other institutions. In this connection, identified below are some of the myths and misconceptions used by government bureaucrats and politicians, school administrators and academics, editors, journalists and publishers, and police to support their resistance to hiring and promoting people of colour, Aboriginal peoples, women, and persons living with disabilities. The rhetorical themes embedded in these myths reflect the discourses of democratic racism.

"Employment Equity Is Reverse Discrimination."

According to this view, employment equity requires employers to discriminate against better-qualified Whites and gives an unfair advantage to people of colour. However, what is required is not reverse discrimination but the end of a long history of employment practices that result in preferential treatment for White males.

"Employment Equity Ignores the Merit Principle."

This myth is perhaps the most widely believed and promoted. It is based on the assumption that employment equity and other anti-discrimination measures will result in the hiring and appointment of "unqualified" individuals and bring an end to the merit principle. In reality, however, equity programs do not require the abandonment of standards and qualifications. Rather, they eliminate irrelevant criteria such as the colour of one's skin, cultural background, disability, and gender.

"Employment Equity Stigmatizes Minorities."

According to this misconception, minorities will never know if the have been selected on the basis of their qualifications or because of their group membership. One could argue that White males for two-hundred years should have been asking themselves the same question. Implicit in the above assumption is the perception that there are no meritorious persons in this group. As one observer noted,

> Do all those corporate directors, bankers, etc., who got their job for extraneous reasons—first because they were somebody's son, second, because they were male, third because they were Protestant, and fourth because they were White—feel demeaned thereby? It would be interesting to ask them—or to ask the same question of those doctors who managed to get into good medical schools because there were quotas keeping out Jews, the skilled tradesmen who were admitted to the union because two members of their families recommended them and so on. (Green, 1981:79)

Clearly implicit in the standard critique of these proactive measures is the notion that being rewarded is the natural result of being part of the majority or the elite. Rewards are only demeaning when one is a member of a minority or a marginalized group (e.g., women).

"Fairness Is Best Achieved by Treating Everyone in the Same Way."

Opponents of equity measures argue that in a democratic society, treating everyone equally is sufficient to ensure fairness in the workplace. However, as Abella pointed out, "We now know that to treat everyone in the same way may offend the notion of equality" (1984:3). She suggested that ignoring differences and refusing to accommodate them is a denial of equal access and opportunity; it is discrimination.

"Employment Equity Means Hiring by Quotas."

Although mandatory quotas are not required in employment equity programs, a widespread perception exists that governments require specific numbers of racial minorities to be hired, promoted, or appointed in specific organizations. Instead, employment equity in Canada requires employers to set goals for their organization, taking into account the number of qualified individuals from target groups that are available in the potential workforce as well as the composition of the internal workforce. Flexible goals and timetables are used to establish benchmarks toward representative hiring and promotion.

THE STATE AS PURCHASER OF GOODS AND SERVICES: CONTRACT COMPLIANCE

Contract compliance is a method of influencing private companies to implement an employment equity program. Under contract compliance, a vendor's contract is contingent on the existence of an equity program. The penalty for non-compliance is loss of the contract with the federal government.

The federal government has specific criteria governing contract compliance. A company must design and implement a program that will identify and take steps to remove barriers in the selection, hiring, promotion, and training of select minority groups. As in employment equity, compliance is largely voluntary; each company is expected to establish special programs in areas where imbalances exist (Jain, 1988).

This program can be criticized on many grounds. For example, the equity programs are not legislated and operate at the discretion of government, and government conducts only random audits of companies that have promised to implement employment equity programs. This allows companies to avoid developing a program until an audit occurs. What develops, therefore, is a cycle of delay in which federal bureaucrats and their corporate counterparts negotiate to delay the implementation of employment equity.

Some companies that supply highly specialized services are not audited. Corporate clients are permitted to set their own goals and timetables to match what is considered reasonable for their peculiar settings, but anti-racism is not a high priority for some companies. Since the government has not applied a criterion of success to this process, companies proceed at their own, often slow, pace.

THE STATE AS PROVIDER AND FUNDER OF SERVICES

Communities affected by racial discrimination continue to feel excluded from public services. They also perceive little support in their efforts to develop the community infrastructures and support systems required to meet the needs of their communities (*Equality Now*, 1984). Studies carried out in the past decade on access to government services (Mock and Masemann, 1987) suggest that racial minorities and Aboriginal peoples do not have equal access to or participate adequately in government programs and services.

Notwithstanding this concern, all levels of government for many years have provided some support to racial-minority, community-based service organizations. Generally, the funding is in the form of short-term project support; but despite varying levels of support and varying criteria, this support is at least a recognition by the state that racial-minority organizations play a critical role in ensuring that the community derives equal benefit from public service. Such support is an appropriate bridging strategy, until public structures and programs adequately serve all communities. However, on the other hand, it encourages the development of parallel services that existing public institutions should be offering.

The question of whether satisfactory service can be effected only by separate provision continues to be debated. Would separate service agencies along racial lines meet minority needs, or would they further fragment the state's delivery system? Would matching rather than mixing racial clients provide more emphatic help to people in need? Is "separate services" a euphemism for segregation?

Another concern is that in providing support to racial-minority organizations, the level of public support is grossly deficient. Separate services have become synonymous with inferior services. Expecting far too much for far too little, minority organizations have been exploited by the state as an expedient way to deliver public services to people of colour.

Arrangements made within public organizations determine to whom services are provided, to whom facilities are made available, and to whom resources are allocated. The attitudes and actions of those who direct these institutions determine who gets what, where, and when. Since people of colour are generally absent from the key decision-making processes in these organizations and in the delivery of services, to what extent do public institutions treat people of colour less favourably?

Although considerable resources have been expended on the training of public servants in "multiculturalism," "race relations," or "managing diversity," to what extent have the special needs of minorities been identified? And to what extent have they been adequately considered and provided for? To what extent have the resources of the state been reallocated in favour of minorities as part of a commitment to equity? Within a framework of genuine, equal sharing of public resources, it appears that the scale on which this has been done by any state agency or institution in Canada is insignificant in addressing the imbalances caused by racially discriminatory policies, programs, and practices. Initiatives taken to measure and address racial minorities' inequalities of access to public programs and services in Canada have been tentative and piecemeal.

CONCLUSION

In many ways, the role of the state in responding to issues of racism and inequality has had to fit within the context of an imagined national culture consisting of a unique blend of English and French cultures, and an identity built on English and French values. As a result, three categories of citizens were recognized: English Canadians, French Canadians, and "others" (Fleras and Elliot, 1992). Only the first two groups had constitutional rights. The construction of undesirable "otherness" has persisted as Canadians have continued to struggle for a national identity (Mackey, 1999). This notion of "otherness" can be seen from three interlocking perspectives:

- "Otherness" provides the dominant White culture with unmarked, invisible privilege and power.
- Issues are deflected in a way that suggests these "others" threaten the democratic fabric of Canadian society.
- There is a reassertion of individual rights and identity over collective identity and group rights

Those positioned within the privileged discourses of democratic racism and Whiteness and intent on maintaining their power assert their claim on the liberal values of individualism, equal opportunities, tolerance, and so on. In so doing, they construct a view of ethno-racial and Aboriginal peoples who do not share these values and therefore are outside the boundaries of the common culture of the state. Anglo-European culture dominance asserts its entitlement and authority within the very policies of the Canadian state, defining all others as "ethnics," "minorities," "immigrants," and "visible minorities," who are then marginalized and rendered subordinate to its unmarked centre. The power elite determines which differences and which similarities are allowed in the public domain (Suvendrini and Pugliese, 1997; Mackey, 1999). That authority "to define crucial homogeneities and differences" is defended within the liberal discourse of equality and progress (Asad, 1979:627).

For many Canadians, the increasing pluralism of Canadian society poses a threat to the way they have imagined and constructed Canadian identity. They hold on to an image of Canada distinguished from other countries, particularly the United States, by its French-English duality. Many Anglo-Canadians and others fear that multiculturalism will never provide a solution to the issue of national identity. Canadians want to resolve French-English tensions without having to address the multicultural issue of identity. One scholar argued that Canada is a nation in which "state-sanctioned proliferation of cultural difference itself is seen to be its defining characteristic" (Mackey, 1996:11). Multiculturalism as state policy embraces, in theory, the notion of cultural and racial diversity. Ethno-racial minorities are declared to be part of the "imagined" community of Canada. Anderson used the term "imagined community" to define the concept of "nation." A nation is "imagined" because the members of even a very small state do not know each other, yet in "the minds of each lives the image of their communion" (1983:15). However, in reality, the policy and practice of multiculturalism continues to position certain ethno-racial groups at the margins rather than in the mainstream of public culture and national identity.

The "symbolic multiculturalism" of state policy does not consider the necessity of restructuring or the need for a reconceptualization of the power relations between cultural and racial communities based on the premise that communities and societies do not exist autonomously but are deeply woven together in a web of interrelationships. Liberal pluralist discourse is unable to move beyond "tolerance," "sensitivity," or "understanding" of the "others." The state's construction of symbolic multiculturalism as a mechanism for maintaining the status quo can be seen in many forms of public discourse around issues of race, culture, difference, politics, and identity. This discourse is not restricted to the public declarations of policy-makers, legislators, and bureaucrats. It is also reflected in the language and practices employed by the state through its institutions and systems, including justice and law enforcement, print and electronic media, cultural and educational institutions, and public-sector corporations (Tator, Henry, and Mattis, 1998).

Visano (2002) argues that Whiteness as a plurality of assumptions, beliefs, and discourses continues to influence public policies that support the status quo. He states, "Law is a set of spectacular and performative moments that promote a neutral response to racial injustices, thereby escaping its complicities. . . . Legal illusions pacify and legitimate the articulation of a culture of Whiteness that guides the everyday behavior of law" (210).

People of colour and Aboriginal peoples are seldom invited into the mainstream discourse of Canadian dominant White culture. The select few are considered to be models and are imagined as being different from others of their kind. In contrast, the airing of diverse perspectives by people of colour on issues related to fundamental human rights (e.g., the Anti-Terrorism Act), Aboriginal rights, multiculturalism, racism, and employment equity are commonly dismissed, deflected, or ignored. In a liberal democracy, justice and equality are already assumed to exist. Therefore, Aboriginal and ethno-racial minorities' demands for access and inclusion are seen as "radical," "unreasonable," "undemocratic," and a threat to cherished democratic, liberal values. The small gains made by minorities and women are seen by the mainstream culture as being "too expensive" economically and ideologically. Dissent by the oppressed is considered disruptive and dangerous.

The "symbolic multiculturalism" of state policies holds to a paradigm of pluralism premised on a hierarchical order of cultures that under certain conditions "allows" or "tolerates" non-dominant cultures to participate in the dominant culture. Such an approach imagines minority communities as "special-interest groups," not as active and full participants in the state and part of its shared history. This paradigm represents notions of tolerance and accommodation, but not of equity and justice. It holds to a unified and static concept of identities and communities as fixed sets of experiences, meanings, and practices rather than of identities as dynamic, fluid, multiple, and historically situated. In summary, state policies continue to be largely centred on the maintenance of the status quo.

SUMMARY

This chapter has analyzed the roles and functions of the state as lawgiver, policy-maker, employer, purchaser, and provider of goods and services. It has argued that public policies, including the Canadian Multiculturalism Act, the Canadian Charter of Rights and Freedoms, the Employment Equity Act, the Anti-Terrorism Act, and human rights codes and commissions have been inadequate instruments to address racial inequality in Canadian society. The discourse of liberalism underlying these policies—the rhetoric of rights, reasonableness, freedoms, equality, standards, tolerance, understanding, and so forth that is incorporated into all of these state responses to some extent—is in itself a limiting factor in the struggle to control racism. Moreover, new public policies in relation to immigration, and more specifically, the Anti-Terrorism Act may function to deprive ethno-racial individuals and groups, particularly those of Arab and/or Muslim descent of their fundamental civil rights.

Although these state responses are vastly different from the more overtly racist and assimilationist policies of earlier governments, the ideology and discourse of racism that are deeply embedded in the collective belief, value, and normative system of Canadian society affect the way laws are interpreted and implemented. The aforementioned public policies may be significant steps on the path to racial equality, but they are not instruments of societal transformation. Despite the government's recognition of ethno-racial diversity through public policies on multiculturalism, minority rights, and equity, social inequality not only continues to operate but is actually reproduced and legitimated through the state.

The notions of tolerance, accommodation, diversity, and equality are woven into the rhetoric of federal, provincial, and municipal politicians and other public authorities. At the same time, their discourses, policies, and practices reflect a deep ambivalence to undertaking the tasks required to attain racial equity. This conflict has shaped the way state policies have been constituted and the way they have been implemented.

This analysis points up a central paradox of modern liberal societies, first identified by Goldberg (1993): as modernity increasingly commits itself to the ideals and principles of equality, tolerance, and freedom, new racial identities and new forms of racism and exclusion proliferate.

Racialized public discourses concerning multiculturalism, rights and freedoms, and employment equity tend to employ ill-defined ideas and implicit notions regarding culture, differences, race, and racism, which when operationalized function socially and politically to marginalize ethno-racial minorities. Democratic racism is manifested in both subtle and overt forms of the discourse that shapes the very policies and practices designed to ameliorate racial inequality.

NOTES

1. Most provinces in Canada have appointed an ombudsperson to handle the complaints of citizens who believe they have been treated unjustly by an agency of government. The office of the ombudsperson cannot deal with individual complaints of racism, but it has investigated complaints against provincial human-rights commissions, primarily regarding long delays in their management of cases.
2. The Charter has not been extensively used in race cases. For a complete review, see Mendes (1997).
3. Because of the far-reaching significance of the equality rights guaranteed in section 15, the federal and provincial governments gave themselves three years to change their laws and policies to comply with the Charter. Thus, while the rest of the Charter came into effect in 1982, section 15 came into effect three years later.
4. It is indeed paradoxical that the Charter, with all its identifiable weaknesses, has nevertheless been perceived negatively by people who fear its power. Judge Rosalie Abella, for example, notes that "in less than a generation, this remedy for discrimination has been seen to be sufficiently powerful that people struggle urgently to find a remedy from equality. How ironic that 'equality-seeker' has become a pejorative term, denoting someone whose claim to fairness is a menace to the nation's economy and psyche" (*Toronto Star*, October 26:A20).
5. The Charter Challenges Program, supported by the federal government, provides some financial assistance to disadvantaged groups who wish to pursue Charter cases.
6. Equitable representation depends on the following factors: the number of designated group members in the working-age population in a certain geographical area, the number of trained or skilled members who are employable or can be readily available, and the equal opportunities for change that exist in each workplace.
7. In March 1987, visible minorities made up 2.7 percent of the public service and 6.3 percent of the Canadian population. By March 1996, their share of the government employment had risen to 4.5 percent, but their representation in the population had jumped to 13 percent. In management, minorities held only 2.3 percent of the executive positions in the public service (Samuel and Karam, 1996).

REFERENCES

Abella, R. (1984). *Equality in Employment: The Report of the Commission on Equality in Employment. Ottawa: Supply and Services Canada.*

Agocs, C., Burr C., and Somerset, F. (1992). Employment Equity: Cooperative Strategies for Organizational Change. New York: Prentice Hall.

Anderson, B. (1983). *Imagined Communities.* London: Verso.

Andreychuk, R. Senator. (2003). *Case Study: Canada Anti-Terrorism Legislation.* Presented at Commonwealth Human Rights Initiative Seminar: Human Rights and Anti-Terrorism Legislation in the Commonwealth. June 5–6.

Angel, S. (1988). "The Multiculturalism Act of 1988." *Multiculturalism* 11(3): 25–27.

Annual Report. (2003). Ottawa: Canadian Human Rights Commission. Available <http://www.chrc-ccdp.ca/publications/reports-en.asp>, accessed August 30, 2004.

Asad, T. (1979). "Anthropology and the Analysis of Ideology." *Man* 14:607–27.

Bannerji, H. (2000). *The Dark Side of the Nation: Essays on Multiculturalism, Nationalism and Gender*. Toronto: Canadian Scholars' Press.

Bhabha, H.K. (1990). "The Third Space." In Jonathan Rutherford (ed.), *Identity: Community Culture and Difference*. London: Lawrence and Wishart. 207–21.

Black, B. (1994). *Report on Human Rights in British Columbia*. B.C. Human Rights Review, Communications Branch, Ministry Responsible for Multiculturalism and Human Rights. Government of British Columbia.

Bourrie, M. (2002). "Tough Anti-Terrorism Legislation Called Threat to Civil Liberties." *Law Times*. Ottawa: Canadian Law Book Inc. Available <http://www.canadalawbook.ca>, accessed August 30, 2004.

Canadian Heritage—Multiculturalism Program. (1990). *The Canadian Multiculturalism Act: A Guide for Canadians*. Ottawa: Multiculturalism and Citizenship Canada. Excerpt reproduced with the permission of the Minister of Public Works and Government Services Canada and the Minister of Canadian Heritage, 1999.

Corporate Policy Branch of Multiculturalism and Citizenship Canada. (1988). "Canadian Multiculturalism Act Briefing Book: Clause by Clause Analysis." Unpublished document released under Access to Information Act.

Creese, G. (1993–94). "The Sociology of British Columbia." *BC Studies* Special Issue 100 (Winter).

Day, S. (1990). *Human Rights in Canada: Into the 1990s and Beyond*. Ottawa: Human Rights Research and Education Centre.

Diène, D. (2004). "Addendum: Mission to Canada." Report for the United Nations Commission on Human Rights, Sixtieth Session, Item 6 of provisional agenda. March. Commission on Human Rights. Geneva, Switzerland: Office of the High Commissioner for Human Rights. Available <http://www.ohchr.org/english/issues/racism/rapporteur/annual.htm>.

Duclos, N. (1990). "Lessons of Difference: Feminist Theory on Cultural Diversity." *Buffalo Law Review* 38:325.

Employment and Immigration Canada. (1988). *Annual Report, Employment Equity Act*. Ottawa: Minister of Supply and Services.

Equality Now! Report of the Special Committee on Participation of Visible Minorities in Canadian Society. (1984). Ottawa: Queen's Printer.

Fleras, A., and J.L. Elliott. (1992). *Multiculturalism in Canada: The Challenge of Diversity*. Scarborough, ON: Nelson.

Gibson, D. (1985). "Protection of Minority Rights Under the Canadian Charter of Rights and Freedoms." In N. Nelville and A. Kornberg (eds.), *Minorities and the Canadian State*. Oakville, ON: Mosaic Press.

Goldberg. D. (1993). *Racist Culture: Philosophy and the Politics of Meaning*. Oxford: Blackwell.

Government's Anti-Terrorism Act, The. (2001). *Former Prime Minister's Newsroom Archive (1995–2003)*. October 15. Privy Council Office site <http://www.pco-bcp.gc.ca>, accessed August 29, 2004.

Green, P. (1981). *The Pursuit of Inequality*. New York: Pantheon.

Harding, S. (1995). "Multiculturalism in Australia: Moving Race/Ethnic Relations from Extermination to Celebration?" *Race, Gender, and Class* 3(1)(Fall):7–26.

Itwaru, A., and N. Ksonzek. (1994). *Closed Entrances: Canadian Culture and Imperialism.* Toronto: TSAR.

Jain, H. (1988). "Affirmative Action Employment Equity Programs and Visible Minorities in Canada." *Currents: Readings in Race Relations* 5(1)(4):3.

James, C. (1995). "Multiculturalism and Anti-Racism Education in Canada." *Race, Gender and Class: An Interdisciplinary and Multicultural Journal* 2(3)(Spring):31–48.

Kallen, E. (1982). "Ethnicity and Human Rights in Canada." In P. Li (ed.), *Race and Ethnic Relations in Canada.* Toronto: Oxford University Press.

Kobayashi, A. (1990). "Racism and the Law." *Urban Geography* 11(5):447–73.

Li, Peter (ed.). (1999). "Race and Ethnicity." *Race and Ethnic Relations in Canada.* Toronto: Oxford University Press. 3–20.

Mackey, E. (1996). "Managing and Imagining Diversity: Multiculturalism and the Construction of National Identity in Canada." Unpublished doctoral dissertation, Social Anthropology, University of Sussex, UK.

—. (1999). *The House of Difference: Cultural Politics and National Identity in Canada.* London: Routledge.

Macklin, A. (2002). "Borderline Security." In R. Daniels, P. Macklem, and K. Roach (eds.), *The Security of Freedom: Essays on Canada's Anti-Terrorism Bill.* Toronto: University of Toronto Press.

Matas, D. (1990). "The Charter and Racism." *Constitutional Forum* 2:82.

Mendes, E. (ed.). (1997). *Racial Discrimination and the Law: Law and Practice.* Toronto: Carswell.

Mirchandani, K., and E. Tastsoglou. (2000). "Toward a Diversity Beyond Tolerance." *Journal of Status in Political Economy* (Spring):49–78.

Mock, K.R., and V.L. Masemann. (1987). Access to Government Services by Racial Minorities. Toronto: Ontario Race Relations Directorate. Available. <http://ceris.metropolis.net>.

Moodley, K. (1983). "Canadian Multiculturalism as Ideology." *Ethnic and Racial Studies* 6(3):320–31.

Mullard, C. (1982). "Multiracial Education in Britain: From Assimilation to Cultural Pluralism." In J. Tierney (ed.), *Race Migration and Schooling.* London: Holt, Rinehart and Winston.

Multiculturalism and Citizenship Canada. (1989–90). *Annual Report of the Operation of the Canadian Multiculturalism Act.* Ottawa.

Ontario Human Rights Code Review Task Force. (1992). *Achieving Equality: A Report on Human Rights Reform.* Toronto: Ministry of Citizenship.

Samuel, J., and A. Karam. (1996). "Employment Equity and Visible Minorities in the Federal Workforce." Paper presented at the Symposium on Immigration and Integration, October 25–27. Winnipeg: University of Manitoba.

Smith, C. (2004). "Borders and Exclusions: Racial Profiling and Canada's Immigration, Refugee and Security Laws." Paper commissioned for annual meeting of Canadian Court Challenges Program, October 2003.

St. Lewis, J. (1996). "Race, Racism and the Justice System." In C. James (ed.), *Perspectives on Racism and the Human Services Sector.* Toronto: University of Toronto Press. 104–19.

Suvendrini, P., and J. Pugliese. (1997). "Racial Suicide: The Re-licensing of Racism in Australia." *Race and Class* 39(2)(October–December):1–19.

Tator, C., F. Henry, and W. Mattis. (1998). *Racism in the Arts: Case Studies of Controversy and Conflict.* Toronto: University of Toronto.

Tepper, E. (1988). "Changing Canada: The Institutional Response to Polyethnicity." In *Review of Demography and Its Implications for Economic and Social Policy.* Ottawa: Carleton University.

Visano, L. (2002). "The Impact of Whiteness on the Culture of Law: From Theory to Practice." In C. Levine-Rasky (ed.), *Working Through Whiteness: International Perspectives.* Albany: State University Press: 209–237.

Walcott, R. (1993). "Critiquing Canadian Multiculturalism: Towards an Anti-Racist Agenda." Master's thesis, Graduate Department of Education, York University.

Wallace, M. (1994). "The Search for the 'Good Enough' Mammy: Multiculturalism, Popular Culture and Psychoanalysis." In D. Goldberg (ed.), *Multiculturalism: A Critical Reader.* Cambridge, MA: Blackwell. 259–68.

Yalden, M. (1990). "Canadian Human Rights and Multiculturalism." *Currents: Readings in Race Relations* (Toronto) 6(1):2.

7

Protecting Equality in the Face of Terror: Ethnic and Racial Profiling and s. 15 of the *Charter*

SUJIT CHOUDHRY

A. INTRODUCTION

The anti-terrorism omnibus bill currently before Parliament is a lengthy and complex statute whose provisions, taken together, have the twin goals of enhancing the effectiveness of existing instruments, and creating new instruments, for law enforcement officials to prevent terror both in Canada and abroad. What is striking about the legislation is its silence on one of the central issues in the public debate over how Canada and other liberal democracies should respond to September 11: ethnic and racial profiling (which I shall refer to simply as profiling in this paper).

Profiling has burst onto the national agenda in the wake of the horrific events of September 11. Long criticized by academics and public interest organizations examining the workings of the criminal justice and immigration systems, profiling has now, as a result of September 11, attained renewed prominence. The reason is clear—the hijackers identified by American law enforcement officials all appear to have been Arab, and the argument made by proponents of ethnic and racial profiling is that had airport security officials engaged in profiling, the terrorist acts of September 11 could have been prevented. It was no doubt this sort of reasoning that led retired Major-General Lewis MacKenzie, now security advisor to Premier Mike Harris, to suggest that profiling would be an acceptable law enforcement strategy to fight terror.[1] Indeed, an editorial in the *National Post* went so far as to state that 'it would be criminally negligent if Air Canada did not engage in racial profiling.'[2] Advocates of profiling have not made clear who it is who would be profiled—Arabs, persons of Middle Eastern appearance, or Muslims, three groups whose membership overlaps, but is not at all identical. I will assume that it is the first group which is at issue.

Despite the lack of clarity over who would be profiled, the mere prospect of profiling has met with a chorus of disapproval. Federal Fisheries Minister Herb Dhaliwal has denounced the practice, having once been profiled himself,[3] as have organizations such as the Canadian Arab Federation, the National Council on Canadian Arab Relations, and the Canadian Muslim Civil Liberties Association. And a controversial directive issued to port-of-entry immigration and customs officers, published in the *Globe and Mail*,[4] prompted federal Immigration Minister Elinor Caplan to state '[t]here is no racial profiling, not by gender or religion.'[5]

What is extraordinary about the debate over profiling is the absence, for the most part, of any analysis of whether it would be constitutional. This is all the more extraordinary, since the constitutional concerns raised by the omnibus bill have already generated considerable interest in the legal community, and will propel various provisions of the statute to court in the weeks, months, and years to follow. In my view, the rather minimal public attention devoted to the constitutional challenges raised by profiling is a direct function of the form that such a policy would likely take. If immigration and law enforcement agencies begin to engage in the profiling of persons of Arab background or appearance, they will do so through means—ranging from internally distributed departmental memoranda, to informal word-of-mouth directives issued by superior officers—which are less visible and hence less susceptible to public scrutiny and democratic debate than publicly promulgated legal texts such as statutes and regulations. Civil libertarians must therefore ensure that in focusing so closely on the text of the omnibus bill, they do not overlook the threat posed by other components of the war against terrorism to the very values that that war seeks to defend. This is particularly true in a multiracial and multiethnic democracy such as Canada, which is constitutionally committed to equality and non-discrimination.

B. WHAT IS RACIAL AND ETHNIC PROFILING?

What is racial and ethnic profiling? It is important to be absolutely clear here, because the debate over profiling has not yielded a precise definition of what that practice is. As Randall Kennedy of the Harvard Law School has suggested, there seem to be two definitions of profiling.[6] The broad definition holds that profiling consists of a decision to detain or arrest an individual, or to subject an individual to further investigation, '*solely* on the basis of his or her' race or ethnicity.[7] The narrow definition is that profiling consists of the use of race or ethnicity *along with other factors*, such as suspicious behaviour. The narrow definition has been seized upon by advocates of profiling, not only because it may better fit the actual practice of law enforcement (a disputed point), but also because it appears to dilute the importance of race and ethnicity, and hence gives the impression that race or ethnicity would not drive law enforcement decisions. However, Kennedy correctly argues that this inference is based on faulty reasoning, because allowing the use of race or ethnicity even as one factor must mean that it can play a 'decisive' role in whether to subject an individual to further investigation.[8] That is, any use of race and ethnicity may serve to distinguish two individuals who otherwise manifest identical suspect behaviour, subjecting one to heightened scrutiny, while letting the other walk free. And if this is true, then decisions to target law enforcement will still be made on the basis of race or ethnicity, even if that factor is one among many.

There are two additional points to note here. First, to the best of my knowledge, profiling has been advocated in order to target investigative efforts, but not as a reason for final decisions, such as to prove guilt in a criminal proceeding, or to deny persons admission to Canada. This is important, because at times, profiling has played this role. During the Second World War, for example, the internment of Japanese Canadians amounted to the use of profiling to conclusively deprive persons of their liberty without due process of law, not merely to subject them to more probing investigation.

Second, Kennedy's definitions are very much framed in the context of the United States, where a significant body of evidence documents the use of profiling in law enforcement against African Americans and Latinos. In recent years, the tool of choice for law enforcement has been the pretext stop, in which a police officer stops a driver, ostensibly because of a traffic violation, for the purpose of searching his vehicle and its passengers in connection with non-traffic related offences, usually related to drugs.[9] The evidence suggests that pretext stops have been employed disproportionately against African Americans and Latinos, because law enforcement officials believe those groups are more likely to commit drug-related offences. Many will be familiar with the recent controversy surrounding racial profiling in New Jersey, where state troopers had a policy of pulling over African American and Latino drivers, while the state governor denied the very existence of that policy for over a decade.[10] What is striking about the American experience is that prior to Sept. 11, a broad consensus had emerged that racial profiling in policing was unacceptable and should be banned. According to a Gallup poll taken in 1999, 81% of respondents, including 80% of white respondents, opposed racial profiling in policing.[11] During the 2000 presidential campaign, both candidates pledged to ban racial profiling in policing, and President Bush reiterated his position as late as July, 2001.[12] Indeed, a bill that would ban profiling—the *End Racial Profiling Act*—was introduced in the United States Senate in June, 2001.[13]

The dominance of the American experience should not lull Canadians into thinking that racial and ethnic profiling is not a problem that we face. It most certainly is. Let me cite two pieces of evidence. The first is a study initially prepared by Scot Wortley of the University of Toronto's Centre for Criminology for the commission on Systemic Racism in the Ontario Criminal Justice System.[14] The study surveyed Torontonians to establish whether the likelihood of people being stopped and searched by police officers differed on the basis of race. The study found that African Canadians were twice as likely as whites to be stopped once, and four times as likely to be stopped more than once. In particular, African Canadian men seem to be the targets of policing activity. The second piece of evidence is an analysis by Wortley of data collected by the African Canadian Legal Clinic that strongly suggests that African Canadians are subject to higher levels of scrutiny than white individuals by customs and immigration officers at Pearson International Airport.[15]

For the sake of simplicity, when I talk about profiling in Canada, I would like to focus on two situations that involve profiling by the state or its agents, and hence that engage the *Charter*. First, there is the use of profiling at the border, with respect to the degree of scrutiny that travellers —be they citizens, permanent residents, or visitors—receive by customs and immigration officers as they enter Canada. Second, there is the use of profiling by airport security, prior to boarding, on both domestic and international flights. Needless to say, there are other examples I could use—such as the profiling of applicants for immigration, or of candidates for civil service jobs—both of which have been suggested in the wake of September 11, and both of which are extremely problematic from a constitutional perspective. But I have chosen these two examples because they are at the forefront of public debate.

C. THE CONSTITUTIONALITY OF RACIAL AND ETHNIC PROFILING UNDER S. 15 OF THE *CHARTER*

Is profiling constitutional? I will confine my constitutional analysis to the *Charter's* equality rights provision, s. 15, and its reasonable limits clause, s. 1. For this reason, I will not be touching upon ss. 7 through 10, which confer procedural rights on persons who come into contact with the criminal justice system. However, as I will show, there is some crossover between the procedural fairness and equality arguments.

The test for determining whether there has been a s. 15 violation was recently rearticulated by the Supreme Court in *Law v. Canada*,[16] and has three parts: (a) that a distinction be drawn, (b) that it be drawn on the basis of a prohibited ground, and (c) that it be a discriminatory distinction. The first two steps would be easily met by a policy of profiling, because the whole point of profiling would be to target individuals for heightened scrutiny on the basis of their race and ethnicity, both of which are enumerated in s. 15 as prohibited grounds of discrimination. The heart of the analysis, as in all recent s. 15 cases, would be whether this differential treatment is discriminatory. To answer this question, *Law* states that we must ask whether the distinction demeans the dignity of the rights-claimant, where dignity 'means that an individual or group feels self-respect and self-worth,' understood in terms of both 'physical and psychological integrity and empowerment.'[17]

Several strands of the Court's equality jurisprudence suggest that profiling would be found to be discriminatory under s. 15. One of these is the centrality of stereotyping to a finding of discrimination. In the context of s. 15, *Law* explained that a stereotype can mean one of two different things. It can mean that a distinction drawn on the basis of a prohibited ground reflects the view that all members of a group who share that characteristic—race, religion, gender, and so on—possess certain undesirable traits that in fact none of them do. The most virulent forms of anti–Semitism, for example, rely on stereotypes of this sort. Far more common, though, is a stereotype that is an over-generalization—that is, an assumption that all members of a group possess certain undesirable traits that some members of those groups possess, when in fact some, or many, do not. The harm to human dignity—what transforms the use of stereotypes into discrimination—is that doing so has the effect of stigmatizing all members of that group, by promoting the view that they are somehow less worthy of respect and consideration, because they all possess the undesirable trait in question. The Court suggested that these claims would be easiest made out if the group at issue experienced pre-existing prejudice, because the use of stereotypes in framing government action can interact with and reinforce that existing disadvantage. Finally, the perspective to be adopted in this inquiry is subjective and objective: 'that of the reasonable person, dispassionate and fully apprised of the circumstances, possessed of similar attributes to, and under similar circumstances as, the claimant.'[18]

Front and center in *Law*, then, are the *social meaning* of government practices and the *social context* within which those practices occur. In *M. v. H.*, for example, the Court found that the exclusion of same sex couples from the spousal support provisions of the Ontario *Family Law Act*, in the context of widespread prejudice against same sex couples, meant that the exclusion promoted the view that same sex relationships were 'less worthy of recognition and protection.'[19] In the context of September 11, profiling would take the fact that some Arabs committed terrorist acts as a reason to subject *all* Arabs to heightened scrutiny. In other words, profiling employs race and ethnicity as a *proxy* for the risk of committing terrorist or criminal acts. But to profile in this way raises the serious danger of tarring an entire group with the crimes of the few, by giving rise to the myth that being an Arab reflects a propensity to engage in terrorist activity.

An aid to the success of this argument would be the demonstration of pre-existing disadvantage faced by Arabs. But the Court also made it very clear in *Law* that proof of pre-existing social disadvantage was not a *sine qua non*. What would be very significant to the s. 15 inquiry, in my view, is the kind of treatment that profiling entails. In both of my examples, profiling would involve the differential imposition of the burden of law enforcement on Arab travellers by agents of the state, in public and in full view of other passengers. The image here is stark: persons of Arab appearance being taken aside before boarding for more intensive questioning, or being steered at customs and immigration for secondary questioning, not because of any evidence that could reasonably give rise to suspicion of their links to terrorist organizations, but rather because of their physical appearance. It would also be significant that profiling would occur in the context of law enforcement, where the goal is to identify those who pose a threat of engaging in criminal activity. The mere fact of being subject to heightened scrutiny in the criminal context carries with it a stigma. Subjecting members of a racial or ethnic group to more probing investigation threatens to stigmatize an entire community.

To be sure, this kind of targeting would be rightly viewed as revolting by many Canadians. Indeed, images of the coercive power of the state directed at visible minorities in a public manner—at blacks in South Africa or in the American South, at Muslims in Bosnia—stand for many Canadians as paradigmatic examples of the grossest violations of human rights. In all of these cases, what was often involved was the selective enforcement of laws that themselves did not distinguish upon the basis of race or ethnicity. And that same history has taught us the chilling lesson that the targeting of state power on minorities can socially label those persons as deviant or inferior. It can lead otherwise reasonable and fair-minded people to rationalize or explain away the discriminatory application of state power, and even to engage in discrimination themselves. This prospect is all the more likely in a climate of fear.

Defenders of profiling might counter that the extent to which profiling stigmatizes turns on *how* profiling is handled in practice. The frame of reference here is the use of pretext stops in the United States, where what is objectionable is not simply the use of racial profiling, but the way in which those stops and the subsequent searches have not infrequently been conducted with guns drawn, with verbal abuse, with no explanation as to why individuals have been stopped, and with no apology afterward.[20] A defender of profiling would argue that if handled professionally and politely, the more intrusive investigation of Arabs by airport security and immigration officials would minimize the indignity and stigmatization experienced by those passengers.

However, I reiterate that the relevant perspective under s. 15 is that of the reasonable person in the position of the rights-claimant. Even if the heightened scrutiny is handled professionally and politely, for the persons who are subject to profiling, the indignity is real. Those travellers would essentially be asked to establish their legitimacy; they would be placed in the position of having to state and justify their reasons for travelling, an entirely legal activity, while other travellers would face no such burden. They would face this burden of persuasion every time they fly, or pass through customs and immigration. The cumulative effect on individuals of bearing this burden, simply because of one's looks, in a historical and social context where the differential imposition of the burden of law enforcement on the basis of race and ethnicity have been identified with the most odious forms of discrimination, would be enormously damaging on their self-respect and self-worth. And it is a cost that advocates of profiling altogether ignore.

Another factor identified by *Law* as relevant to a finding of discrimination is the nature of the interest at stake, the idea being that the more important the interest, the more likely a finding of discrimination. In the two examples I have chosen, the interests at stake are nothing less than physical liberty and privacy. It is important to be absolutely clear on how physical liberty and privacy are implicated in these two situations. I am not suggesting that anyone has a constitutional right to travel by air, or that non-citizens have a constitutional right to enter the country. However, in both of these cases, profiling would entail additional questioning by agents of the state regarding a variety of personal matters. It might also entail the involuntary, physical redirection of travellers to areas where they can be questioned further. The significant point here is that both privacy and physical liberty attract the protection of several rights in the *Charter*, unlike many of the interests which the court has found sufficiently important to count as reasons in favour of finding distinctions discriminatory e.g., the protection of a human rights code,[21] or the right to seek spousal support payments.[22] The constitutional status of these interests should count as a reason to view with deep suspicion distinctions drawn with respect to their enjoyment.

Although *Law* strongly suggests that racial profiling would be found to violate s. 15, it contains one significant holding that cuts the other way. The Court in *Law* stated that a distinction is not discriminatory if it treats a rights-claimant 'differently on the basis of actual personal differences between individuals.'[23] The Court made this point by reference to an earlier decision upholding the separate placement of a child with a learning disability, which, the Court held, did not discriminate against her because it was made in her best interests.[24] This argument, if anything, could cut against profiling, because profiling treats people differently regardless of anything about them *per se*, but rather because of behavioural inferences drawn from their race and ethnicity. However, the Court in *Law* then went on to say that distinctions made on the basis of 'informed statistical generalizations' would not be discriminatory, because they corresponded closely enough with actual need.[25] What does this statement mean? Look at the holding of *Law* itself: the Court upheld age-based restrictions on survivors' benefits under the Canada Pension Plan on the ground that younger individuals were more likely to be able to re-enter the workforce than older persons—using 'age as a proxy for long-term need'[26]—knowing full well that this generalization did not hold true for many individuals. What the Court seemed to be saying is that relying on generalizations is acceptable in some circumstances, even if those generalizations are stereotypes. This is an apparent contradiction to what the Court said a few paragraphs earlier in its judgment.

The implications of permitting statistical generalizations for the constitutionality of profiling under s. 15 are clear and dramatic. But *Law* contained narrowing language, suggesting that this derogation from equality principles was only permissible in the context of remedial social benefits legislation which is under-inclusive, and stating that generalizations would be impermissible 'where the individual or group which is excluded by the legislation is already disadvantaged or vulnerable within Canadian society.'[27] These caveats should have been enough to put to rest any suggestion that the use of statistical generalizations would be acceptable in applying the burden of law enforcement upon a racial and ethnic minority.

What gives me pause for thought, though, is the majority judgment of the Supreme Court in *Little Sisters v. Canada.*[28] In that case, the Court held that Customs agents had discriminated on the basis of sexual orientation by targeting their enforcement efforts at homosexual erotic materials being imported by a lesbian bookshop, while adopting a relatively lax attitude to the importation of heterosexual materials by mainstream booksellers. *Little Sisters*, on its face, would thus seem to set a helpful precedent, because it found profiling based on sexual orientation to be discriminatory under s. 15. But the majority went on to say that the targeting would have been constitutional had it been based on 'evidence that homosexual erotica is proportionately more likely to be obscene than heterosexual erotica,' because in those circumstances, there would be a 'legitimate correspondence between the ground of alleged discrimination (sexual orientation) and the reality of the appellant's circumstances (importers of books and other publications including, but by no means limited to, gay and lesbian erotica).'[29] So even in a law enforcement context involving the differential application of burdens on a vulnerable minority, *Little Sisters* suggests that an informed statistical generalization would count as a reason for holding that profiling is not discriminatory, despite what was said in *Law*.

Little Sisters forces us to consider the rationale behind the use of, and the constitutional discipline to be imposed on, informed statistical generalizations. Amazingly, both *Law* and *Little Sisters* say little to nothing about either of these issues. This is a glaring oversight, given the obvious tension between permitting statistical generalizations and the Court's own emphasis in *Law* on the harm to equality caused by over-generalization. What are likely at work here are efficiency concerns of different sorts. In the social benefits context, the alternative to rules determining eligibility on the basis of personal characteristics such as age is a process of case-by-case decision-making that directly assesses need. Such a situation would be time-consuming, expensive, and potentially open to abuse because it would require that individual decision-makers be vested with considerable discretion. In the law enforcement context, the Court in *Little Sisters* suggested that targeting is permissible because Customs 'is obliged to use its limited resources in the most cost-effective way,' an explicitly utilitarian line of argument that gives priority to maximizing the yield of finite law enforcement activities.[30] Proponents of profiling at airports and at immigration will no doubt point to this latter comment in *Little Sisters* in defence of the constitutionality of such policies.

However, there are three large problems with permitting informed statistical generalizations to defeat claims of discrimination. First, statistical generalizations may often be unreliable, or, even worse, reflective of discrimination themselves. To permit governments to rely on them as justification for differential treatment may do no more than to revictimize the victims of discrimination. This is a well-known argument in the context of racial crime statistics, which are notoriously unreliable because they may reflect discriminatory policing practices in the process of investigation. A proponent of profiling would respond that the events of September 11 are different, because the perpetrators of those terrorist acts are clearly drawn from one ethnic group. But here we encounter the second problem—the question of fit. The justification for using personal characteristics such as race or ethnicity as proxies for risk is that they have some sort of predictive value. *Law* says nothing, though, about *how much* predictive value they must have to suffice for the purposes of s. 15. More precisely, profiling generates both false positives and false negatives, because they subject to scrutiny people who pose no risk whatsoever, and omit from scrutiny people who do in fact pose some risk. *Law's* silence on this issue is deeply problematic, because both false positives and false negatives generate equality concerns. For the former, innocent persons are subject to the unequal burden of law enforcement because of their race and ethnicity, and for the latter, the law is enforced unequally against persons who pose a risk because of their race and ethnicity.

This leads me to the final and most important point: that statistical generalizations cut against the grain of a conception of equality itself—that is, that individuals not be judged on the basis of presumed group characteristics, but rather on the basis of their individual traits.[31] This theme has been very important in the Court's equality jurisprudence, and comes from its case-law interpreting human rights codes. That jurisprudence has been shaped by the impulse that access to the goods whose distribution is regulated by those statutes—principally employment, but also housing and services—should not be governed by criteria that erect arbitrary and irrational barriers, but instead by criteria that are relevant, such as talent, ability to pay, and need. The difficulty with profiling is that its subjects persons to heightened scrutiny not on the basis of criteria that are tied to their conduct, which is the legitimate concern of law enforcement, but instead to the colour of their skin or appearance, which is not.

What is the way forward? As constitutional lawyers know, concerns regarding administrability are classically considered under s. 1, where the burden of proof is on the governments who possess the relevant information, not rights-claimants, who do not. Because they are motivated by efficiency concerns, the justification of statistical generalizations should therefore be relegated to s. 1. But even there the Court must soon develop an account of the legitimate and illegitimate use of statistical generalizations in equality rights case. Left unchecked, the combination of *Little Sisters* and *Law* is dangerous.

As a tentative suggestion, I would propose the courts distinguish on the basis of the nature of the policy at issue. In cases involving under-inclusive social benefits schemes, a good case can be made for the constitutionality of distinctions drawn on prohibited grounds of discrimination if the statistical generalizations meet some threshold level of fit. But with respect to the burden of law enforcement, which by necessity entails deprivations of liberty and privacy, generalizations of this sort are deeply problematic. It is here that procedural fairness concerns can inform the interpretation of s. 15. Section 8 of the Charter prohibits 'unreasonable' search and seizures, and s. 9 prohibits 'arbitrary' detention and arrest, which has been interpreted to impose a requirement of 'articulable cause.' Although a majority of the Supreme Court has not spoken to this issue directly, the dissenting judgments of Sopinka J. in *R. v. Ladouceur*,[32] and LaForest J. in *R. v. Belnavis*,[33] as well as the Ontario Court of Appeal's judgment in *R. v. Simpson*,[34] all indicate that detaining an individual simply because of her race constitutes an arbitrary detention or arrest under s. 9 (*Ladouceur, Simpson*) or makes a search unreasonable under s. 8 (*Belnavis*). In my view, the interpretation of 'arbitrary' under s. 9 and 'unreasonable' under s. 8 should inform the interpretation of 'discrimination' under s. 15, so as to render race and ethnic-based deprivations of liberty and privacy based on statistical generalizations *per se* discriminatory. What is required instead is a process of individual decision-making on a case-by-case basis.

D. CONCLUSION: SECTION 1, ALTERNATIVE MEANS AND THE ANTI-TERRORISM BILL

If a policy of profiling were found to violate s. 15, many difficult issues would arise under s. 1, such as whether that policy would be prescribed by law, the appropriate level of deference owed by the courts, and whether profiling minimally impairs the right to equality. By way of conclusion, I want to focus on the last of these three issues. The obvious alternative to race- or ethnic-conscious policies for airport security and immigration is the use of other criteria that are not prohibited grounds of discrimination, nor thinly veiled proxies for them. Indeed, I want to argue in favour of one provocative alternative to profiling, which is to subject *everyone* to intrusive investigation both by airport security personnel and immigration officers. This policy would be extremely effective, and would comport entirely with the equality guarantee. But amazingly, not a single proponent of profiling has even considered it, even if only to reject it.

If we were to take this proposal seriously, as we should, what would be the principal arguments against it? One argument is that it would be extremely costly, and that in a world of scarce resources, governments cannot be expected to adopt the absolutely least intrusive means for securing their public policies. However, we should be extremely sceptical of this claim. The same voices that are calling for racial and ethnic profiling also claim that in the war on terrorism, money is no object, and that significant resources should now be devoted to Canada's military, intelligence services, and law enforcement agencies. And the expectation is that significant resources will be made available. If this is true, the plea of poverty rings hollow. The true question is not whether moneys are available, but the relative priority to be attached to different kinks of expenditures prompted by September 11. At the very least, in tallying up the costs of the war on terror, the costs of complying with s. 15 must be taken into account. Indeed, I would go even further, and argue that in the allocation of scarce resources, compliance with the *Charter* should presumptively take priority.

The other argument against a policy of blanket scrutiny is that it would exact enormous costs in terms of liberty and privacy. No doubt, the infringements on liberty and privacy of a blanket policy would be severe, and would be a significant cost to be weighed. However, the policy would also have an enormous benefit, because it would eliminate one of the principal costs of profiling: the stigma born by those who are singled out for heightened investigation. What this means is that a blanket policy would *redistribute* the costs of the fight against terrorism, and ensure that they are borne by everyone, not just those who through no choice of their own share the race and ethnicity of those responsible for September 11. Indeed, distributing the costs in this way might lead to a better social valuation of the war on terror, because those who advocate racial and ethnic profiling are not the ones who will bear the costs of that policy. It is deeply ironic that the same voices who call for racial and ethnic profiling are precisely those who now call for solidarity across ethnic and racial lines, and proclaim that we should all be willing to surrender some freedom in favour of security. But if solidarity is truly their guiding principle, and their willingness to surrender freedom is genuine, then their policy proposals should match their rhetoric. Profiling does not.

NOTES

* I would like to thank Ron Daniels for his invitation to speak at the conference, Rebecca Jones for excellent research assistance, Ira Parghi for perceptive comments on an earlier draft, and Kent Roach, Julian Roy and Scot Wortley for extremely helpful discussions and invaluable background material. None of them necessarily shares the views that I express here. I acknowledge the financial support of the Faculty of Law, University of Toronto, in preparing this manuscript.

1. S. Schmidt, 'Ontario denies anti-terror policy is racist: Retired general says checking Arabs "common sense"' *National Post* (5 October 2001) (available on-line at: http://www.nationalpost.com/search/story.html?f=/stories/20011005/721871.html&qs=MacKenzie%) (dated accessed: 7 November 2001).
2. 'Profiles in Prudence (editorial)' *National Post* (20 September 2001) A17.
3. D. Leblanc & S. McCarthy, 'The War on Terror: Terror bill worries Dhaliwal' *The Globe and Mail* (30 October 2001) A11.
4. E. Oziewicz, 'The Brink of War: Border alert targets pilots, Canadian guards told to watch for men with technical training and links to 16 "conflict" countries' *The Globe and Mail* (19 September 2001) A1.
5. S. McCarthy & E. Oziewicz, 'The Brink of War: CANADA'S RESPONSE—Ministers defend new border alert' *The Globe and Mail* (20 September 2001) A6.
6. R. Kennedy, 'Suspect policy: racial profiling usually isn't racist. It can help stop crime. And it should be abolished' (1999) 221:11 *New Republic* 31 (available on-line at: http://www.thenewrepublic.com/archive/0999/091399/coverstory091399.html) (date accessed: 7 November 2001). Kennedy's definitions actually refer only to racial profiling, but in fact encompass profiling on the basis of both race and ethnicity.
7. *Ibid.* (emphasis mine).
8. R.L. Kennedy, *Race, Crime and the Law* (New York: Vintage Books, 1998) at 148.
9. The constitutionality of this practice was upheld by the United States Supreme Court in *Whren v. United States,* 116 S.Ct. 1769 (1996).
10. P. Verniero & P.H. Zoubek, *Interim Report of the State Police Review Team Regarding Allegations of Racial Profiling* (20 April 1999).
11. F. Newport, 'Racial Profiling is Seen as Widespread, Particularly Among Young Black Men' *Gallup News Service* (9 December 1999) (available on-line at: www.gallup.com/poll/releases/pr991209.asp) (date accessed: 7 November 2001).
12. 'Remarks by the President to National Organization of Black Law Enforcement Executives' (30 July 2001) (available on-line at: http://www.whitehouse.gov/news/releases/2001/07/20010730-5.html) (date accessed: 7 November 2001).
13. S 989.
14. S. Wortley, 'The Usual Suspects: Race, Police Stops and Perceptions of Criminal Injustice,' *Criminology* (forthcoming); Commission on Systemic Racism in the Ontario Criminal Justice System, *Report of the Commission on Systemic Racism in the Ontario Criminal Justice System* (Toronto: Queen's Printer for Ontario, 1995) at 352 to 360.
15. R. James, 'Black Passengers Targeted in Pearson Searches? Lawyers plan court fight over "racial profiling" by customs officials at airport,' *The Toronto Star* (29 Sunday 1998) A1.

16. *Law v. Canada (Minister of Employment and Immigration),* [1999] 1 S.C.R. 497 [hereinafter *Law*]
17. *Ibid.* at para. 53
18. *Ibid.* at para. 60
19. *M. v. H.,* [1999] 2 S.C.R. 3 at para. 73.
20. D.A. Harris, *Driving While Black: Racial Profiling On Our Nation's Highways* (American Civil Liberties Union, 1999) (available on-line at: http://www.aclu.org/profiling/report/index.html) (date accessed: 7 November 2001). For those who think that demeaning police conduct toward members of racial minorities is strictly an American problem, we need only remind ourselves of the public strip search conducted of Ian Vincent Golden, an African Canadian, the constitutionality of which is currently before the Supreme Court of Canada: *R. v. Golden,* [1999] S.C.C.A. No. 498 (hearing date: Feb. 15, 2001).
21. *Vriend v. Alberta,* [1998] 1 S.C.R. 493.
22. *M. v. H.*, supra note.
23. *Law, supra* note at para. 71.
24. *Eaton v. Brant (County) Board of Education,* [1997] 1 S.C.R. 241.
25. *Law, supra* note at para. 106.
26. *Ibid.* at para. 104.
27. *Ibid.* at para. 106.
28. *Little Sisters Book and Art Emporium v. Canada (Minister of Justice),* [2000] 2 S.C.R. 1120 [hereinafter *Little Sisters*].
29. *Ibid.* para 121.
30. *Ibid.* at para. 120.
31. *Miron v. Trudel,* [1995] 2 S.C.R. 418 at para. 131.
32. [1990] 1 S.C.R. 1257 at 1267.
33. [1997] 3 S.C.R. 341 at 376–7.
34. (1993), 12 O.R. (3d) 182 (C.A.). Also see *Brown v. Durham Regional Police Force (1998),* 131 C.C.C. (3d) 1 at 17 (Ont. C.A.) (stating that detention under Ontario *Highway Traffic Act* for reasons of race would be illegal because of lack of articulable cause).

PART 3

EVERYDAY AND CULTURAL RACISM

So far, we have provided the reader with a Canadian historical review of racism and colonialism and articles that demonstrate that since the early period of the nation-building project, racism has been an integral part of Canada's reality. But State policies and other institutionalized forms of racism are not separate from the intellectual discursive realms. Ideas, images, stereotypes, and other forms of everyday and cultural racism are just as powerful as they are dangerous. They shape our ideas of ourselves and others and our interactions with each other. Discourses of the "other" can breed feelings of fear and dislike toward a socially constructed "other" and consequently the latter can never feel secure enough to call this land home no matter how long this individual might have been here or whether this individual was born here. As the postcolonial theorist Edward Said[1] pointed out, the West has relied on a binary oppositional construction of an "Us" (the West) and an "Other" (what Said referred to as the Orient) for the origin and continuance of imperialism and colonialism. In this construction, the two opposites have not been treated as equals because the West has been constructed as more civilized, refined, and intellectually superior. Said termed "Orientalism" a discourse that deals with, makes statements of, and analyzes the "Orient." The Orient, then, is a constructed idea whose discourse has the function to dominate and control those defined as inferior. Because of the ability of such construction to produce particular "knowledge" of the "other" that shapes the social relations that have been created in places like Canada where the "Us" and "Them" have shared the land for a few hundred years, it is important that we dedicate a specific section on the ways in which cultural and ideological structures interact with everyday encounters and help justify the colonization of Indigenous peoples and the racist treatment of them and of various "others."

Literature, media, and sport cultures are interesting places to begin our critical examination of constructed ideas of a racialized "other" because they have negative impacts on the everyday lives of Indigenous groups and other racialized "others." These ideas provide the ideological contexts under which forms of exclusion/inclusion are lived by everyday. As the various articles show here, a racialized body is never fully accepted as equal, and a tolerance of him/her can be easily interrupted, as the case of Ben Johnson in Abdel-Shehid's essay so lucidly demonstrates. As we were reviewing the article for the purpose of completing this collection of essays, the case of NHL Indigenous player Chris Simon reminded us of how such racist, everyday, discursive practices still linger in 2008. Simon was severely punished and suspended for his aggressive behaviour against another player during a game. Though aggression and violence are not foreign to the sporting culture of hockey and other sports, we do not question here the punishment, as

1 E. Said, *Orientalism* (London & Henley: Routledge & Kegan Paul, 1978).

we do not condone such acts and we acknowledge this was not his first act of aggression during a game. Yet many other similar acts performed by other players have not received such consequences and, more importantly, such popular outcry and media exposure. Most disturbing have been the comments of one NHL official that Simon's time off might be good for him so that he can deal with his anger and alcohol issue. Such comments are rooted in historical colonial and racist assumptions, ideas, and images of the angry, lazy, drunk, Indian male, and hence did not escape the attention of many Indigenous peoples who are wondering why reference to his ancestry, and the stereotypical ideas associated with it, had to be made. What impact can such comments have on ordinary, contemporary, Indigenous peoples? How can they ever be assured that they are ever accepted as equals? How are Indigenous youth who look up to Simon made to feel when someone of such a status can be given this disrespect? And can they expect any more for themselves? The articles in this section examine the impacts that everyday and cultural racism continue to have on those subjected to it, and yes, words, just as other acts of racism, do hurt. Following Abdel-Shehid's examination of the interconnection between race, sports culture, and national identity, Janice Acoose's article analyzes how literature and other imagery texts have historically created an imaginary inferior "Indian" woman and ample negative images of "Indian" men. Such knowledge production has fed readers and viewers with stereotypes such as the promiscuous, savage "Squaw" or the romanticized and sexualized "Indian Princess." Both stereotypes do not accurately reflect the realities of Indigenous women's lives, and, as Acoose demonstrates, have had many negative consequences for Indigenous women, such as the over-representation and reality of racialized sexual violence against Indigenous women in Canada and in other parts of the world.

Both Deer's and Aujla's articles discuss the historically constructed discourses of the Asian and South Asian "Other" in Canada. Such discourses have been consistently filled with anxiety, fear, stereotypes, and exoticization. Moreover, they have not been reserved only to recent immigrants, but also, as Aujla illustrates, to second-generation South Asians who have been subjected to this treatment of otherness when, for example, they are told to "Go home" or "Go back to your own country." Everyday encounters like these can lead them to feel like "outsiders in their own land," and can only serve to fuel intolerance, racism, and tension between groups in Canada.

CRITICAL THINKING QUESTIONS

1. What does Abdel-Shehid mean by "Criminalization of Black Masculinity"? How is it linked to Canada's nationalist representations of itself and its "others"?

2. Do you think that the case of Ben Johnson is a good example of both criminalization of the Black male body and of forgetful remembering that happens in a racist construction of nationalist imaginary? If so, please explain how.

3. Discuss how literature and other imagery texts influence one's own ways of being, knowing, and understanding of Indigenous women. What impact does this influence have on the everyday lives of contemporary Indigenous women?

4. Locate and watch a recently released popular movie that deals with Indigenous peoples of North America. Discuss how you think the movie portrays Indigenous men and women: in a stereotypical way; in a positive or negative way? What impact do you feel the movie could have on non-Indigenous viewers? What impact could it have on Indigenous viewers?

5. What does Deer mean by the process of "othering"? How is this process connected to a construction of moral panic about Chinese and other Asian immigrants to BC? Can you think of other similar experiences of moral panic discourses about other non-white racialized groups that have recently occurred in Canada?

6. Drawing from the various articles of this section, how do you see racist and colonial discourses of the past to continue to influence dominant discourses and perceptions of non-white racialized populations in Canada? What ideas, images, and stereotypes have contextualized these discourses? How do you think the media has influenced the construction or perpetuation of these discourses?

FURTHER RESOURCES

Abdel-Shehid, Gamal. *Who Da Man? Black Masculinities and Sporting Cultures*. Toronto: Canadian Scholars' Press, 2005.

Acoose, Janice. *Iskwewa-kah'ki Yaw Ni Wahkomakanak: Neither Indian Princesses nor Easy Squaws*. Toronto: Women's Press, 1995.

Collins, Peter C. and John E. O'Connor, eds. *Hollywood's Indian: The Portrayal of the Native American in Film*. Lexington: The University Press of Kentucky, 1998.

Henry, Frances and Carol Tator, eds. *Discourses of Domination: Racial Bias in the Canadian English-Language Press*. Toronto: University of Toronto Press, 2002.

James, Carl E., ed. *Seeing Ourselves: Exploring Ethnicity, Race and Culture*. Toronto: Thompson Educational Publication, 2003.

Teelucksingh, Cheryl, ed. *Claiming Space: Racialization in Canadian Cities*. Waterloo: Wilfrid Laurier Press, 2006.

FILMS

Race is a Four Letter Word. 2006. Director: Sobaz Benjamin. Producer: Annette Clarke. National Film Board.

Description: While examining conflicts concerning race in Canada, director Sobaz Benjamin also reveals much about himself and his own struggles to appreciate the meaning of his heritage as an "Afro-Saxon" Briton, then Grenadian and now Haligonian–Nova Scotian-Canadian.

In the Name of the Mother and the Son. 2005. Director: Maryse Legagneur. Producer: Yves Bisaillon. National Film Board.

Description: Takes us to the heart of Montreal's Saint-Michel neighbourhood for an intimate look at two young Quebecers of Haitian origin as they search for hope and freedom. Like their parents who left Haiti to settle in Canada where they had to start over from scratch, James and Le Voyou try to make a life for themselves, each in his own way. It is a declaration of love by two young men for their mothers and a sensitive account of the prejudices faced by young people of Haitian origin in the neighbourhood of Saint-Michel.

Crash. 2004. Director: Paul Haggis. Producer: Paul Haggis. Lions Gate.

Description: For two days in Los Angeles, a racially and economically diverse group of people collide with one another in unexpected ways. Examines fear and bigotry from multiple perspectives as characters careen in and out of each other's lives. For a critical review of the popular movie *Crash*, we refer the reader to Robert Jensen & Robert Wosnitzer's article "Think Piece: 'Crash' and the Self-Indulgence of White America" in the Black Commentator, Issue 176 of March 23, 2006. (www.blackcommentator.com/176/176_think_crash_jensen_wosnitzer.html)

Redskins, Tricksters and Puppy Stew. 2000. Director: Drew Hayden Taylor. Producer: Silva Basmajian. National Film Board.

Description: Takes complex issues like Native identity, politics, and racism and wraps them up with one-liners, guffaws, and comedic performances. Overturns the conventional notion of the "stoic Indian" and shines a light on an overlooked element of Native culture—humour and its healing powers.

For Angela. 1993. Director: Nancy Trites Botkin and Daniel Prouty. Producer: Daniel Prouty. National Film Board.

Description: A story of racism inspired by the experience of Rhonda Gordon and her daughter Angela who were victims of racism. The film is aimed at encouraging change with regard to how we view racism.

KEY CONCEPTS/TERMS

Criminalization of Black Masculinity

Forgetful Remembering

Indian Princess/Easy Squaw Images

Literature as an Ideological Apparatus

"Othering" Process

Moral Panic

Running Clean: Ben Johnson and the Unmaking of Canada

GAMAL ABDEL-SHEHID

8

INTRODUCTION

> As an apparatus of symbolic power, (the nation) produces a continual slippage of categories, like sexuality, class affiliation, territorial paranoia, or "cultural difference" in the act of writing the nation (Homi Bhabha, 1994: 140).

We can now begin to explore these processes in recent events in black sporting cultures in Canada. I will begin with a discussion of track and field, a sport in which Canada is rapidly becoming an important world player. The rise to prominence of Canadian track athletes presents a challenge to conventional ways of representing Canada. As we will see, the kinds of narratives used to represent hockey (and Canada by extension) are quite different than those in track and field, with perhaps the greatest difference in the representations of temporality in the two sports. For example, while hockey is represented as timelessly Canadian, track and field is inflected with an itinerant temporality. In addition to this difference, the demographic nature of the two sports is worth noting. Here, the difference between track and field and hockey is great: the second part of this chapter will explore how forms of blackness are split and often used to police or expurgate other ones.

In this chapter, I suggest that Canadian multiculturalism relies on racialized tropes of cleanliness and uncleanliness that are mapped onto bodies. Specifically, I argue that in sports, the body of Ben Johnson is a place where these practices achieve very poignant meaning in the national imaginary. I suggest that one of the reasons why Johnson's body is so evocative is via a linking of his black body to steroid use, crime, and uncleanliness. Moreover, there is a parallel between the negative treatment of black bodies in nation-state practices (such as immigration and citizenship) and political economy (such as providing a cheap and flexible labour supply), and the treatment of these same bodies as dispensable within sport, as in Johnson's case. In other words, Canadian nationalist practices of representing, and more often exploiting, non-white and First Nations peoples reverberate in the realms of political economy, cultural production, and sports. In addition, I suggest that representations of Ben Johnson can be read as elements of a national tragedy: we will see the key role played by "race" in the construction of this narrative.

Racism and State Practices

In thinking about how blackness works in the national imaginary, we must contextualize the role that the state plays in reproducing racist and capitalist agendas. Without trying to suggest a one-to-one connection between state practices and the nationalist imaginary, the parallels are worth underlining. One way to think about this connection is through the slogan that I hear (and chant myself) at demonstrations in support of undocumented immigrant workers in Canada: "Good enough to work, good enough to stay." This slogan highlights how a significant number of people of colour come to Canada from other countries as a cheap and flexible labour source. In turn, the slogan highlights the role of the state in often deeming this labour as being at once unnecessary *and* disposable. Li and Singh Bolaria (1988: 14) suggest that "[t]he oppression of racial groups is by no means a historical accident, but is rooted in the social and economic development of Canadian society."

The authors suggest that economic racism is a feature of colonialism in the Americas, and point to the historical example of plantation slavery as evidence: "Theories that attempt to account for racial domination have to begin with how an apparently irrational concept such as 'race' becomes rational in the process of reproducing cheap labour" (ibid.: 27). This argument is compelling if somewhat limited. In what follows I will accept its major tenets with a few qualifications. The first is that this argument does not account for other forms of identity such as sexuality, gender, and ethnicity. For example, "race" and racism also act as a form of sexual subjugation, as well as acting as the marker that makes certain bodies deemed more sexually available than others.[1] In addition, this argument pays little attention to black and non-white performances in the field of politics and semiotics, which alter the basic script of racism (as I discussed previously). These interventions do not mean that racism's overall framework recedes; they do, however, suggest that victims of racism have a role to play in either resisting or perpetuating it.

Singh Bolaria and Li offer a way to think about how the iconography of blackness, crime, and nation in Canada are not random or even merely rhetorical. Media representations of the Dubin Inquiry, Ben Johnson, and Donovan Bailey must be seen in line with other circumscriptions of non-white identities as an imperative of the racist/colonialist foundations of the Canadian State. These circumscriptions are particularly useful in upholding a discursive entity called "Canada," defined in opposition to First Nations, black, etc. For example, the inability to "forget" Ben Johnson, which I discuss below, can be contextualized if we look at two recent historical phenomena that underline the role of the state in reproducing racism.

This latest manifestation of intense anti-immigrant racism emerges out of the late 1980s in Canada and corresponds to an economic crisis precipitated by globalization; severe economic dislocation due to the North American Free Trade Agreement; and increased neo-liberal attempts by governments both federal and provincial in Canada to "attack the deficit." This has caused a shift in the political language in Canada that is becoming increasingly anti-immigrant, evident in the re-emergence of neo-Nazi hate groups like the Heritage Front, and the rise of the far-right Reform (now Alliance) Party. For example, anti-immigrant racism made up one of the strongest planks of the Reform Party, as well as that of the Liberal and Progressive Conservative Parties throughout the 1990s. This is seen through some of the comments of Reform's one time immigration critic, MP Art Hanger, who has "made a national name for himself" by attacking federal immigration policies as being lenient (*Globe and Mail,* 31 Oct. 1994: A1).

Throughout the 1990s, Reform Party policy explicitly linked immigrants with economic impoverishment of the nation. The party's policy on immigration held that the rate of new immigrants should be cut to 150,000 per year as long as the unemployment rate is above 10 per cent (ibid.). This logic implies that capitalist economic crises such as unemployment can be solved by reducing the number of immigrants. This perpetuates the racist stereotype, contrary to documented evidence (see Noorani and Wright, 1995) that maintains that immigrants are a "drain" on the national economy; they come here and take "our" jobs, and rip off welfare.

Another example of this paranoia is seen in the "McLeod Report." This report, written by a federal immigration intelligence officer, was apparently commissioned in 1993 by then Ontario Liberal Party Leader, and Leader of the Opposition of the Ontario Legislature, Lyn McLeod. The report called Somalis "masters of deceit and corruption" and stated that "*Our Western and primarily Christian based way of life* has little meaning or relevance to *these people*" (*Share,* 18 Nov. 1993; emphasis added). These events support Singh Bolaria and Li's argument that capitalist crises demand a kind of scapegoating of immigrants or those marked as such. This logic equates immigrants and refugees with crime, evident in calls for strict changes to the immigration regulations in 1994 by the Federal Government with the scaling down of the number of immigrants to 200,000 from 250,000 per year (*Globe and Mail,* 31 Oct. 1994: A1).

Within this context emerges the criminalization of black masculinity as a response to national "crisis." The criminalization of black masculinity as a response to economic crisis has been demonstrated by Stuart Hall et al. in *Policing the Crisis,* as well as more recently in the work of Tricia Rose (1994).[2] While Hall et al. discuss Britain, and Rose discusses the United States, Canadian racism is no better. It deploys tropes of black masculine criminality in similar ways. In other words, Canadian preoccupations with law and order have historically meant preoccupations with whatever the Canadian nation-state has deemed to be the "immigrant" or "native population."[3]

This criminalization is exemplified in the recent paranoia over the deportation of two young black criminals. The first is O'Neil Grant, who, along with Gary Francis and Lawrence Brown (both of whom are also black), was charged with the murder of Georgina "Vivi" Leimonis in the "Just Desserts" robbery. The second is Clinton Gayle, who was recently convicted to two life sentences for shooting and killing Todd Baylis, a Metro police officer. There are a few important points worth noting here. The first is that both incidents occurred in 1994, and were followed by extreme paranoia, whipped up by the mainstream media, the government, and the Metro police. The fact that both men were of Jamaican origin and had previous criminal records only heightened the hysteria. There was a rise in calls for a "crackdown" on deportations and a rise in the rhetoric of immigrants as criminals. For example, Sergio Marchi, then federal immigration minister, claimed in the aftermath of the Baylis killing (August 1994) that it was time that the government improved its enforcement of deportation orders of suspected criminals (*Calgary Herald,* 25 Aug. 1994: A14).

These incidents provide a sense of the political and economic context of Canada in the late 1980s and 1990s and will inform both this discussion and that of the following chapter. These factors provide some explanation for how and why Ben Johnson becomes the nation's "ghost" in media representations. In short, the primarily "itinerant" presence of black bodies in Canada—often felt to have been brought here as disposable labour—means that their right to be Canadian is always in question. Therefore, calls for deportation are often the first response to a whole series of factors that may or may not involve wrongdoing by black folks. In that sense, the reading of black masculinity as disposable and itinerant is indispensable to an understanding of Ben Johnson's place in the national imaginary.

Can't Forget Ben

Until his positive test for the use of anabolic steroids (stanozolol) in September 1988, a performance-enhancing substance banned by the International Track and Field Association, Ben Johnson was the most successful track athlete in Canadian history, and one of the most successful in the world in recent history. He had established several world records, both for indoor and outdoor competitions, and had established a worldwide persona that was reflected by his sizable endorsement contracts with several companies based in Western Europe, Japan, and North America. Johnson achieved his ultimate success at the 1988 Summer Olympics in Seoul, South Korea. There, in front of over seventy thousand fans and a worldwide television audience in the billions, Johnson ran the 100-metre sprint in a world record time of 9.79 seconds, beating his competitor an American rival Carl Lewis by 13/100ths of a second, an unprecedented margin of victory (*Montreal Gazette*, 24 Sept. 1988: A1, 2). This victory, Canada's first Olympic sprinting gold medal since the 1930s, was met with euphoria by many Canadians at the time. This euphoria was shattered when, two days later, it was revealed that Johnson tested positive in a compulsory urine test after the race for stanazolol (*Montreal Gazette,* 27 Sept. 1988: A1, 2). Afterwards, according to the *Montreal Gazette,* "retribution was swift and severe." This included the retraction of the gold medal and a statement by Jean Charest, then Canada's sports minister, that Johnson would never be able to compete for Canada again. In addition, Johnson's world record was stricken from the record books. Third, there was a statement from then Prime Minister Brian Mulroney acknowledging that this was "a moment of great sorrow for all Canadians" (*Montreal Gazette*, 27 Sept. 1988: A1, 2).

But perhaps the most lasting and insidious form of "retribution" regarded Johnson's "citizenship." In what became known as a national infamy, there was a progression in the representation of Ben Johnson from one of a "Canadian hero" in victory to one of a "Jamaican" after disqualification. This process was encapsulated by the editorial cartoon of 13 October 1988, in the *Globe and Mail,* which featured three identical images of Ben Johnson but a different subtitle under each one. The first read: "Canadian Wins Gold Medal"; the second read "Jamaican-Canadian Accused of Steroid Use"; and the third read "Jamaican Stripped of Gold Medal" (A7). This cartoon reflected how Johnson's Jamaican heritage had not been mentioned until he tested positive for steroid use. In coverage afterwards, Johnson becomes progressively less Canadian as his status as a "lawbreaker" is revealed.

The following passage from the *Globe and Mail* also reveals the change in attitudes toward Johnson following his disqualification:

> Angella Issajenko, an Olympic (black Canadian) sprinter from Mr. Johnson's home club, Mazda Optimists, told reporters in Seoul, "A white (Canadian) team member came up to where the Jamaicans were sitting and said, 'You can have Ben back now, he is not Canadian now' " (*Globe and Mail*, 13 Oct. 1988: A7).

This movement, or the discursive passage of Johnson's body from "our own" to "Jamaican," mirrors the kind of troubles that black folks often receive in relation to the Canadian nation-state, whether in the form of harassment at borders, police harassment, or in employment situations. This context is central to understanding the portrayal of Donovan Bailey's 100-metre sprint victory at the World Championships of Track and Field, in August 1995. Moreover, another incident of drugs and sport, specifically those of Ross Rebagliati, is also impossible to think about without referencing what happened to Ben Johnson, as I discuss below.

Tragedy, Sports Writing, and Nation

While narratives of cleanliness and purity are indispensable to understanding the place Ben Johnson occupies in the national imaginary, the role of tragedy in such narratives is also crucial. It helps here to consider the nature of tragedy, and how it works through the representations of nationalism in sports.

Before considering tragedy specifically, it is helpful to consider the role of sports writers within such a constellation, given that sports writing is highly influential to how we think about sport. While no major study exists on sports writing *per se*, it is important to make some preliminary claims about how it operates as a genre and what that means for understanding "race," sport, and nation. Sports writing within the print media is highly hyperbolic. It attempts to make human beings larger than life, with a tendency toward extreme opinions and high drama. Moreover, sports writing often promotes a very Manichean view of the world, with clear demarcations between good and evil, rich and poor, and so on. Another important point is that the biggest stories are the ones about big games, but also the ones that are controversial. An example of this is how immediately following his failed drug test, Ben Johnson left Seoul to return to his home in Toronto, via New York's La Guardia Airport. A security officer, who had worked at the airport for over thirty years, swore that he had never seen such a massive throng of reporters and onlookers as had gathered for Johnson's arrival in New York. As a figure involved in a major sporting controversy, Johnson's appeal was larger than that of such major world figures as Mikhail Gorbachev, Fidel Castro, etc. While it is true that this interest was to some extent organic to the events, we cannot overlook the role that sports writers played in hyping this event.

Often, the kinds of stories sports writers attempt to represent, in addition to those that grip the sporting public, are those containing the most pathos and drama. One of the most common of these stories is tragedy. Lewis Gordon summarizes the relation between the tragic protagonist and the community as follows:

> For the community's demands to emerge, the kind of rightful action that must emerge is the reconstitution of justice. In other words, regardless of the characters' points of view, the world must be restored to a certain order. *The tragedy in tragedies is therefore that the "innocence" of the characters who occupy a wrongful place in the drama is ultimately irrelevant.* Thus, the tragic protagonist finds himself [sic] guilty by virtue of deed and circumstance, not

> intent, and finds himself [sic] suffering, ironically, for the sake of justice. The tragic drama cleanses the community of its own evasions. Justice is tragically restored (1996: 302; emphasis added).

Gordon's reflections are helpful in thinking about how Ben Johnson was initially received and treated. In fact, this description almost literally explains what happened to him: although he clearly did commit a "transgression," as we know, he was not alone. However, it was Johnson who became the scapegoat, not his coaches, etc.

What is important to stress here is how the components of tragedy are racialized, i.e., what are the meanings of "justice" and "guilt" within a *racist culture*[4] that ultimately makes these categories illegible or highly distorted? For example, the community's demand for justice, as well as the desire to turn the hero or heroine into a scapegoat, have a double resonance in relation to blackness, whereby black folks are always already read as guilty. In that sense, recalling the racialized foundations of tragedy helps us make sense of the gravity of what happened to Johnson. This backdrop is important to remember when we consider how Johnson's name appears in nationalist discourses after Seoul. For while the immediacy of the event has waned, there remains a persistent desire to raise Johnson's name as a "ghost" that haunts the nation, as we will see.

Ben Johnson and the Boundaries of Canada

In the wake of Ross Rebagliati's recent positive test for marijuana use at the Nagano Olympics, Canadian Olympic Association (COA) officials were bombarded with questions about drug use, role models, and the possible appeal of the decision to strip Rebagliati of his gold medal. One of the more prominent questions was: "Is this incident comparable to Ben Johnson?" In other words, was Rebagliati's transgression in 1998 linked to Johnson's a decade earlier?[5] The answer that came from various COA brass and from former Olympians such as Silken Laumann (athletes asked for their opinions by sports reporters) was an emphatic "No." Richard (Dick) Pound, the Canadian representative on the International Olympic Committee, looked incredulously at media types who dared to suggest a connection. *Toronto Star* columnist Dave Perkins wrote, "This isn't steroids. This isn't cheating. This wasn't Ben Johnson—although the initial punishment pending appeal, medal stripped and banishment from the Olympics—is the same" (*Toronto Star*, 11 Feb. 1998: C1). Federal sports officials ruled out a link between the two events and in fact lobbied to have Rebagliati's gold medal reinstated. Reaction in the federal Parliament was the same. While in 1988, as we saw above, "retribution was swift and severe" for Ben, Rebagliati received different treatment.[6] In fact, he was defended by prominent MPs like Sheila Copps, who in effect argued that smoking a little pot was not so bad (*Toronto Star*, 12 Feb. 1998: A6).

While the double standard meted out to Rebagliati and Johnson is worth noting, there is something else at play that betrays something far more pernicious about Canadian nationalism. In addition to the bizarre way that Liberal cabinet ministers defended, or even trivialized pot smoking, another interesting facet of this event comes to mind that suggests something crucial about Canadian identity. It points to the role that *invoking* Ben Johnson, or rather the "affair" of Ben Johnson, plays in constructing a Canadian identity. In other words, it shows the importance of the construction of the "other" as necessary to make Canada one. Both the questions asked by reporters and the responses, all of which were negative, are an exercise in how to construct a "proper" Canadian identity. These examples suggest how important invoking negative invocations of Ben Johnson are to the construction of the interior of Canada. Thus, this strategic use of Ben Johnson may point to the marker or limit of what is "Canada" and "Canadian."

Canada and Its Ghosts

In the Rebagliati case, involving Ben Johnson negatively is what makes "us" Canadian; Johnson's body is the black backdrop against which other athletes often get illuminated. Bhabha suggests that in constructing the nation, the nation constructs its double, or its *ghost.*[7] While Bhabha does not emphasize the pun, I cannot help but think of the historical confluence between black folks and ghosts in the Enlightenment/colonialist imagination. In the Canadian context, the ghost is often First Nations peoples or non-white Canadians. In Canadian sport and popular discourse, this process, I want to argue, is mapped onto Ben Johnson's body. Much like the racist assaults on immigrants as the cause of all social ills, this doubling or ghost-making process shows what has to be kept out of bounds in order to keep Canada *"clean."*

In the questions surrounding Rebagliati, Johnson's name is a flashpoint of memory. It is brought up, then summarily dropped. All comparisons between Johnson and Rebagliati were dismissed out of hand. Dick Pound's response suggests a process of writing the nation that has no room for "ghosts" and/or "cheaters." In other words, Pound's reaction may be evidence of the threat that Johnson's name poses if it is allowed to stick around. The demand to distance oneself from Ben Johnson is proof of the performative acts required in the drawing of Canadian boundaries to exclude the "outsider" or minority.

This example suggests a paradox of Canadian myth-making. He is not forgotten as such, but rather what happens, in addition to his body being marked as exterior, is a *forgetful remembering*. Johnson is only remembered enough to be forgotten, or displaced to the border or beyond.[8] Forgetful remembering cannot be read in isolation against other readings of black masculinity, such as that of O'Neil Grant and Clinton Gayle. Locating

Black Canadian Cultural Studies (Part 1)

Two of the key figures in this growing field are Dionne Brand and Rinaldo Walcott. In his 1997 book, *Black Like Who?,* Walcott suggested that as a result of racism, discriminatory immigration and employment policies, and police harassment, a diasporic sensibility emerges among black Canadians. As a way of rooting themselves, New World black peoples continue to make identifications that are beyond the national boundaries, and beyond what the state suggests.

this process together within the current crisis of Canadian nationalism offers clues to discerning how racist nationalism works in the aftermath of post-colonial shifts. In addition, it is indispensable to thinking about representations of Donovan Bailey.

Donovan Bailey, Ben Johnson, and Forgetful Remembering

Since 1995, when he won the World Outdoor Track and Field Championships in Göteborg, Sweden, Bailey has been hounded by reporters wanting to compare/link him to Johnson.[9] The following quotes come from the August 1995 media coverage of Bailey's 100-metre sprint victory. There are three features to this coverage. First, Bailey is not understood by himself, but represented vis-à-vis the "forgetting" of Johnson; second, Bailey and Johnson are represented as oppositional; third, this opposition is framed within a Manichean rhetoric of cleanliness, whereby Bailey is represented as "clean," to Johnson, who is represented as "unclean." Throughout the coverage, Donovan Bailey, the gold medal winner, is unable to stand alone in the representation; he is rhetorically linked to Ben Johnson. The fact that Johnson's career had been over for quite some time at that point (he retired in 1993) is somehow not enough to remove him from the representations. Moreover, in reading the headlines and the quotations below, it is important to keep in mind Bhabha's discussion of the ghost of nationalism. The spectral quality of the representations is shocking.

Major headlines from Toronto dailies represented Bailey and his victory in relation to Johnson. The *Toronto Star* headline on the cover of the sports page reads: "Canada's Bailey outruns world" yet the subtitle of the headline reads: "But fastest human still *can't escape Johnson's shadow*" (*Toronto Star*, 7 Aug. 1995: B1; emphasis added). The *Toronto Sun* cover headline reads: "Canada Can Finally *Forget Ben Johnson*" (*Toronto Sun*, 7 Aug. 1995; emphasis added). In the first page of the sports section of the same newspaper, the headline reads: "1–2 punch to *Ben's* legacy" (ibid.; emphasis added). An article in the *Globe and Mail* is entitled: "Echo from *past* chases new heroes" (*Globe and Mail*, 7 Aug. 1995: D1; emphasis added). Moreover, only in the *Toronto Star* sports page is Bailey's name part of the headline, yet even there, Johnson's name is mentioned in the subtitle.

Second, Bailey and Johnson are represented oppositionally. In one instance, Bailey is upheld as a more amicable athlete than Johnson. The *Toronto Sun* reports, "Bailey, like the *disgraced* Johnson, was born in Jamaica. Like Johnson, he is big, strong, and muscular. But *in contrast* to the *sullen Johnson, Bailey is personable*" (*Toronto Sun*, 7 Aug. 1995: Sports section; emphasis added).

The *Globe and Mail* piece establishes the opposition in a rhetoric of sprinting styles: "His [Bailey's] race at the world championships in Göteborg, Sweden, was in many ways the *polar opposite* of that other fellow's [Johnson], he of the famous flying start (*Globe and Mail*, 7 Aug. 1995: D1; emphasis added). The *Toronto Star* piece does not explicitly position the sprinters as oppositional, but it does establish throughout that Bailey is tired of the constant comparisons with Johnson: "The question about Johnson's legacy delivered to Bailey at the post-race news conference, in Göteborg, Sweden, won't be the last" (*Toronto Star*, 7 Aug. 1995: B1).

Randy Starkman's comment in the *Toronto Star* illustrates the process of forgetful remembering. In contrast to Rebagliati, and due to a racist imaginary that links all black bodies, Bailey is consistently linked to Johnson. Starkman's quote, in an ironic note of self-fulfilling prophecy, asserts that these questions will never go away. In some respects, it can be read as a warning to Bailey that, in fact, the nation will constantly invoke Ben's name as a racist attempt to denigrate or expel him.

Oppositional positioning and the inability to mention Bailey unto himself leads to a third feature of the reporting. This is that the tests rely on a narrative of redemption and a rhetoric of cleanliness. The desire to link blacks to uncleanliness is a longstanding figure of colonial discourse, or representing blackness in colonialism. Frantz Fanon's work has been very influential[10]:

> In Europe the Negro has one function: that of symbolizing the lower emotions, the baser inclinations, the dark side of the soul. In the collective unconscious of *homo occidentalis,* the *Negro—or, if one prefers, the color black—symbolizes evil, sin, wretchedness, death, war, famine* (1967: 191; emphasis added).

Print media usage of this discourse is what helps to establish Ben Johnson as the epitome of evil and uncleanliness. The effect of such devices renders Bailey's body as not that of a sprinter but rather one frozen within the category of eraser or cleansing agent. His victory is understood as "erasing the shame" of the "tarnished" image of Canadian sprinting and, by extension, Canada. This is done through language that links Johnson to the Seoul Olympics and drug use. Starkman stresses that Bailey is a clean runner, quoting Bailey's manager (not Bailey, incidentally): " 'He has to run clean and he does run clean,' [Bailey Manager Adrian] Keith said. 'He's probably facing more [drug] testing than anyone because of Ben Johnson' " (*Toronto Star,* 7 Aug. 1995: B1). In the *Toronto Sun,* Bailey's victory is described as "erasing the bitter aftertaste of Johnson's *steroid-tainted* run at Seoul in 1988" (*Toronto Sun,* 7 Aug. 1995: Sports section; emphasis added). The *Globe and Mail* piece achieves the linking of Johnson with drug use by neologistic means: "Mention that other fellow [Johnson], he whose name will not appear in this column, and they know instantly as well. It has become like one long word that turns up in every language: *Canadadrugcheatben.*" (*Globe and Mail,* 7 Aug. 1995: D1). If Bailey's body is an eraser, by contrast, Johnson's body and legacy are an impurity. Johnson is that which must be forgotten or disavowed; he is referred to as "shame," a "spectre," and a "long, dark shadow."[11]

In addition, this myth-making process is not merely the preserve of print media reporters. Witness the following definition of Ben Johnson in a recent article by Steven Jackson:

> Furthermore, the *tarnished* legacy of Ben Johnson continues to influence the lives of Canadians, especially those black athletes who are following in the former sprinter's footsteps. . . . It would appear that, despite all the attempts at damage control including the Dubin inquiry, Canada continues to be *haunted* by the Ben Johnson saga (1998: 23; emphasis added).

In sum, linking Johnson's body to impurity, or ghostliness, is necessary in order to establish the nation's interior, or "Canada," as clean. It is a necessary outcome of the whitening of the nation.[12] As cited above, this practice accords with linking blackness to uncleanliness in the racist imaginary and, moreover, such a myth betrays the extent

Black Canadian Cultural Studies (Part II)

Dionne Brand, in *Bread out of Stone* (1998) suggests that Canadian identity relies upon a forgetting, a forgetting of the past. This process, she argues, is necessary because immigration, especially non-white immigration, is seen as folks leaving what must be a worse place, in order to come to Canada, which is always read as undeniably better.

to which notions of Canadian nationalism are linked to white supremacy. Furthermore, given the oppositional way that Bailey's body is represented, it illustrates the colonialist practice of using some black bodies to discipline other ones. In addition, it is worthwhile to consider how the concept of cleanliness is linked to drugs, "race," and crime.

Commissioning Canada: The Dubin Inquiry, Innocence, and "Race"

> I still get asked in interviews, "Is there racism in this country?" Unlike the United States, where there is at least an admission of the fact that racism exists and has a history, in this country one is faced with a *stupefying innocence* (1994: 178: emphasis added).

The above quote from Dionne Brand links notions of innocence, racism, and whiteness in Canada. It allows us to think about how a racialized concept of innocence is used in addition to forgetful remembering as a device for nation-making in Canada. In what we could call "exhibit A" to Brand's quote, I offer the following quote by Stephen Brunt of the *Globe and Mail* on the morning after Bailey's 1995 victory at Göteborg: "But what happened seven years ago [the Johnson disqualification] was like the *unbearding of Santa Claus.* Try as hard as you can to make it the way it was, and *it's still impossible to reclaim innocence*" (*Globe and Mail*, 7 Aug. 1995: D1; emphasis added). I offer Brunt's quote as "exhibit A" in support of Brand's comments because Brunt's deployment of "Canadian innocence" works the same way as denials of racism. Brand's quote shows to what extent a constructed notion of *innocence*, or innocence as a form of identity, is a hallmark of Canadiana.[13] In order to sustain Canadian nationalism, there must be an innocence about racism in this country, an innocence about its horrors, foundations, and legacies.

Brand notes how Canadian "innocence" is built in contrast to the comparative "guilt" of the United States. Brunt's quotation is an example of how the projection of guilt works in Canada. However, here the guilty party is not the United States but Ben Johnson, or in Brunt's words, "that other fellow." Such a move enables "Canada" to equal pure, pristine, etc. By extension then, the non-innocent, or guilty ones, exist outside of "Canada," i.e., outside of the law. This is clearly how Brunt imagines Ben Johnson—as an outlaw, i.e., someone who has transgressed and impurified the Canadian body politic. Thus, Brunt's linking of Johnson's body to the fictional *end of innocence* for Canada remaps racism onto the bodies of its victims.

Moreover, there is a temporal element as well that involves the claim that the period before 1988, before stanazolol, was the Golden Age of Canadian sport, or track and field specifically. It allows for an erasure of histories of racism and their replacement by mythologies of black invasions.[14] In addition, there is a gendered element to the construction of innocence that is central to the nation-making project. Brunt's highly sexualized rhetoric and his linking of Johnson to defilement help to name Johnson as an example of what Davis (1983) calls "the black rapist." However, unlike Davis, Brunt is unaware of the mythical quality of this designation, which was the claim that black rapists were an invention of white racists in the United States as a response to a loss of political and economic power.[15] This pattern continues if we briefly consider the Dubin Inquiry into steroid use among athletes in Canada that was convened after Johnson's positive drug test in Seoul in 1988. I suggest we read the Dubin Inquiry as an exercise in contemporary re-drawings of nation in Canada to discipline immigrant bodies and to construct Canada as an innocent, crime-free zone.

While little scholarly work has been done on the Dubin Inquiry, John McAloon's (1990) work provides a place to begin. McAloon argues that the Dubin Inquiry, while not being a court of law *per se*, had the effect of "trying" certain individuals and "passing judgment" upon them in the realm of popular culture. McAloon also claims that such an organization had the effect of making steroid use in Canadian amateur sport a problem of certain incorrigible individuals, as opposed to being the systemic problem that it is. He notes that the focus on individuals is worth stressing given that it established the innocence of larger sports institutions, such as the federal Sports Ministry and SportCanada. Such institutions, which should have taken their fair share of the apportioned blame, emerged as "innocent" given the Dubin Inquiry's character as a "Real Life Soap Opera" (1990: 59) and its personalizing of systematic crises. Further, McAloon argues that the Dubin Inquiry was an exercise in the production of melodrama as a device to achieve this:

> . . . the Dubin Inquiry created and offered the public an eagerly consumed melodrama. . . . Rather than some straightforward exercise of unequal power relations by political authorities anxious to protect themselves, . . . melodramatization as a specific process and effect is here claimed to be the crucial means by which the *state went free* (1990: 43).

McAloon correctly points out that this is not merely a trial about individuals; it is more precisely about an individual. Despite all the hype, pomp, and circumstance, the Dubin Inquiry was a public way of making Ben Johnson accountable for his "sins." He writes:

> The frame of the judicial trial insisted that there yet be one principal defendant and no matter how long he was held off-stage, no matter how many times it was insisted that Johnson was not the sole target, no matter how much more dramatic other stories turned out to be, the public and the media would not let him abandon his central role (1990: 53).

Once again, there is another paradox at work here involving Johnson, blackness, and the Canadian "public."[16] Similar to the process of forgetful remembering, this paradox involves the Dubin Inquiry. In spite of its official narration as an inquiry into the use of performance-enhancing drugs, that is not what it was. This spectacle is about Ben Johnson. McAloon's analysis of the Dubin Inquiry reflects the obsession many Canadians had with Ben Johnson.

While McAloon's argument is sound, he does not consider the question of "race" and how it works in the making of Canada's self-image. Without a discussion of "race," McAloon's analysis is guilty of the same shortcomings that he correctly attributes to the Dubin Inquiry: it prioritizes individuals as opposed to systems and structures in describing the Johnson phenomenon. Moreover, the obsession with Ben Johnson, and why he appears as the nation's ghost, has no systematic theory, and it makes us unable historically and politically to locate why the Dubin Inquiry took the form that it did.

In contrast, my reading of the Dubin Inquiry suggests that it was a place not only to let the state appear blameless, but it was possible only through the construction of Johnson as a black man who was guilty of failing to live up to his role as either a "Canadian role model" or "law abiding immigrant." My claim that the Dubin Inquiry was more about policing black bodies than about the question of steroid use is backed up by the Inquiry's redundancy. The reason I call the Dubin Inquiry redundant is because the only proof revealed in the Inquiry was the fact of rampant steroid use among athletes in high-level athletics. However, this is not news. For years, many sporting insiders, athletes, and coaches have claimed that at high levels of competition, all athletes are on a regimen of performance-enhancing drugs of different kinds. These performance-enhancing drugs are often illegal. While this knowledge has existed mostly as rumours circulating among people "in the know," there are plenty of instances where hardcore "evidence" was presented.

One such example noted in the Dubin Inquiry was the testimony by Desai Williams. Williams, a former Canadian Olympian who ran with Ben Johnson at the Mazda Optimist Track Club and who trained under the same coach, Charlie Francis, admitted during the Inquiry to using steroids. Williams's justification was that steroid use was/is a regular component of world-class track and field (*Toronto Star*, 13 Apr. 1989: A17). In addition to Williams, Francis and Johnson, among others, testified something to the effect that "everybody" was doing it, and the only way for Canadian athletes to be competitive internationally was to "join the club."[17]

What is interesting and telling about the Inquiry's role in the production of Canada, and the production of innocence as a form of Canadian identity, is that in spite of testimony from people like Williams and others, there remains a persistent belief that world-class athletes can run clean and win, which helped to establish the supposed "necessity" of the Inquiry. In other words, the legitimacy of the Dubin Inquiry relied on a *forgetting* of the fact that it is almost impossible to run clean and win at world-class levels and, moreover, that Canadian athletes are not alone in taking steroids.[18]

Such a wilful denial of contemporary realities is seen in the testimony provided by Abbie Hoffman. At the time of the Inquiry, Hoffman was the director of SportCanada, the body overseeing amateur athletics in Canada. Responding to questions about the use of drugs among world-class athletes, Hoffman stated that "[p]eople are too quick to condemn athletes as steroid users when hard work and training might account for size gains and performance improvements" (McAloon, 1990: 58). Hoffman's naivete, or should we say innocence, in the face of the preponderance of steroids in international track and field is astonishing.[19]

Hoffman's insistence on innocence has little meaning outside of the larger politics of "race" and nation in Canada. In her statements to the inquiry, she noted that officials at SportCanada were "only vaguely aware of rampant rumours of steroid use among Canadian athletes" (McAloon, 1990: 58). Whether her ignorance is feigned or not, her comments are crucial here. If the official position is that you can run clean and win, in spite of evidence to the contrary, then the Dubin Inquiry can have only one conclusion: punish the "cheaters." The "innocence" of Hoffman, Charest, and others establishes the frame of the Inquiry and its ultimate cultural meaning. What this amounts to is the redundancy of the Dubin Inquiry. More to the point, given that the Inquiry told us little that we did not know, we are left with no option but to conclude that the Dubin Inquiry was designed for nothing else but the following: to name Johnson as a "scapegoat."

Thus, the Dubin Inquiry is an important marker in the history of "race" and nation in Canada. Its effect on the national imaginary is to once again name the "immigrant" (in this case, black masculinity) as being outside the law. Within the current context of racism in Canada, the Dubin Inquiry helps institutionalize a discourse of the cheating immigrant, more specifically the cheating Jamaican, in sporting cultures and beyond. Given how Ben Johnson has become the "ghost" of the paranoid nationalist, the Inquiry marks the institutional and archival justification for the tropes of spectrality that have come to define Ben Johnson: "shame," "spectre," "ghost," etc. It acts to reinforce longstanding beliefs that immigrants haunt and pollute the body politic and it acts to inform the work of media and, as we will see, other athletes.

Thus, the Inquiry institutionalizes another temporality of nation that the figure of Johnson represents. The threat of this temporality is what must be kept in mind as we move to the next section, since it is here that I discuss Donovan Bailey's role within this particular iconography. In other words, how is Donovan Bailey, represented as he has been, performatively dealing with these representations? For while it is true that the media have a great deal of power in shaping the discourse of nation, the fact is that this is not a hermetic process. In fact, what Bhabha would call *performative* re-mappings of nation can come from a number of sources: one of these sources is the people on the ground. But superstars like Bailey also have an impact with his popularity as well as the predominance of sports media within popular culture more generally. The following section deals with these issues and tries to suggest what it means to the performance of black masculinities in Canada.

Bailey's Performance in Making (Other) Canadas

Not surprisingly, these actions, and the use of Johnson as a negative image against which to write the nation, are not limited to white folks. In this sense whiteness becomes an identity—which is to say, it is not reducible to skin colour. Whiteness, as many have argued, is a set of cultural practices designed to reproduce racist notions of superiority and inferiority. This is seen if we take a look at Donovan Bailey's subjective actions regarding media positioning of Ben Johnson and Bailey's own status as a hero.

What I will show in this section is that Bailey himself appears to be performing this Canadian racism in order to secure his own identity. Bailey's performance tells us about the ability of Canadian racism to inscribe itself on black and white subjects and the way that construction of a kind of macho Canadian identity is predicated on beating up on black folks. Bailey's post-Atlanta performance suggests a partial desire to distance himself from Johnson, and, in turn, the limits of the nation.

Bailey's original attitude when asked repeatedly about Ben was often solidarity toward him and a defiant Jamaican pride (which took some of us back to another era). For example, immediately after winning the gold medal in the men's 100-metre sprint in Atlanta in 1996, Bailey spoke live to National Broadcasting Corporation reporters and reminded everyone that he was a Jamaican and proud of it. Further, when asked about how he identifies nationally by the *Toronto Star* immediately after the victory, his response was: " 'I'm Jamaican, man,' said Bailey in a post-race conference. 'I'm Jamaican first. You've got to understand that. That's where I was born. That's home. You can't take that away from me. I'm a Jamaican-born Canadian sprinter' " (28 July 1996: D1). In addition to this exuberance, recall that in the lead-up to the Atlanta Games, in a *Sports Illustrated* article by Michael Farber, Bailey openly criticized Canadian racism and defended Ben Johnson by claiming that Canada was "as blatantly racist as the United States" (*Sports Illustrated*, 22 July 1996).

Bailey's post-race performance is a significant moment in Canadian track and field and in the history of post-colonial Canada. His proud display of Jamaican identity and his insistence on being "Jamaican first" shows the extent to which Bailey was, in fact, identifying with Ben Johnson in opposition to the racist press, government, and public who scapegoated Johnson. Moreover, his final sentence, as reported in the *Toronto Star*, suggests he is signifyin(g)[20] on the racism meted out to Johnson and the illusion of citizenship and heroic status that was bestowed upon Johnson. Recall that in the aftermath of Ben Johnson's positive drug test in Seoul, his identity moved from being "Canadian" to "Jamaican-Canadian" to "Jamaican-born."[21] Bailey's dis-identification—a reclaiming of the pejorative term "Jamaican-born—is a gesture aimed at solidarity between himself and Ben Johnson.[22] Moreover, it is a way of taking power away from state officials to name and identify blackness at their whims.

However, in the aftermath of such strident anti-racism, it appears that Bailey may have changed his tune. At the Canadian outdoor track and field championships in Vancouver in 1997, Bailey responded to questions about Johnson's possible return to track with a dismissal of his country-man. Bailey told a news conference: "Ben should just get a real job." Such a response differs from the previous gestures of solidarity or of the gestures of ambivalence regarding Johnson, which Bailey felt were necessary in order to stake out his name to a media and public continually linking him to Johnson. In the post-race news conference at Atlanta, Bailey noted, "I did it for myself, for my family, for my country—so I wouldn't say it was just to revoke the past" *Toronto Star*, 28 July 1996: D1). In addition to disparaging Johnson, Bailey has been wont to express brash anti-American sentiments toward his rivals. After losing the 100-meter sprint in the World Championship in Athens to American Maurice Greene in the summer of 1998, Bailey was far from gracious in defeat. According to news reports, Bailey rolled his eyes at Maurice Greene's repeated gestures to thank God for his victory. And when it was Bailey's turn to speak, he mocked Greene and told reporters that he was thankful to God for being allowed the opportunity to answer questions.

What is noteworthy about Bailey's shift is that it betrays the traces of a classic Canadian nationalism—bashing Americans and bashing immigrants (or those who "look" like them). Far from being different, what is disappointing in Bailey's case is that what seemed like a possible "multiculturalist" masculinity, where he rejected the terms of official Canadian nationalism and tried to do something different, has given way to the same old clichés. Bailey's quasi-secure status as a Canadian hero has translated into distancing himself from Ben (or just dissing Ben), and a crass and superficial anti-Americanism (which may be read as anti-black as well). What is instructive here is that Bailey seems to be responding to the fear that occurs if Ben's name sticks around. If Ben's name sticks, you may become less "Canadian." Ironically, or perhaps not, Bailey is responding to the same panic that motivates conservative nationalists such as Dick Pound and others.

Conclusion

Racism continues to inscribe itself in the national sporting imaginary. In track and field, where many big stars are black and male, the way that track is represented is very different from hockey. Whereas hockey is represented as the timeless and essentially Canadian core, track and field suggests a different temporality. This temporality is influenced by a number of factors, not the least of which is the influence of nation-state practices on blackness and its demand that it exist as a flexible labour supply, as well as on Canadian fantasies of cleanliness, as witnessed through the analysis of the Dubin Inquiry.

These discourses manifest themselves in a continued (re)making of Ben Johnson as the "ghost" that haunts the nation. The inability to forget Ben, or the injunction to forgetfully remember him, marks the limits of the nation. Remembering/forgetting Ben marks the nation's boundaries and separates good from evil, pure from impure, "Canada" from "immigrant." Donovan Bailey's performance is therefore a very important piece of the puzzle. Bailey's performance does not mitigate the racism of the press and public, nor does it in any way lessen Ben Johnson's status as a "ghost." In fact, Bailey's dissing of Ben is what reproduces a Manicheism between a good blackness and an un-Canadian one.

Once again there is an attempt to incorporate blackness into the nation or the national landscape, if you will. However, what happens is that one kind of blackness becomes Canadian at the same time as another is pushed to the border, deported. Injunctions by black Canadian sprinters to run clean are a gesture at erasing Ben Johnson. Donovan Bailey's performance is evidence that to be Canadian is to diss Ben or to forget him. In a sense, forgetting Ben is similar to forgetting social problems and the possibility of resistance.

But from an anti-racist perspective, this does not have to be the way it goes. To begin, one way to resist racism is to remember Ben differently; that is, to remember the hypocrisy and maliciousness of state officials. In addition, this would also entail a refusal of the racist demand to diss Ben in order to be granted citizenship or rights to the nation. Refusing to invoke Ben in such a fashion opens the door to imagining Canada differently, as was witnessed by Bailey's brief period of solidarity. Such an imagination opens the door to a sustained critique of Canadian racism and capitalism. Moreover, such an imagination offers the chance to develop creative spaces from where we can understand blackness differently, which is the second point in my anti-racist critique. In other words, it must allow us to develop strategies that continually resist the conflation of all black bodies into one—in this case the conflation of Bailey into Johnson and vice versa. Writing the nation differently means developing a way of conceiving and figuring blackness, or black subjectivity, which would resist monosyllabic or Manichean versions of blackness. Resisting Manicheism, the good-evil dichotomy, is crucial, since the demand to be Canadian often rests on such binaries in order to secure itself.

NOTES

1. For more on the way that "race" acts to construct black women's bodies as accessible, please see Angela Davis (1983, chapter one). In terms of a discussion on the confluence of the sexual and economic subjugation of black women, and the role of the Canadian state in such, please see Silvera (1989); and Brand, in Bannerji (1993).
2. Rose's work is not as detailed in terms of linking racism and economic crisis.
3. There are many sources to cite regarding this; in terms of an overview, please see Li and Singh Bolaria (1988).
4. See Goldberg (1993).
5. Note that Ben Johnson's opinion was not sought.
6. For a longer discussion of the actions of Charest and Mulroney, see McAloon (1990).
7. Bhabha (1994: 143). Emphasis on "historical" in original; other emphases added.
8. There is a resonance between my concept of forgetful remembering and Bhabha's claim that nation and its unanimism are in fact challenged. While I argue that remembering to forget is a necessary task in Canadian nationalism, Bhabha (1994: 160) suggests that its opposite, "forgetting to remember," opens up ways to imagine the nation differently.
9. While I am writing the revisions to this chapter in September 1999, this is no longer the case. However, I cannot go into detail here about the current representations.
10. Please see also Anne McClintoch (1995, chapter five).
11. *Toronto Sun* (7 Aug. 1995: front page); *Toronto Star* (7 Aug. 1995: B1); *Globe and Mail* (7 Aug. 1995: D1).
12. The obsession with cleanliness as necessary for healthy nationhood has been a hallmark of Canadian nationalism for quite some time, and perhaps had one of its most fervent periods in the early twentieth century. For more on this, please see Valverde (1991).
13. Tess Chakkalakal (1999) has shown to what extent being "humanitarian" is central to Canadian nationalism, especially in the area of foreign policy.
14. Such inversions are common in what Butler (1993) calls "the racist imaginary." Borrowing on Fanon, Butler shows how the Los Angeles police justified their continued beating of Rodney King through a belief that King, while almost unconscious, was attacking them.

15. With specific reference to nationalism, this is also a longstanding practice, the most poignant example being the D.W. Griffiths's film *Birth of a Nation*, which features a white woman being raped by a black man (a white actor in blackface) to symbolize the defilement of the nation.
16. The reason I have placed *public* in quotation marks is because of its dubious reality. For me, and many others with an anti-racist conscience, what we were watching was a completely different spectacle, one that was clearly about pinning the blame on Ben Johnson while other guilty parties escaped without blame. While McAloon is for the most part aware of this, he does often refer to public in a white way. The same is true for Steven Jackson's (1998) use of "Canada."
17. In the wake of the Tour de France steroid scandal, Ben Johnson argued in an interview with the *Daily Mail*, an English newspaper, that this proves he was "not the only bad guy." See *Toronto Star* (5 Aug. 1998: E9).
18. While there is insufficient space to discuss this here, I suggest that many fans refuse to accept athletes' "impurity" because of persistent desires for our "superheroes" to be superhuman.
19. Note also that in 1997, when Johnson applied to be reinstated by the International Amateur Athletic Federation, he and his agent, Morris Chrobotek, were arguing that effectively Johnson did not do anything wrong because everyone else was on steroids as well.
20. "Signifyin(g)" is an African-American term. It usually involves a form of wordplay among competing individuals. Sometimes it also refers to artists building on and revising the work of their predecessors. For more on signifyin(g), see H.L. Gates (1998), *The Signifying Monkey*.
21. For more on this, please see the *Globe and Mail* (13 Oct. 1988: A7). See also "Responding to the Crisis," Nation of Immigrants Project, Phase II, 1996, Ontario Council of Agencies and Serving Immigrants (OCASI).
22. Munoz (1997: 353) suggests that, "Disidentification . . . is the way that a subject looks at an image constructed to exploit and deny identity and instead finds pleasure, both erotic and self-affirming."

SELECT BIBLIOGRAPHY

Bhabha, H.K. 1994. "The Other Question." In H.K. Bhabha, *The Location of Culture*. New York: Routledge.

Brand, D. 1993. "A Working Paper on Black Women in Toronto: Gender, Race and Class." In *Returning the Gaze*, edited by H. Bannerji. Toronto: Sister Vision.

Brand, D. 1994. *Bread out of Stone*. Toronto: Coach House.

Butler, J. 1993. "Endangered/Endangering: Schematic Representations and White Paranoia." In *Reading Rodney King, Reading Urban Uprising*, edited by R. Gooding-Williams. London: Routledge.

Davis, A. 1983. *Women, Race, & Class*. New York: Vintage.

Fanon, F. 1967. *Black Skin, White Masks*. New York: Grove Press.

Gates, H.L., Jr. 1988. *The Signifying Monkey*. New York: Oxford.

Goldberg, D.T. 1993. *Racist Culture.* Oxford: Blackwell.

Gordon, L.R., T.D. Sharpley-Whiting, and R.T. White. eds. 1996. *Fanon: A Critical Reader.* Oxford: Blackwell.

Hall, S., et al. 1978. *Policing the Crisis: Mugging, the State and Law and Order.* London: Macmillan.

Jackson, S.J. 1998. "A Twist of Race: Ben Johnson and the Canadian Crisis of Cultural and National Identity." *Sociology of Sport Journal*, vol. 15: 21–40.

Li, P., and B. Singh Bolaria. 1988. *Racial Oppression in Canada.* Toronto: Garamond.

McAloon, J. 1990. "Steroids and the State: Dublin, Melodrama, and the Accomplishment of Innocence." *Public Culture*, vol 2: 41–64.

McClintoch, A. 1995. *Imperial Leather.* London: Routledge.

Noorani, A., and C. Wright. 1995–6. "About the Hype." *This* December/January 1995–6: 232–238.

Rose, T. 1994. "Rap Music and the Demonization of Young Black Males." In *The Black Male. Representations of Masculinity in Contemporary American Art*, edited by T. Golden. New York: Whitney Museum of American Art.

Silvera, M. 1989. *Silenced.* Toronto: Sister Vision.

Valverde, M. 1991. *The Age of Light, Soap and Water: Moral Reform in English Canada, 1885–1925.* Toronto: McClelland & Stewart.

LITERATURE, IMAGE, AND SOCIETAL VALUES

JANICE ACOOSE

> Racism becomes a destructive force of cultural supremacy in its abject justification of aggressive ethnocentricity. There can be no doubt that racism is a destructive force if one racial type can act in concert to justify conquest and subjugation of another racial type towards genocide. . . . Cultural imperialism, whether by physical means or psychological means, achieves the same destructive ends—even while such cultural blindness might be perceived from inside only as a necessity to bring to knowledge those who are perceived to have no knowledge.
>
> (Jeannette Armstrong, *Give Back*)

Prior to the publication of Maria Campbell's autobiographically-based *Halfbreed*[1] and the numerous texts by Turtle Island writers like Emma LaRocque, Beatrice Culleton, Jeannette Armstrong, Lee Maracle, Ruby Slipperjack, Marie Anneharte Baker, Beth Cuthand, Louise Halfe, and others, Indigenous women were (and continue to be, albeit in a more subtle way through texts authored by non-Indigenous writers) misrepresented. They were generally represented in canadian literature somewhere between the polemical stereotypical images of the Indian princess, an extension of the noble savage, and the easy squaw, a more contemporary distortion of the squaw drudge. Such representations create very powerful images that perpetuate stereotypes, and perhaps more importantly, foster dangerous cultural attitudes that affect human relations and inform institutional ideology. This chapter challenges readers first to put aside all their assumptions about literature and particularly, the frequently encountered idea of its apolitical aesthetic character. This chapter also encourages readers to think critically about the relationship between text and reader with a view to understanding how one's own way of being, knowing, seeing, and understanding the world has been shaped and informed by literature, an apparatus of the prevailing ideology which I have previously characterized as white-eurocanadian-christian-patriarchal. Lastly, I implore readers to look to their own ideological foundation as a way of understanding personal perceptions and cultural attitudes towards Indigenous women. Delineating the relationship between literature, society, and image, this chapter explores the social history of the Indian princess and easy squaw stereotypes and offers an explanation for their existence by looking to some of the European traditions in which they were constructed.

During the early history of this developing nation now known as canada, christian fundamentals informed ideology, or the way the early settlers viewed, knew, understood, and existed in the world. By extension, the slowly evolving canadian literary canon had absorbed a european-christian (inherently patriarchal) ideological base; generally in relation to french-canadians it was catholicism and far more often in relation to english-canadians, it was protestantism. Influenced by weakening colonial ties to great britain and france, canadian literature took on a nationalistic character that spoke to the nation's political, social, and economic evolution. But despite loosening relations from metropolitan cultures, it continued to draw on fundamental European christian patriarchal ideology for cultural and literary values.

Although eventually secular liberalism modified and to some extent replaced the earlier christian dogmatism, the new ideology continued to justify imperialist practices, including the construction of literary texts. Liberalism, born out of ongoing struggles with church authority, was transplanted into the developing canadian nation and manifested in state policies and practices, including literature, which is at the core of this discussion. Liberalism, which Kenyan writer Ngugi Wa Thiong'o refers to as "the sugary ideology of imperialism," dominated canadian academia and continues to strongly influence writers' textual constructions. Liberalism, he argues, "blurs all antagonistic class contradictions, all the contradictions between imperialistic domination and the struggle for national liberation, seeing in the revolutionary violence of the former, the degradation of humanity . . ."[2]

Ideology, as a very basic way of understanding, reflects a particular group's way of being, knowing, seeing, and understanding the world and is of primary importance in the search for the roots of the Indian princess and easy squaw images. Robert Berkhofer posits that when the european explorers, fuelled by the adventurous dreams of expansionism and potential mercantile profits, landed in what they subsequently referred to as the americas, they comprehended the "New World and its peoples in terms of their own familiar conceptual categories and values."[3] These concepts, as indicated previously, were at that time based on a european-christian-partriarchal fifteenth century vision of the world. That world, previous to Columbus's historic voyages, was extremely limited and thus, as Berkhofer argues, narrowly defined the New World inhabitants. In fact, as most scholars are now aware, the first image of the Indian, like the term itself, came from Columbus's erroneous cartography.

As Berkhofer explains, the most detailed New World ethnography came from Amerigo Vespucci's *Mundus Novus* (published about 1504–1505), which provided europeans with information about so-called New World peoples. His writings mark the origin of Indigenous women's imprisonment in a white-christian-male biased ideological paradigm. Observing Indigenous women from this very ethnocentric male-biased point of view, he maintains that when they "had the opportunity of copulating with Christians, urged by excessive lust, they defiled and prostituted themselves."[4] Overall, Vespucci was astonished to learn that he and his countrymen could not distinguish women "virgins" from women who had previously born children "by the shape and shrinking of the womb; [or] in the other parts of the body."[5] Vespucci's words "women go about naked [and] . . . their bodies . . . are barely tolerably beautiful and clean" are useful for this discussion because he expressed an obvious physical attraction for the Indigenous women he encountered, although his christian sense of morality (like that of numerous other subsequent white males) conflicted with his desires. Seemingly, he must therefore reject his desirous inclinations, project them onto the women he desires, and suggest that it is the women who are lustful and given to prostituting themselves. Vespucci's ethnocentric observations, as well as subsequent early writers' observations, informed early white-european-christian-patriarchal ideology, and thus encouraged the perpetuation of stereotypic images of Indigenous women as promiscuous, and later as either Indian princesses or easy squaws. Our lives, as Indigenous women, are still constructed within this very male-centred white-european-christian, and now a white-eurocanadian, ideology. This ideology informs canadian institutions which construct and reproduce stereotypical images of Indigenous women that are based on binary opposites: good and bad.

In the historical context, Indigenous women were stereotyped as good when european interests were furthered by some sort of liaison. Before a so-called good christian whiteman could have relations with an "Indian" woman, however, she had to be elevated beyond an ordinary Indigenous woman's status. In most historical references, such Indian women were thus accorded the status of royalty. For example, Dona Marina, the Aztec who had a liaison with Hernando Cortez, is described as the daughter of a "native" nobleman. Pochahontas, who saved John Smith from a supposedly tortuous death, is described as Princess Pochahontas. Raymond William Stedman explains that the princess legend was necessarily born with the mythical representation of the New World sixteenth-century artist Stradanus' illustration. Stradanus' drawing exhibited a scantily dressed New World "maiden ris[ing] from her hammock to greet Americus Vespucci."[6] To reinforce this point, Stedman's text carries various images of barely clothed Indigenous women which appeared in illustrations symbolizing the New World: an Amazonian-type woman with spear and severed head, dated 1581; a regal looking woman with bow and arrow sitting on top of an armadillo, dated 1594; and a majestic-type woman in a carriage pulled by two armadillos, dated 1644. A more contemporary discussion by Angelika Maeser-Lemieux which draws on Native writer Rayna Green's "The Pochahontas Perplex: The Image of Indian Women in American Culture," points out that "the Indian Queen was presented as a symbol of the Americas which, after the colonial period in the United States, became divided into the figures of Princess of Squaw."[7] The bad Indigenous woman or squaw (the shadowy lustful archetype) provided justification for imperialistic expansion and the subsequent explorers', fur traders', and christian missionaries' specific agendas. Stedman maintains that these women were, to men like Columbus, barely human.

The good and bad—princess and squaw—images appear throughout the pages of eurocanadian history and literature. For information about pre-canadian Indigenous peoples, many writers looked to the Hudson's Bay Company (HBC) documents and the *Jesuit Relations and Other Documents*. While the HBC documents are first and foremost records of business transactions between the company and "Indian" fur traders, and the *Jesuit Relations and Other Documents,* correspondence from the colonial outpost to the Jesuits' home office (usually written to request more financial and human resources, and thus missionaries' conditions were greatly exaggerated), these two primary sources are inappropriately referred to as historically authoritative. Not unlike the discussions of many other previous and contemporary historians, Peter C. Newman's discussion of relations between "Indian" men, company men, and "Indian" women in *The Company of Adventurers* relies heavily on the fur trade records. Depending primarily on the records, Newman refers to George Nelson as an observant Nor' Wester who quoted "the bitter lament of an Indian who hoped to pass off his second wife to a whiteman because he considered her to have been debauched by past associations with them."[8] According to Newman's selective use of quotations, Indigenous women are represented as "Sluts—to satisfy the animal lust, and when they [the men] are satiated, they cast them off, and another one takes her . . . until she becomes an old woman, soiled by everyone who chuses to use her.[9] While many contemporary scholars are aware that primary sources, early recordings, and textbooks are full of ethnocentrism, they continue to use these fallacious records without much critical thought. Incorrectly informed canadian history books and creative literature consequently continue to encourage the construction of stereotypical representations and foster attitudes which view Indigenous women as easy squaws, or whores whose only purpose is a sexual one.

Historically, relative to their own cultures, Indigenous women exercised autonomous control over their bodies and relations with others. Although this chapter does not engage in a discussion of Indigenous women's history, an overview of weccp influence on Indigenous peoples generally, and Indigenous women specifically, is warranted here. Relying heavily on Howard Adams's *Prison of Grass*[10] and Emma LaRocque's *Defeathering the Indian*,[11] I want to offer an overview of the social, political, economic, and spiritual conditions which led up to the shift in ideological norms from an Indigenous woman-centred one to a white-eurocanadian-christian patriarchal one. While scholars offer numerous different theories about the political, economic, social, and spiritual subjugation of Indigenous peoples, there is a general consensus that a shift in power did indeed transpire. Few scholars, however, are insightful enough to challenge existing patriarchal interpretations of history which exclude discussions of shifts in power from an Indigenous woman-centred way to a white-eurocanadian-christian patriarchy. Such shifts inevitably erased Indigenous women's meaningful social, economic, political, and spiritual participation in the leadership of their communities, as well as the exercise of and control over their bodies and relations with others. As Adams's comprehensive discussion of the colonization of prairie Indigenous peoples and LaRocque's succinct analysis of stereotypical representations of Native peoples make clear, colonial institutions, as the disseminators of white-eurocanadian-christian patriarchy, functioned as cultural destroyers.

In the 1800s, by way of colonial laws enacted through treaties and various legislative acts, "Indians" and "Half-breeds" were segregated to small areas of land called reserves where they could be observed and to a large extent controlled by the various agents of colonialism. Once isolated from other Indigenous groups, Indigenous peoples were encouraged to believe that their social, political, and economic disparity was peculiar and of their own creation. Adams maintains that any development of a political nature or of consciousness was immediately repressed. To make this point he draws attention to the way that certain kinds of ceremonies and practices that strengthened Indigenous peoples were outlawed, while those that posed no threat to the settlers' ideology were encouraged. This point becomes abundantly clear when one considers the erasure of Indigenous women as pipe carriers, drummers, singers, and medicine/spiritual people, as well as participants in political, social, and economic matters.

As both LaRocque's and Adams's texts illustrate, our cultures became closed and static, albeit they were once inclusive of and supported and strengthened by Indigenous women's full participation; progressively adaptive to new ideas; flexible in terms of social, political, and economic evolution; and distinguished from one another. Our once community- and consensually-based ways of governance, social organization, and economic practices were stripped of their legitimacy and authority by white christian males, who imposed an ideologically contrasting hierarchical structure. Of specific importance to this discussion is the removal of women from all significant social, political, economic, and spiritual processes. Where women once participated and contributed in meaningful ways as part of clan, tribal, and council consensus governments, under the colonial regime (and in a more subtle way today) they were generally excluded. A perusal of some of the historical records of the treaty negotiations reflects the blatant exclusion of women's participation. Indeed, the "Indian" headmen's signed Xs to the various treaty documents confirms the whites' insistence on including males only in this process. Subsequently, Indigenous women's political powers (including and especially the freedom to exercise control over their bodies and relations with others) were almost completely eradicated while their energies were channeled into less threatening activities, such as ladies' auxiliary groups, church rituals (marriage, baptism, confession, communion, confirmations, and funerals), as well as far less important social activities such as church sponsored teas, bake sales, and bazaars. All activities in the communities were at least on the surface, completely controlled from birth until death by those previously identified white christian male authorities (and to a lesser extent by white-christian females). This kind of control was apparent in the organization of personal relations, the home, and community, in which women were strongly encouraged to adhere and conform to patriarchy. To understand the effective methods used to indoctrinate Indigenous women, one need only refer to the christian-operated "Indian" residential schools and the "halfbreed" day schools. In those schools, women were primarily taught domestic skills which would prepare them for their future roles as good farm wives and christian ladies. Any woman who demurred from the stringently imposed ideological norms underlying such teaching was viewed as an archaic squaw or a promiscuous, immoral, and scandalous woman. While a great majority of women no doubt gave in to the overwhelming pressures, some women deliberately chose to maintain their independence and autonomy. These women, however, were subjected to ridicule, banishment and alienation, and eventually represented in stereotypic and derogatory images by white christian (male and female) authorities, and by submissive Indigenous men and women who were coerced by christian patriarchy. Adams's retelling of a frightening ordeal with some Mounties who picked him up on the highway outside his community is a shocking reminder of prevailing attitudes during the 1950s and 1970s. Adams maintains that the Mounties tormented and insulted him by referring to "halfbreed babes [who] liked to have their fun lying down . . . [who] liked it better from a whiteman . . . redskin hotboxes who didn't wear any pants at all . . . [and who went] to bed with anyone for a beer." Although Adams appears somewhat perplexed by their fetishistic interest in "halfbreed" girls, he says "they seemed to have an obsessive interest in native girls [although they implied] . . . that Metis girls were little more than sluts and too dirty for Mounties."[12]

Beatrice Culleton's *In Search of April Raintree* calls attention to this ethnocentricism and stereotypic representation, which has blinded too many euro-canadians and fostered dangerous cultural attitudes about Indigenous women. In her novel Culleton creates the literary character Mrs. Semple—who functions as April Raintree's social worker—to give voice to that ethnocentricism. Mrs. Semple, who refers to the "native girl syndrome," comments spitefully to April that when Native girls get out on their own they become pregnant, indulge in alcohol and drugs, engage in shoplifting and prostitution, end up in jail and skid row, live with abusive men, and ultimately live off society.[13] While Culleton's Indigenous characters reflect feelings of powerlessness and oppression compounded with disillusionment and frustration, which encourage behaviours that sometimes appear to mirror stereotypes, her novel subverts the "native girl syndrome," the easy squaw, and to a much lesser extent, the "Indian" princess, by contextualizing Indigenous women's lives in a social, political, and economic reality that is too often imposed by the power structures.

As critical readers we must ask ourselves how stereotypical images like the Indian princess or easy squaw affect our values, beliefs, and attitudes. In "Sexism and the Social Construction of Knowledge" Margaret Anderson explains that there is a relationship between images and reality, "either because images reflect social values . . . or because images create social ideals upon which people model their behaviour and attitudes."[14] Delineating this point, Anderson writes: "Moreover, these images are produced by working people; even if we see them as social myths, they are connected to the social systems in which they are created."[15] Ngugi Wa Thiong'o suggests that while many people naively look upon literature as merely belonging to a surreal or metaphysical realm, it powerfully shapes our attitudes and beliefs towards life. Thiong'o strongly argues that

> the product of a writer's pen both reflects reality and also attempts to persuade us to take a certain attitude to that reality. The persuasion can be a direct appeal on behalf of a writer's open doctrine or it can be an indirect appeal through "influencing the imagination, feelings, and actions of the recipient" in a certain way toward certain way toward certain goals and a set of values, consciously or unconsciously held. . . .[16]

Examining literature in relation to colonization, Thiong'o connects literature to that system of oppression and genocide. He argues that literature is a "subtle weapon because literature works through influencing emotions, the imagination, the consciousness of a people."[17] The western ruling classes, he explains, reflected themselves, their images, and their history in the literature, while the colonized saw only "distorted image[s] of themselves and of their history" in the colonizer's constructions.[18]

Metis-Canadian Howard Adams writes that "as soon as native children enter school they are surrounded with white-supremacist ideas and stories—every image glorifies white success" because "white supremacy, which had been propagated since the beginning of European imperialism, became woven into Canadian institutions such as the church, the schools, and the courts, and it has remained the working ideology of these institutions."[19] He insists that Native people "cannot avoid seeing the cultural images and symbols of white supremacy, because they are everywhere in society, especially in movies, television, comic books, and textbooks."[20]

Joan Rockwell insists that what we read does affect us. Using the universal existence of censorship as an example, she explains that policy makers apparently must believe that what we read (even when fictionalized) "may have some potentially dangerous influence on people's beliefs, and consequently (possibly) on their social behaviour."[21]

Bearing this in mind one must seriously consider the influence of literature, particularly because for a great majority of people, knowledge about Indigenous peoples' cultures and history is very limited. One must also keep in mind that literature is much more than finely crafted words in the form of poems, essays, short stories, and novels, preserved over time and subsequently selectively offered to readers as the best of a particular era. More than literary art belonging to an unreal or metaphysical realm whose aesthetic qualities please the reader, literature is powerfully political because it persuades and influences the oftentimes unsuspecting reader. Because it absorbs and conforms, in varying degrees, to place of origin, political and nationalistic agenda, literature manifests ideology and expresses the dominant group's economic, philosophic, religious, and political codes and conventions. Canadian literature as an expression of the nation's prevailing ideological structures continues to erode the ethos of Indigenous peoples generally, and relative to this discussion, Indigenous women specifically.

When readers approach texts, they carry with them culture, language(s), experiences, economic status, and social/political agendas which inevitably influence interpretation. As such, these things function as ideological filters through which meaning is derived. Relative to the reader, texts similarly laden with culture, experiences, economic status, and social/political agendas strongly influence interpretation. In contrast to the reader's participation, texts can subliminally encourage readers to adopt specific ideological norms and values. Specifically, literary texts constructed outside an Indigenous ideological paradigm imprison Indigenous women in stereotypes which obscure and distort their very real and lived experiences. Literature, in this context, functions as an ideological apparatus. In "Art as Ideology"[22] Janet Wolff reinforces this point, drawing attention specifically to ideas, cultural values, and religious beliefs whose embodiment in cultural institutions strongly encourages ideological confirmation. To exemplify this point Wolff refers to schools, churches, art galleries, legal systems, and political parties, as well as cultural artifacts like texts, paintings, and buildings, as examples of cultural institutions. Perhaps feelings of powerlessness, oppression, confusion, and frustration encourage some Indigenous women to mirror stereotypical representations and consequently these women encourage the perpetuation of stereotypes. But do these few women's weaknesses justify representing all or most Indigenous women in the same way? Are we not still individuals with unique characters and spirits? Is it fair to dismiss significant and relevant cultural attributes of Metis, Saulteaux, Cree, Dakota, Dene, Blackfoot, Huron, Sushwap, Mohawk, and many other cultures' women because of a few who may encourage stereotypical representations? I often wonder why our very real contemporary lives as professors, doctors, lawyers, politicians, artists, university students, and nurturing responsible mothers are excluded from the pages of contemporary literary texts authored by many non-Indigenous writers.

If people were to refer to our mythologies and sacred stories they would find references to very important female deities and spirits. Some of these stories hold our sacred knowledge of Creation, and although many of our Elders protectively guard them because they fear that others might misrepresent them (much the same way prophets from other major religions were sceptical about having their words recorded and subsequently misrepresented), others have taken risks and recorded some stories in texts written in English.

For example, Paula Gunn Allen's *The Sacred Hoop, Spider Woman's Granddaughters,* and *Grandmothers of the Light* provide comprehensive reviews of specific Indigenous peoples' myths of female deities and superhuman spirits.[23] Also, Sam Gill's and Irene Sullivan's *Dictionary of Native American Mythology* refers to some female deities and spirits, contextualized within a dictionary type reference book.[24] Marla Powers's "Sex Roles and Social Structures" discusses the White Buffalo Calf Woman who, according to the author, represents Lakota culture and all humanity, and who brought the sacred pipe and rites which encourages faith in and longevity with all the relations.[25] Native-american anthropologist Dr. Beatrice Medicine's "Warrior Woman—Sex Role Alternatives for Plains Indian Women" maintains that among the Plains Indian cultures, "there was considerable variation in the roles of women and men."[26] Citing numerous studies, she refers to the Lakota, Navajo, Blackfoot, Ketenai, Tlingit, Ottawa, Peigan, Gros Ventres, and Cheyenne cultures in which some women's roles as warriors challenge prevailing myths about Indigenous women. Lorraine Littlefield's "Women Traders in the Maritime Fur Trade" challenges existing stereotypic images of Indigenous women by explaining that among the Haida, Nootka, Tlingit, and Tsimshian the role of women as traders was not anomalous in relation to, but in fact an extension of, women's hunting, fishing, and trading practices.[27] Priscilla Buffalohead's "Farmers Warriors Traders: A Fresh Look at Ojibway Women" provides an interesting comparison to the better known missionary accounts which encouraged the depiction of women in accord with the settlers' ideology of women as submissive to patriarchal authority.[28]

In its images of women, Indigenous peoples' literature is very different from european and eurocanadian literature. The latter grew out of the so-called book of god, wherein the social ordering of christian members is based on the very stringent and powerfully protected god-king-man-boy-woman-girl hierarchy. In such a christian dogmatic tradition, women are merely appendages to men and are generally represented as either innocently pure virgins, virtuous mothers, or fallen women. Even the foremost mother, according to a very narrow christian biblical interpretation, immaculately conceived the child Jesus and thus maintained her virginal state. This kind of christian fundamentalist obsession with purity and virginal innocence has created extremely problematic situations for all women, and when sexism is compounded with racism, such obsession creates overwhelmingly intolerable situations for Indigenous women.

NOTES

1. Maria Campbell, *Halfbreed* (Toronto: McClelland and Stewart, 1973).
2. Ngugi Wa Thiong'o, "Literature and Society," *Writers in Politics* (London: Heinemann Educational Books, 1981), 20.
3. Robert Berkhofer, *The Whiteman's Indian: Images of the American Indian from Columbus to the Present* (New York: Vintage Books, 1979), 4.
4. Ibid.
5. Ibid.
6. Raymond William Stedman, *Shadow of the Indians: Stereotypes in American Culture* (London: University Press of Oklahoma, 1989), 32.

7. Angelika Maeser-Lemieux, "The Metis in the Fiction of Margaret Laurence: From Outcast to Consort," in *The Native in Literature: Canadian and Comparative Perspectives,* eds. Thomas King, Cheryl Calver, and Helen Hoy (Winnipeg: ECW Press, 1987), 125.
8. Peter C. Newman, *Company of Adventurers* (Ontario: Penguin Books Canada, 1985), 271.
9. Ibid.
10. Howard Adams, *Prison of Grass: Canada From a Native Point of View* (Saskatoon: Fifth House Publishers, 1975).
11. Emma LaRocque, *Defeathering the Indian* (Agincourt, Canada: Book Society of Canada, 1975).
12. Adams, 38–39.
13. Beatrice Culleton, *In Search of April Raintree* (Winnipeg: Pemmican Publications, 1983), 66.
14. Margaret Anderson, "Sexism and the Social Construction of Knowledge," 29.
15. Ibid.
16. Thiong'o, 7.
17. Ibid., 15.
18. Ibid., 36.
19. Adams, 14.
20. Ibid.
21. Joan Rockwell, "A Theory of Literature in Society: The Hermeneutic Approach," *Sociological Review Monograph* 25 (August 1977): 32–42.
22. Janet Wolff, "Art as Ideology," in *The Social Production of Art* (London: Macmillan Education Ltd., 1981).
23. Paula Gunn Allen, *The Sacred Hoop: Recovering the Feminine in American Indian Traditions* (Boston: Beacon Press, 1986); Paula Gunn Allen, ed., *Spider Woman's Granddaughters: Traditional Tales and Contemporary Writing by Native American Women* (New York: Ballantine Books, 1989); Paula Gunn Allen, *Grandmothers of the Light: A Medicine Woman's Sourcebook* (Boston: Beacon Press, 1991).
24. Sam Gill and Irene Sullivan, *Dictionary of Native American Mythology* (Toronto: Oxford University Press, 1992).
25. Marla Powers, "Sex Roles and Social Structures," in *Myth, Ritual, and Reality* (Chicago: Chicago University Press, 1983), 203.
26. Beatrice Medicine, "Warrior Woman—Sex Role Alternatives for Plains Indian Women," in *The Hidden Half: Studies of Plains Indian Women.* (Lanham: University Press of America, 1983), 267.
27. Lorraine Littlefield, "Women Traders in the Maritime Fur Trade," in *Native People, Native Lands: Canadian Indians, Inuit and Metis,* ed. Bruce Alden Cox (Ottawa: Carleton University Press, 1987), 173.
28. Priscilla Buffalohead, "Farmers Warriors Traders: A Fresh Look at Ojibway Women," *Minnesota History* 48 (1983): 236–44.

10

The New Yellow Peril: The Rhetorical Construction of Asian Canadian Identity and Cultural Anxiety in Richmond

GLENN DEER

INTRODUCTION

On October 21, 1995, the following advertisement for the CBC Evening News appeared in the *Richmond Review* (p. 5), a community newspaper that serves a suburban readership situated south of the city of Vancouver, British Columbia:

> LET'S TALK. In the last year, the Chinese Canadian population grew by 26 percent. It's the biggest social transformation Greater Vancouver has ever experienced. As the pace of change increases, so do racial tensions.
>
> We need to talk. **Monday at 6:30**, Kevin Evans moderates a discussion on race relations in a special **CBC Evening News Forum.**
>
> Watch FACING FORWARD: RACE RELATIONS IN VANCOUVER

The race relations "News Forum" (subtitled "Neighbours: Beyond Political Correctness" when it was broadcast on the evening of October 23, 1995), and the events that precipitated it, provide an exemplary starting point for the analysis of discursive struggles over community identities, urban environments, and race in British Columbia. Televisual discourse, with its invitation to "talk" and its deployment of the promise of "moderation," sought to intervene in a series of race relations crises. These crises, as I will demonstrate, were forms of "moral panic" (Hier & Greenberg, 2002) that were both conspicuously constructed by the news media themselves and linked to a locally established discourse of Anglo-European Canadian entitlement to space. These discourses were not only present in the local news but were long preceded by the locally published "official" history of Richmond, written by Leslie Ross and entitled *Richmond: Child of the Fraser* (1979), a historical text that not only reinforces the character of Anglo-European spatial entitlement in Richmond but simultaneously places Asian Canadians in an abject, outsider space. This historical text provides evidence of how Richmond officially regarded itself in the 1970s, not as the multicultural and dynamic zone of development that it would become thirty years later, but as a semi-rural, Euro-Canadian community on the road to steady urban transformation. The cultural homogeneity and assumptions of Euro-Canadian spatial priority found in this history provide an important context for understanding the tensions between long-time residents and immigrant newcomers in the mid-1990s, and these tensions would erupt as a series of significant media events. Such cultural contexts paved the way, in 1995, for local newspapers to provide an increasingly dramatic narrative of racialized conflicts over immigrant-driven increases in urban change, and the participation of prominent print journalists and citizen correspondents became an integral part of the CBC forum held on October 23 of that year. This study will consider these cultural contexts, local histories from the 1970s like *Richmond: Child of the Fraser*, and the media discourses of 1995 that culminated in the October CBC forum entitled "Neighbours: Beyond Political Correctness."

RACE AND HISTORICAL CONTEXTS

Vancouver and Richmond have shared a long, but often overlooked, history of significant conflicts between dominant white communities and minority communities of colour, mainly of Asian background, over the ownership and use of space in British Columbia's lower mainland. From the earliest race riots of 1887 and 1907 by anti-oriental leagues that destroyed Chinese and Japanese homes and businesses on Vancouver's Hastings and Powell streets (Adachi, 1976; Ward, 1978; Anderson, 1991), to the appropriation of Japanese-Canadian properties and boats in Richmond's fishing village, Steveston, and the forced internment of 22,000 Canadians of Japanese ancestry in 1942 (Adachi, 1976; Miki & Kobayashi, 1991), to the more recent publicity in 2004 over the strategic positioning of Indo-Canadian and Chinese-Canadian political candidates in particular Vancouver and Richmond constituencies, these intertwinings of group allegiances, competition, and racialized spaces are central in understanding the past and present cultural dynamics of the Greater Vancouver region.

The ethnic-social character of the city of Richmond, surrounded by the arms of the Fraser river and located south of Vancouver, has been described by Ray, Halseth, and Johnson (1997) as being historically a "steadfastly European space within Greater Vancouver" (p. 88), although the fishing village of Steveston, with its long association with Japanese fishermen and some Chinese cannery workers in the early 1900s, is the "exception." Ray, Halsesth, and Johnson also note that as of 1971, Richmond had a "homogenous British/European identity," and that Asians were a small portion of the population: they comprised only 5.5 percent of the total (and these were mainly Japanese), with 36 percent residing in Steveston. By 1986, they note, the Chinese-Canadian presence grew to 8.3 percent of the population, and the proportion of total immigrants grew to a significant 31.3 percent of the total Richmond population (Ray et al., 1997, pp. 88–89). Ray, Halseth, and Johnson's analysis contends, however, that negative reactions in the popular press to the newly increased Asian Canadian presence had "little to do with physical change per se, and instead is reflective of a long history of ideas about immigrants, race and place in the suburbs" (1997, p. 83).

RACE AND SPACE IN 1995

The change in the annual immigrant population in Vancouver and Richmond, nevertheless, had an impact on local resources, material conditions, the perceptions of residents, and the production of discourse. Racialized tensions over space in Vancouver and Richmond came to a head in the fall of 1995, not only because of sheer population changes but also because of the perception of these changes by media editorialists and local citizens. British Columbia's annual incoming immigrant population doubled from 24,474 in 1980 to a dramatic 48,529 in 1994 (Hutton, 1997, pp. 300–301). The surge in the population of new immigrants, many from Hong Kong and Taiwan who chose to settle in either Vancouver or Richmond, put obvious pressures on real estate prices, school capacities, hospital services, transportation corridors, political constituencies, rezoning of residential properties, and the use of parklands.

Public controversies in the local press that responded to these pressured zones in the fall of 1995 were focused on three prominent spaces: the activity of Asian shellfish harvesters at the waterfront around Stanley Park; the perceived decline of English language use in the Commercial Drive area of East Vancouver; and the sprawling Asian business developments and malls, with Chinese signage, along Richmond's Number 3 Road. First, John Nightingale, executive director of the Vancouver Aquarium, bitterly accused Asian immigrants of "strip mining" the marine resources in the Stanley Park foreshore (Pynn, 18 September 1995). Second, preceding the Nightingale accusation by several weeks were a series of cynical reports during August 1995 by *Vancouver Sun* columnist Elizabeth Aird, who wrote about the deterioration of public harmony, the proliferating crime and violence, and the resulting "white flight" from Vancouver to outlying "little Rhodesias" like Tsawwassen (Aird, 15 Aurgust 1995; 17 August 1995; 22 August 1995; 21 September 1995). Third, adding to Aird's complaints of white flight, Dave McCullough, publisher of the *Richmond Review*, described the Asian "cultural ghettoization" of Richmond (McCullough, 9 September 1995; 21 October 1995; 27 January 1996). The threat of the new "Yellow Peril" (see Ward, 1978, p. 6) extended even beyond British Columbia's Hong Kong diasporic community to Markham, Ontario, where the deputy mayor, Carole Bell, inveighed against Asian theme malls and the flight of long-established residents and businesses away from the growing Chinese-Canadian communities ("'Fleeing the Asians' remark rebounds," 22 September 1995). Generally, the press noted the ironic historical turn-about in the fact that migration from Hong Kong, whose political life as a British colony ended in 1997, was turning Asians into the new "colonizers" of British Columbia (Aird, 17 August 1995). Yasmin Jiwani, in a 1995 North Vancouver anti-racism forum, confirmed that the media were consistently representing "Asian Canadians as opportunistic businessmen buying out Vancouver" (p. 14).

DISCOURSE STRATEGIES: MORAL PANIC AND UNMAPPING HISTORY

The print journalists cited above constructed public anxiety and moral panic over territorial threats by Asian immigrants through *spatial tropes*. These spatialized tropes of the Asian presence were persistently used in relation to prominent and popular spaces in the cultural imagination of British Columbians: the moral panic engendered by the notion of "strip mining" or irresponsible consumption collocated with Stanley Park, for example, triggered significant public outcries over this renowned tourist attraction and public green space. Threats to the supposed integration of communities on Commercial Drive were lent greater urgency by the juxtaposition of alienated schoolchildren, threats of linguistic isolation, and increasing crime in the news: the trope of "white flight" reinforced this anxiety over increasingly tension-filled and racially demarcated urban spaces. Chinese-language signs were translated into the broader and ominous effects of the trope of "cultural ghettoization." The moral panics of these three tropes—"strip mining," "white flight," and "cultural ghettoization"—served both to articulate the anxiety of white British Columbians and to wage a discursive war against the threat to Anglo-European privilege.

It should be noted that even though many residents from a Euro-Canadian background might have felt no threat at all from these changes, the prominence of the newspaper columns and letters, the advertisements of the televised discussion, and the actual CBC television forum certainly conjured up the conditions of a necessary moral panic that divided the population into an imagined dichotomy of long-time residents versus immigrant newcomers.

The concept of moral panic is outlined by Sean Hier and Joshua Greenberg (2002), who analyze the news media's coverage of the Fujian boat migrants who arrived on the coast of Vancouver Island in 1999. They employ Stanley Cohen's *Folk Devils and Moral Panics* (1972) to illustrate how the Fujian migrants were constructed as a greater threat than was warranted by the actual circumstances. Moral panic can be defined as "a tendency for a large part of society to consolidate in response to a threat, which can be real or imagined" (Hier & Greenberg, 2002, p. 140). Not only does the threat increase the solidarity of the threatened community but "this threat is believed to be so dangerous to the social body and the 'moral order' that 'something must be done,'—that is, some regulatory process must be mobilized. 'Doing something' usually involves reconsidering or amending existing mechanisms of social control" (Hier & Greenberg, 2002, p. 140). Hier and Greenberg demonstrate how the moral panic that evolved during the media coverage of the Fujian migrants drew upon racist stereotypes, and combined "racialization…with a discourse of illegality" (Hier & Greenberg, 2002, p. 147).

The analysis pursued here will demonstrate that the type of moral panic described by Hier and Greenberg can apply to the events of 1995 in Vancouver and Richmond: news readers were drawn into the controversies through the panic-evoking tropes of "strip mining," "white flight," and "cultural ghettoization." The letters to the *Richmond Review,* and the editorials by Dave McCullough and Elizabeth Aird, show that "media discourses have the capacity to recruit and mobilize news readers as active participants in the social construction of moral panic" (Hier & Greenberg, 2002, p. 139). The solidarity of the threatened Anglo-European communities was doubly evoked by the racialized spatial tropes of territorial threat, and by the rhetorical stances taken by the writers, who implied that their audience was monoculturally of Anglo-European descent, monolingually English, and from social classes who would sympathize with the urban professionals described by Elizabeth Aird. In other words, these writers did not compose their discourse in a manner that assumed their readers might be a diverse mixture of multilingual social classes, ethnicities, and communities: if they had assumed different stances and appealed to a greater variety of interests, their own positions would not have reinforced an insider/outsider dichotomy. The print journalists were drawn into the audience-forming dynamics of moral panic by assuming that their ideal readership was an ethnically bounded one, one conterminous with each of the areas threatened by the Asian "Others."

In identifying these journalists and writers as contributing to the process of "Othering," this study is drawing attention to the privileged power they have in the social construction of panic: the "Others," those Canadians who are not part of the mainstream, certainly do not have the same influence over the discourse of moral panic as do the journalists cited above. "Others" obviously do not have the same level of power to turn the tables on the news pundits, or would not have the power to "Other" or minoritize these privileged wielders of public discourse who have held prominent positions as producers of editorial rhetoric for large televisual and newspaper audiences. The CBC Evening News advertisement cited at the beginning of this study exemplifies the mobilization of an authoritative and powerful public "regulatory" process that attempts to exert some mitigating social control over the elements of panic, but simultaneously reinforces the importance of that very form of panic it is attempting to defuse. If one were to pose the question, Who initiated and developed this process of moral panic and Othering? one need look no further than the professional producers of public discourse themselves, the journalists, for part of the answer.

My investigation of the dynamics of audience scripting here is partly indebted to the traditions of rhetorical analysis and discourse theory (George Dillon, 1986; Roger Fowler, 1981; Barry Brummett, 1994), and to the combination of these methods with critical race theory as practiced by Frances Henry and Carol Tator (2002), but also to now classic formulations of discursive power in the work of Michel Foucault (1977, 1980) and Edward Said (1978). The construction of Asian presence as a threatening Other reflects the dominant discourse of the newspapers and their control over the elements of national identity. As Henry and Tator (2002) remind us, "Media representations are discursive formations . . . [that] have enormous power not only to represent social groups but also to identify, regulate, and even construct social groups—to establish who is 'we' and who is 'other' in the 'imagined community' of the nation-state" (p. 27). Furthermore, Henry and Tator emphasize that the dominant discourse so often reproduced in the media represents "the ideological positions of their elite owners" (p. 7). I would emphasize again that journalists and television news anchors enjoy special powers, with access to privileged social and economic capital for their public roles in shaping the national imaginary.

While the work of Henry and Tator—which productively combines the anti-racist energy of feminist academic bell hooks with the micro-level scrutiny of Dutch linguist Teun van Dijk and the mass media insights of John Fiske—is an exemplary model for any intervention in mediations of race and space, it is also helpful to connect discourse analysis to Sherene Razack's (2002) concept of "unmapping," a spatially self-conscious process that denaturalizes the ownership of spaces and uncovers the hierarchies of power and violence that are embedded in white representations of territory. In Razack's terms, this critical work also connects the body to a space: "To unmap means to historicize...asking about the relationship between identity and space" (p. 128). For example, in Razack's unmapping of white male sexual crime perpetrated against the First Nations woman Pamela George, she unpacks the raced and gendered elements productive of a male ideology that assumed a privileged and predatory stance against an utterly abjected and sexualized aboriginal body confined to the demeaning streets of Regina's "the stroll." Razack's insistence upon the importance of racialization as a determinative force in situating victims of violence is helpful in countering the charges of those who might insist that spatial competition is separable from race. Unmapping the problem of histories of space and the media construction of social relations in Richmond and Vancouver, therefore, entails restoring the racial link to identity and space, not eliding or naturalizing the occupation of a space by a particular group. Razack's intervention is compatible with both the discursive critique of Henry and Tator, and the views of John Fiske, who points out that while "discourses work to repress, marginalize, and invalidate others" (1996, p. 4), there are "continuous but unequal opportunities for intervention, and discursive guerrillas are key troops in any political or cultural campaign." (p. 6).

Thus, I will proceed to "unmap" some of the repressed elements of power, racialization, and history by using the methods of rhetorical criticism and discourse analysis. I will proceed in this investigation by first performing a rhetorical and racial unmapping of the "official" history of Richmond, a book published in 1979 entitled *Richmond: Child of the Fraser*; I will then confront some of the scholarship that attempts to downplay the significance of race in the space disputes of the 1990s, specifically the work of David Ley; finally, I will undertake a rhetorical analysis of the October 23, 1995 CBC *News Forum.*

UNMAPPING "OFFICIAL" HISTORY: RACING *RICHMOND: CHILD OF THE FRASER*

Carol Tator, Frances Henry, and Winston Mattis, in *Challenging Racism in the Arts*, state that "Canada suffers from historical amnesia" (1998, p. 10), or a "collective denial" that has "obliterated from . . . collective memory the racist laws, policies, practices, and ideologies that have shaped Canadian social, cultural, political, and economic institutions for three hundred years" (ibid.). What is important to note is that despite the racism uncovered by the scholarship of Adachi (1976), Ward (1978), Miki and Kobayashi (1991), or Anderson (1991), cultural and historical amnesia and elisions persist in the officially commissioned history of Richmond, the book titled *Richmond: Child of the Fraser.* A critical unmapping of the racialization of space in this book will reveal some important contexts that inform our understanding of Richmond, and will illustrate a continuity between the assumptions of Euro-Canadian spatial primacy, articulated in the 1970s, and the development of threats to this primacy in the tensions encountered in the mid-1990s. My purpose here is not to read this historical work as a simple racist text—it is clearly not—but it is a book that bears the vestigial, more understated elements of Anglo-European precedent and originary claims to space, claims that are revealingly indicative of racialized power when read with a sensitive eye for the fate of Asian Canadian lives and sensibilities within its pages.

Richmond: Child of the Fraser is a history of Richmond that was published in 1979 to mark the city's centenary. It was written by Leslie Ross, who held an undergraduate degree from Simon Fraser University and an M.A. in American and diplomatic history from Wichita State University at time of the book's publication. Her work was produced, according to the title page, "Under the direction of the Historical Committee of the Richmond '79 Centennial Society" (Ross, 1979). Ross's history is thus directed by collective municipal interests, and her book is a traditional eulogization of the Anglo-European explorers, dike builders, pioneer families, farmers, and patriarchal reeves who imposed environmental order, domesticity, and civic institutions on the fertile Fraser floodplain from 1879 to 1979. Ross's 244-page, oversized book contains 344 black and white photographs, including depictions of First Nations Salish dwellings on the Fraser, Spanish maps, pioneer homes, bearded and waistcoated Victorian patriarchs, clapboard storefronts, bridge construction, lacrosse teams, fishing boats, fire brigades, popular dance hall actress Lulu Sweet, and a prominent gallery of Richmond's past mayors (all men of Anglo-European background).

Ross's history is a typical commissioned municipal history that combines geographical facts, European settler narratives, and chronicles of the building of city infrastructures. The book's metaphoric link to an innocent and naturalistic "Child of the Fraser" River performs the symbolic and mythic task of managing an interesting ontological contradiction: the title's reference to the "child" naturalizes a delta space that was, in fact, cut away from the natural forces of the river through the human-controlled process of dike-building. The photographs of the fishing industry, dairy farming, and horse racing remind the audience of the dramatic transitions in the environment of Richmond, mainly initiated by human industrial means, as marshes and boggy wetlands were turned into neat, cultivated grids of farmland, then redeveloped into the residential tracts that characterize the Richmond of the present day.

This official history of Richmond, however, recirculates and reinforces the city's Anglo-European character: Asian Canadian communities are not only a marginal presence in this history but also a subtly Othered and abjected group. While Ross tries to present an historically objective account of how Japanese and Chinese labourers established themselves in Steveston, their presence is notably collocated with delinquent or violent criminal actions. For example, one of the few Chinese individuals named in the book is an accused murderer, Yip Leck, who is described in the set-off quotation from the *Vancouver Province* as "one of the ugliest specimens of a bad Chinaman ever landed in British Columbia. His face is the blackest of his race, his upper teeth protrude and his hair is like that of a barbarian" (Ross, 1979, p. 63). Ross's intentions in employing this quotation might be to simply convey the attitudes that prevailed in 1900 toward a man who was apparently guilty of killing the chief of police, Alex Main, with a brush hook; but her adding of the phrase, "A reporter from the *Vancouver Province* minced no words to express his verdict" (Ross, 1979, p. 62) seems to validate the propriety of both the guilty charge and the condemnation of the accused's skin colour, race, and "barbarian" appearance.

Earlier in the book there is a brief acknowledgement that Chinese labourers helped to dig some of the nineteenth-century dikes by hand (Ross, 1979, p. 49), but subsequent references are to the competition posed to the local white labour force by the Chinese, especially in the fish canneries. The efficiency of Chinese labour in the canning companies supposedly compelled white owners to invent the "Smith Butchering Machine" (Ross, 1979, pp. 114–15), a machine that Ross tells us was habitually called "the Iron Chink." Ross uses the pejorative phrase at least four times without contextualizing its racist intentions, or distancing herself from its demeaning forcefulness (Ross, 1979, pp. 114, 115, 117, 126). A typical example of Ross's use of the term occurs in the many references to how Chinese contract labourers were thought of as "living machines." The implied attitude throughout this narrative is that the Chinese possessed no inner life, value systems, or emotional integrity. Asian labour and bodies were replaceable with machines because their function in the construction of Richmond was that of a tool without an identity. As Ross writes, "even those employees who eagerly recruited Oriental labour around the turn of the century actively sought to replace them where it was to their advantage. Technological advances such as the 'Iron Chink' is one example of this, as is the hiring of native and East Indian workers who were less organized and more limited in their ability to force higher wages" (1979, p. 126).

Ross thus recirculates the abject metaphor of the Asian labourer as a disposable machine, and both documents Anglo-European discrimination and inscribes herself within its privileged historical perspective. Few Chinese-Canadian writers would be able to use the term "Iron Chink" without discursively distancing themselves from its acceptability or framing it with stronger self-conscious irony. Such distancing or irony do not occur within this history.

Ross's 1979 perspective on racism is therefore a noticeably passive and complicitous one. In writing about white exclusionary labour practices, she appears to validate the economic rationale for racist practices:

> the origin and exact nature of discrimination is not always easy to determine, but its presence is usually clear. From their earliest arrival, the foreign tongued, physically and culturally distinctive peoples were viewed with fear and suspicion, both founded and unfounded. Tensions grew and faded over the years, very often in relation to the availability of work and money. In British Columbia, the Asiatic Exclusion League, Knights of Labour, and the Workingmen's Protective Association (in Victoria) were formed to defend the rights of white workers and to encourage the passage of legislation to stop the immigration of Oriental workers. (Ross, 1979, pp. 126–27)

Ross's quiet nod to the rationalization of past discrimination is extraordinary when viewed from a raced perspective. The above passage, we must note, is located on a page that conspicuously contains four photographs: two of these show the government's confiscation of Japanese-Canadian fishing boats on the Fraser River in 1942. The phrase "the rights of white workers" hovers, though unintentionally, over a photograph of hundreds of the tragically impounded Japanese-Canadian fishing vessels.

Ross's history, as this example shows, consistently articulates the discourse of white privilege, and unapologetically rationalizes the actions of the dominant Euro-Canadian government, never granting ethical, spiritual, or emotional sensibilities to those affected by these policies. She even appropriates population statistics (Ross, 1979, p. 126) from Ken Adachi's history of Japanese Canadians, but mainly to use Adachi's own terms to highlight the negative "clannishness" of Japanese Canadians (ibid.). As well, Ross's characterization of the appropriation of Japanese-Canadian property is not only portrayed as a necessary and consensual wartime sacrifice, but also an inconvenience that is on par with or below the sacrifices made by Richmond school children who gave up their springtime parties:

> World War II was a time of sacrifice for everyone. Families were separated by the enlistment of husbands, fathers, and sons for service elsewhere in Canada and overseas. . . . The leisure time of many was given to the Red Cross and other groups. . . . School children gave up their May Queen celebrations. Clocks reverted to Daylight Saving Time or "War Time" to conserve energy and the Japanese residents gave up their homes. (Ross, 1979, p. 163)

The euphemization of anti-Asian sentiment that served to maintain Anglo-European dominance in Richmond's fishing industry and control over the Steveston docks is symptomatic of how a dominant culture elides or euphemizes what Mona Oikawa calls the "cartographies of violence" (2002). This official centennial history is an example of "sanitized landscapes and hegemonic ideologies of forgetting" (Oikawa, 2002, p. 75). Such historical elisions might be overlooked as a mere product of the limited social and political knowledge of the late seventies, when Japanese-Canadian interventions like Joy Kogawa's *Obasan* (1981) had not yet remapped the repressed or under-reported mistreatment of Japanese Canadians from the testimonial perspective of the survivor. Ross's history does cite Ken Adachi's important social history of Japanese Canadians, *The Enemy That Never Was* (1976), yet does little to grasp Adachi's rightfully indignant exposure of racist practices.

Nor does Ross's "official" and seemingly comprehensive history acknowledge the earlier oral and photographic history project written by Daphne Marlatt, with photos by Robert Minden and Rex Weyler, and oral testimonies by Maya Koizumi: *Steveston Recollected: A Japanese-Canadian History* (1975). The personal testimonies here convey the lived and phenomenologically rich lives of their multiple authors, a webwork of stories that is beautifully evoked by Robert Minden's photographic cover of a fishing net exquisitely and intricately cascading over the weathered, richly grained, and damp deckboards of a pier. Looking at this net, one feels that Ross's *Child of the Fraser* could have enveloped much more of the human side of Richmond had she actually held such a net in her hands and spoken with the fishermen who used it.

I have analyzed these histories of Richmond in detail and *raced* them because they are part of the historical and historicizing "symbolic economy" that has constructed the cultural identity of the city. As the literary critic and theorist Stephen Greenblatt reminds us, "In any culture there is a general symbolic economy made up of the myriad signs that excite human desire, fear, and aggression" (1990, p. 230; also see Bourdieu, 1993). While Greenblatt's investigation of the symbolic economy focuses on how literary artists use language and narrative to manipulate this economy, hisotrical discourses, popular culture, and mass media forms also produce symbols and images that shape this economy: the symbolic economy that is peculiar to Richmond is largely made up of all the Richmond-focused narratives, historical materials, government documents, photographs, and recordings for public audiences, and *Richmond: Child of the Fraser* is certainly a visible part of the symbolic and historical construction of Richmond. For example, four copies of the book are available in the various branches of the current Richmond Public Library system, and it is also retailed by the Richmond Cultural Centre where the main library is housed. The above reading of *Richmond: Child of the Fraser* supports the claims of Ray, Halseth, and Johnson, and their assertion that racial tensions in Richmond are "reflective of a long history of ideas about immigrants, race and place in the suburbs" (1997, p. 83). It is within this type of Anglo-European symbolic economy that the racialized moral panics of 1995 can be situated. But this foregrounding of longstanding racism in the discussion of Richmond race relations has recently been problematized by David Ley, and it is to the challenge of his perspective that I will turn next.

THE RHETORIC OF RACISM?

David Ley (1997) takes issue with Ray, Halseth, and Johnson in his article "The rhetoric of racism and the politics of explanation in the Vancouver housing market." Ley argues that popular and even scholarly identifications of racism in community complaints about change are too simplistic and ignore the complexities of economic pressure and the pre-existing "anti-growth" movement (p. 342) in Vancouver. Ley does not entirely deny "racist motivations as having some part in the protests of the last decade" (p. 344), but hopes to "open up discussion" (p. 344) that has been dominated by a racialized rhetoric. Ley's challenge can also be supplemented by John Rose's more recent 2001 study that considers whether the attitudes of fifty-four Richmond residents drawn from a network of students, acquaintances, and others are indicative of a more general resistance to growth rather than racist resistance to immigrants. Ley usefully draws out the contradictions that are created by commercial interests, and demonstrates that real estate agents, contractors, and developers benefited by supporting liberal anti-racist campaigns to challenge the protests of Anglo-Europeans against the Asian redevelopment of large areas of Vancouver and Richmond. Ley also points out that the many letters sent to the city that complain about the rapid development, tree removal, and so-called "monster" homes in Vancouver's wealthy Shaughnessy and Kerrisdale communities are ironically forgetful: such complaints, Ley emphasizes, show an ironic blindness to the European displacement of First Nations people from their camps and hunting grounds, and also ignores the "free market principles that these conservative homeowners pursue at work while seeking the protection of state regulation at home" (1997, p. 339).

Ley reiterates that there is a lack of "supporting evidence" (1997, p. 344) in the claims of Ray, Halseth, and Johnson (1997), who detect a "reinvented articulation of old racist concepts" in the 1990s popular discourse (Ley, 1997, p. 344).

My defence of Ray, Halseth, and Johnson would be, first, to highlight the term "*reinvented* articulation" in their original formulation, and to grant that the explicit racism of a five-hundred dollar head tax or the withholding of franchise rights are obviously not part of the racializing strategies of the 1990s. The management of racialized space has shifted from these earlier direct tactics for controlling the movements of Asian Canadians to discursive and symbolic modes. However, as my unmapping of the "Iron Chink" and the "barbarian" in Ross's *Richmond: Child of the Fraser* demonstrates and as the symbolic analysis of the tropes of *panic* reveal ("strip mining," "white flight," "cultural ghettoization"), the vestigial elements of the older warning signals of the "yellow peril" (Ward, 1978, p. 6) racializations continue to haunt us in the symbolic economy. While Ley is justified in urging more discussion of the economic and social factors that support community conservatism—and conservationism—the strong arguments made by Ray, Halseth, and Johnson about the symbolic and metaphoric regulation of spaces are unheeded if we fully side with Ley's position.

As a rhetorical critic and discursive guerrilla, I agree with Raymond Breton's theorizing of intergroup competition that balances both the analysis of the tangible or material forms of power that David Ley favours ("jobs, income, education, housing," in Breton's terms, 1999, p. 292), and the unpacking of the "cultural-symbolic": "the conception of collective identity, cultural character of the society, and the distribution of recognition and status in the social order" (Breton, 1999, p. 292). The symbolic realm or the social imaginary is a crucial aspect of national identity formation, and in the managing of a group's powers. Breton's helpful formulations would thus bolster Ray, Halseth, and Johnson's claim, and my own agreement with them, that in "the creation of a marginalized Chinese geography in Richmond…the fences of circumscription between groups are just as powerful if they are presumed or imagined rather than real" (Ray, Halseth, & Johnson, 1997, p. 96).

"LET'S TALK": UNMAPPING "NEIGHBOURS: BEYOND POLITICAL CORRECTNESS"

The imagined fences of circumscription are much in evidence in the discourse that follows. On August 23, 1995, D. Hannem, a Richmond resident, had a letter published in the *Richmond Review* (p. 9), an exerpt from which I have sampled below:

> With all the change and development in the Lower Mainland, were the original westerners ever asked (excluding politicians) if all this change was welcome? If, with this change, we would be willing to pay higher property taxes for these monster bathrooms to be built and located in our area? Or the continuing destruction of our trees? Blocks have been uprooted with no tree left standing to make way, as our good mayor likes to describe it, for a growing progressive community with no traffic problems. Now let's not get into our schools. Why are we paying for adults to learn English as a second language?

The preceding letter by Hannem—and note its unconsciously ironic claim to being from the position of an "original westerner," a claim that certainly demonstrates the type of originary entitlement represented in Ross's book of history—was a typical example of the flurry of letters responding to the pressures of population change and published by the *Richmond Review* and the *Vancouver Sun* in 1995. Nearly one month later, after Elizabeth Aird's columns on "white flight" were published in the *Vancouver Sun* and Dave McCullough published an editorial titled "Welcome to the Ghetto of Richmond" (16 September 1995), a Vancouver advertising executive named Don Fisher had his letter published in the *Richmond Review* on September 23, 1995 (p. 9). He wrote: "Heartfelt thanks to (publisher Dave McCullough) for your courageous and insightful commentary concerning the problems associated with immigration in Richmond (Welcome to the ghetto of Richmond, Sept. 16)…. There is undeniably an enormous attitudinal and cultural gulf between the society in which I have grown…and those of most newcomers." Two days later, Don Fisher had another letter published on the "Opinion" page of the *Vancouver Sun* (25 September 1995) in which he supported John Nightingale's criticism of the "immigrants…for devastating the marine life" and reiterated his view that a "tremendous attitudinal and cultural chasm…exists between long-time Canadian residents and many newcomers from Asia." Fisher's letter appeared on a page that was accompanied by a large graphic depiction of an enormous net engulfing a map of Stanley Park. The dramatic image visually reinforced the all-consuming and unstoppable predatory actions of the Asian-Canadian fishers. Eight letters were published in the September 25,

1995 "Opinion" section: two of these letters were explicitly critical of Nightingale, five in support (including Fisher's), and one tried to emphasize the conservation principles in a racially neutral way.

This section of my investigation will focus on the CBC *News Fourm*, "Neighbours: Beyond Political Correctness," the "Let's Talk" forum that was advertised in the October 21, 1995 issue of the *Richmond Review*. My analysis will pursue two lines of questions: first, how did the televisual discourse confront the spatialized tropes of panic—the "strip mining" of Stanley Park, "white flight," and the "cultural ghettoization" of Richmond? Second, were these spatialized tropes of moral panic defused, revised, or merely recirculated?

First, Kevin Evans, the well-known Vancouver CBC news anchor, introduced the discussion and emphasized the severity of the rising racial tensions. He attempted to rhetorically inscribe an instant sense of collective "neighbourliness" by using the "we" pronoun and employing a metadiscourse of self-conscious politeness; yet Evans also identified and divided the participants and listening audience into the groups of "old neighbours" and "new neighbours." While Evans's seriously amiable address scripted a sense of shared problems, his language also suggested that careful steps were necessary in a potentially risky discursive minefield, a zone where hidden tensions could escalate and problems could spin out of control. This emphasis on sensitive, tactful propriety was in tension with the other goal of the program, expressed in its introductory caption, "Neighbours: Beyond Political Correctness," and the tension between these two goals would hamper the delivery of the moderator throughout the forum:

> We need to talk about a very difficult subject before things get worse. Vancouver has undergone tremendous transformation in the last little while quite suddenly. There are predictable tensions between some of the old neighbours and some of the new neighbours. But there is a reluctance to bring those tensions out into the open to be dealt with. For the next hour I invite you to join us as we risk stepping beyond the limits of political correctness and speak as neighbours—respectfully, carefully, sensitively, and hopefully above all, honestly. (CBC, 1995, October 23)

While Kevin Evans delivered this prologue, the camera provided visual confirmations of his verbal claims, including images of Asian theme malls in Richmond, newly built large homes, and views of identifiably "Asian" Canadians on the street. While the theme music for the program played, captioned statistics on immigration were presented in a dramatic order to emphasize the urgency of the problem: the first caption stated, "a quarter million Hong Kong immigrants have come to Vancouver," while the second one stated, "half in the last five years." The camera then cut to an interview with Eric Wong, an editor for the *Ming Pao*, a Chinese-Canadian newspaper:

> There was one evening and I was walking downtown, near Burrard Street, and I actually got a man shouting over to me, "Go back, Chinaman, go back!" And I was stunned by that kind of shouting in the street. And I did not know what to do. (CBC 1995, October 23)

There is next a visual segue to Don Fisher, the aforementioned letter writer, who is interviewed in the same lobby:

> I am not a victim, I am a participant in the process, and I am ready, willing and able to accommodate my new neighbours. At the same time, I am concerned about the kind of society that is going to evolve. And that's it in a nutshell, because in accommodating them, I feel by definition, I will lose some of the rights and privileges that I've had, and lose some part of the Canada that I've lived in and come to enjoy. (CBC, 1995, October 23)

The forum thus juxtaposes two voices: first, the voice of the Asian newcomer who testifies to his racist Othering on the streets of Vancouver, where he is told to "go back," an experience that reinforces the vulnerability of non–Euro-Canadians to racist threats based on white claims to spatial primacy; and, second, the "accommodating" voice of the white Canadian who fears for the loss of an exclusively Euro-Canadian-dominated space, and who exerts a preceding and proprietary claim to forms of cultural "enjoyment."

Don Fisher's involvement in maintaining the moral panic of the "strip mining" trope is shortly staged in a tense exchange in the opening minutes of the forum between Kevin Evans, Fisher, and Victor Wong, an activist in the Chinese-Canadian community who has often challenged media stereotyping:

> Kevin Evans: Do you think that there are some people who try to shut down that dialogue...by pointing the finger and calling someone a racist?
>
> Don Fisher: Yes, I do, and I think a good example of that was the recent remarks by John Nightingale, the Executive Director of the Vancouver Aquarium, when he was talking about predation of the foreshore around Stanley Park.
>
> Kevin Evans: He said, I believe, that Stanley Park and the foreshore area is being strip mined by newcomers who have no concept of conservation.
>
> Don Fisher: That's correct, and I think...
>
> Kevin Evans: And Victor Wong, who is here as well tonight, who is with the Vancouver Chinese Canadian Association, said, "That's racist!" You pointed the finger at John Nightingale at the Aquarium, and that shut down the discussion.
>
> Victor Wong: No, no it didn't shut down the discussion at all. In fact we had a discussion later on that day when the story broke. (CBC, 1995, October 23)

The above sample of the exchanges between Evans, Fisher, and Wong show how much of Evans's discourse was metadiscursive, or spent on discussing the difference between racist language and open discussion. Victor Wong's subsequent attempt to absolve the Asian "harvesters" of irresponsible practices would not be accepted by Don Fisher, who later asserted his knowledge as an advertising expert to uphold the notion that different cultural groups have predictable habits that are empirically measurable. The trope of "strip mining," neutralized by Wong's retranslation of the term as "harvesting," was thus deflected into Fisher's discussion of "cultural predilections," a view that reinscribes the logic of moral panic by attributing individual behaviors to the predictable beliefs of a group. Wong was unconvinced, however, by Fisher's claims, and the rhetorical interlocutors were visibly unchanged through their "neighbourly" debate. Fisher neither retracted nor modified the trope of strip mining, but asserted that empirical evidence might show that some cultural groups are less respectful of conservation ethics: as Fisher stated, "I think that John Nightingale in the first case identified a group or groups that were responsible, and I think that he was basically saying that because of their cultural predilections or whatever [that] this was, a common acceptable practice for them, and they were not really respectful of the fisheries management program there" (CBC, 1995, October 23).

The second major trope of moral panic, "white flight," was previously circulated in the newspaper columns of Elizabeth Aird, also a participant in the televised forum. She displayed her anxiety about participating in the forum, since she was severely criticized by Chinese Canadians for her columns. "I'll speak frankly as a journalist," she confessed. "I'm not really keen to be sitting here, talking about this. I certainly was attacked." Aird also asserted that reporters are obliged to cover stories that are "under-reported," and that issues of "cultural difference" fall in this area. She continually drew attention to her own lack of comfort.

While Aird asserted the ethics of reportorial obligation and the freedom of the press, her own body language, anxious tone, and defensive posture appeared to physically reproduce the moral panic of "flight" from a feared Asian Other. Aird's newspaper columns had deployed tropes of "white flight" to announce the fear of cultural changes, the increased crime in her neighbourhood, and her victimization: she then doubled this victim status by describing how she "went to a race forum and was really hammered." Exasperated by the critical attention she received, Aird impatiently confessed to the forum audience, "I am considered a racist now by many, many people for putting [the columns] in the paper—hey, and the point I'm at now I'm thinking, gee, maybe I am, maybe I just don't know I am. The whole discussion is just so confusing" (ibid.).

Aird's rhetorical coping strategy in maintaining her ethos involved separating the discursive force of her columns, in which she had juxtaposed crime, immigrants, and the prevalence of the Chinese language, from the verbal criticism that she had received from the Chinese-Canadian community: for Aird, there was no reasonable relationship between her own discursive "violence" and the subsequent "attacks" on her by Chinese Canadians. Tommy Tao, a Chinese-Canadian lawyer at the forum, drew attention to Aird's use of the "white flight" phrase, and how it implied that she was writing for a monolingual, exclusively Euro-Canadian audience: "when your column says that there has been a white flight from the neighbourhood because there are some new immigrants—Chinese immigrants—who don't speak English. The Chinese people who read that column felt that, from whose perspective is she speaking? Does she really understand the situation? How can you say that the Chinese children don't speak English when they try so hard to learn?" (CBC, 1995, October 23). Aird responded to Tao's questions with another evocation of her reportorial authority and objectivity; yet her rationalizing of the trope of white flight was occluded with her description of her role as an uncertain messenger: she stated, in her defence, "I was reporting on two people, simply saying, here's something that happened, what does it mean, I'm not sure: is this one of the reasons people are leaving?" (CBC, 1995, October 23). Aird's televisual defence did little to mitigate the force of her print-disseminated use of the "white flight" trope of moral panic. Her verbal profession of reportorial impersonality and objectivity was contradicted by the obviously personal and impassioned rhetoric of her August 17, 1995 column titled "People are leaving town to find an English-speaking street for their kids": this was a column that surveyed the decline of Vancouver's neighbourhoods due to "prostitution…traffic, noise, drugs," and the "white flight" of families who could no longer find English-speaking friends for their children in the older neighbourhoods. Aird concluded this column with the line, "unfortunately, my friends are starting to leave town." It is notable that Aird's column never quotes the voices of those immigrant Others that seem to threaten her urban spaces, nor does she withdraw or revise the tropes of "white flight" in her televised appearance. She simply reinscribes her role as the victim of urban decline and she implies that the trope of "white flight" remains a legitimate marker of Vancouver's spatialized moral panic.

The third major trope of moral panic that was central to the forum, Richmond's "cultural ghettoization," was authored by *Richmond Review* publisher Dave McCullough, who was also a vocal forum participant. McCullough's rhetorical stance, interestingly, was the most contrite amongst the group who represented the dominant media. McCullough characterized his own language as "clumsy," and revealed that his newspaper actually received many more racist and "ridiculous" letters of complaint than he would ever consider publishing. McCullough drew attention to "the number of Chinese language signs on Number Three Road, where there is an intense conglomeration of commercial activity" (CBC, 1995, October 23), but he averred that it was ridiculous to object to the speaking of Chinese in the parking lots of Richmond's grocery stores, as evidenced in some of the views of Richmond's citizen correspondents. He returned to his original "fear" (a term he used frequently) that the Chinese "affinity" for culture and language would lead to a long-term segregation of the community: "That fear may be a legitimate fear. What are we going to do to make sure that there is an integration and an understanding, rather than this split based on the fact that we can't speak Chinese or read it?" (CBC, 1995, October 23). McCullough's fears were then addressed by Eric Wong and Nancy Li, who tried to turn the fear of segregation into a discussion of how to assist ESL speakers in becoming more comfortable with their second language.

It seemed apparent that the trope of "ghettoization" or segregation was easier to defuse while in the presence of Asian Canadians who were participating in a dialogue rather than in print, though McCullough was never given an opportunity by the moderator to provide answers to their responses. Kevin Evans initially tried to provoke McCullough into identifying himself with Aird's "white flight" by pointing out that McCullough himself had also left Richmond: "You yourself have left Richmond and moved to Ladner. Was not one of the reasons that you did that because you were not feeling as comfortable as you used to feel?" (ibid.). However, McCullough refutes this characterization of his movement, and points out that his first residence in the Lower Mainland was not Richmond at all, but Ladner. McCullough then quickly discounts his own language, and that of Nightingale and Aird, as "clumsy," and hopes for a more "constructive discussion of what are obviously hot-points" (ibid.).

CONCLUSION

In the CBC forum, key reproducers of the discourse of moral panic, including the tropes of "strip mining" (Don Fisher), "white flight" (Elizabeth Aird), and "cultural ghettoization" (Dave McCullough), were matched with Asian-Canadian interlocutors to engage in a dialogue that could act as a "regulatory process" to manage the moral panic (Hier & Greenberg, 2002) over the control of the Anglo-European character of Richmond and Vancouver, a character that has been materially and symbolically present in past history, discourse, and the control of space. The CBC forum participants were compelled to play roles that they had already scripted for themselves through their authorship of newspaper commentaries and letters. While the rhetorical framing of the forum by the CBC promised a neighbourly dialogue that could move beyond "Political Correctness," the panelists could not transcend their roles both as the representatives of the "old neighbours," with primary claims to Canadian space, and as the producers of the key tropes of moral panic. Certainly lost in this discussion was any deep historical context that would have demonstrated that Euro-Canadian claims to space were already founded on the displacement of First Nations people from colonized territories. As the panel was divided into the "old" and the "new" neighbours, with Anglo-European Canadians pitted against non-Western, multi-ethnic Canadians, Don Fisher, Elizabeth Aird, and Dave McCullough were scripted into roles that they were required to perform to confirm their dramatic status as the disseminators and embodiments of the tropes of spatial moral panic. Fisher, and especially Aird, proceeded to act out their roles as the bearers of the signs of "strip mining" and "white flight" in order to dramatize the tangible and lively existence of the moral panic for the moderator and the televisual audience. Indeed, the rhetorical stance of the moderator, Kevin Evans, often displayed the forensic tone of the cross-examiner who was attempting to confirm the victimized identities of Fisher, Aird, and McCullough. Since these participants were compelled to reprise their dramatic roles in the race relations crises during 1995, they were initially unable to move into the idealized and promised dialogue that would move "beyond political correctness." The participants, with the exception of McCullough, reified their media-prescribed roles as representatives of Euro-Canadian anxiety.

The CBC forum constituted an important discursive site that wove together several social anxieties about continuous threats to the established spatial and race-marked norms of Canadian identity. Such social anxieties have continued to circulate in different forms since 1995, and certainly Hier and Greenberg's study of the Fujian migrant issue in 1999 demonstrates how the "old racist phobias are . . . reinvoked" (Hier & Greenberg, 2002, p. 158) four years later.

While Fisher, Aird, and McCullough did not retract their tropes, some partial modification of their force was achieved by the dialogue, especially in the exchanges with the contrite McCullough. The televised forum could only temporarily intercede in the moral panic of racialized competition by providing a ritualized space or a formal setting of moderated "neighborliness," a space wherein the visible and physical proximity of the forum participants belied the disagreements still brooding below the surface.

REFERENCES

Adachi, K. (1976). *The enemy that never was.* Toronto: McClelland and Stewart.

Aird, E. (1995, August 15). The city through bleak-colored glasses: Is it media-fed vision? *Vancouver Sun,* p. B1.

Aird, E. (1995, August 17). People are leaving town to find an English-speaking street for their kids. *Vancouver Sun,* p. B1.

Aird, E. (1995, August 22). "Fairy tale" melting pot has vanished. *Vancouver Sun,* p. B1.

Aird, E. (1995, September 21). Shouting "racist" a surefire way of stifling discussion of cultural dilemmas. *Vancouver Sun,* p. B1.

Anderson, K. (1991). *Vancouver's Chinatown: Racial discourse in Canada, 1875–1980.* Montreal and Kingston, ON: McGill-Queen's University Press.

Bourdieu, P. (1993). *The field of cultural production.* (R. Johnson, Trans.). Cambridge, UK: Polity Press; New York: Columbia University Press.

Breton, R. (1999). Intergroup competition in the symbolic construction of Canadian society. In P. Li (Ed.), *Race and ethnic relations in Canada* (2nd ed., pp. 291–310). Don Mills, ON: Oxford University Press.

Brummett, B. (1994). *Rhetoric in popular culture.* New York: St. Martin's Press.

CBC. (1995, October 23). Neighbours: Beyond political correctness. [CBC Evening News Special Presentation]. Vancouver: CBC.

Dillon, G. (1986). *Rhetoric as social imagination.* Bloomington, IN: Indiana University Press.

Fisher, D. (1995, September 25). Letter, "Opinion" page, "A tide in the affairs of Stanley Park." *Vancouver Sun,* p. A13.

Fisher, D. (1995, September 23). Publisher's column gutsy, insightful [Letter to the Editor]. *Richmond Review,* p. 9.

Fiske, J. (1996). *Media matters: Race and gender in U.S. politics.* Minneapolis, MI: University of Minnesota Press. "Fleeing the Asians" remark rebounds. (1995, September 22). *Vancouver Sun,* p. A15.

Foucault, M. (1980). *Power/knowledge.* New York: Pantheon.

Fowler, R. (1981). *Literature as social discourse.* London: Batsford Academic and Educational.

Greenblatt, S. (1990). Culture. In F. Lentricchia & T.

McLaughlin (Eds.), *Critical terms for literary study.* (pp. 225–49). Chicago: University of Chicago Press.

Hannem, D. (1995, August 23). Respect for Asians must be earned. *Richmond Review,* p. 9.

Henry, F., & Tator, C. (2002). *Discourses of domination: Racial bias in the Canadian English-language press.* Toronto: University of Toronto Press.

Hier, S. & Greenberg, J. (2002). News discourse and the problematization of Chinese migration to Canada. In F. Henry & C. Tator (Eds.), *Discourse of Domination* (pp. 138–62). Toronto: University of Toronto Press.

Hutton, T. (1997). International immigration as a dynamic of metropolitan transformation. In E. Laquian, A. Laquian, & T. McGee (Eds.), *The silent debate: Asian immigration and racism in Canada.* (pp. 285–314). Vancouver: Institute of Asian Research, University of British Columbia.

Jiwani, Y. (1995). The media, race and multiculturalism. B.C. Adisory Council on Multiculturalism: Anti-Racism Forum, North Vancouver, March 17–18, 1995. Victoria: Government of British Columbia.

Kogawa, J. (1981). *Obasan.* Toronto: Penguin. Let's Talk [Advertisement for CBC]. (1995, October 21). *Richmond Review,* p. 8.

Ley, D. (1997). The rhetoric of racism and the politics of explanation in the Vancouver housing market. In E. Laquian, A. Laquian, & T. McGee (Eds.), *The silent debate: Asian immigration and racism in Canada.* (pp. 331–48). Vancouver: Institute of Asian Research, University of British Columbia.

Marlatt, D. (Ed.). (1975). *Steveston recollected: A Japanese-Canadian history.* Victoria, BC: Aural History, Provincial Archives of British Columbia.

McCullough, D. (1995, September 9). Markham whines, so why is Richmond so silent? *Richmond Review,* p. 10.

McCullough, D. (1995, October 21). Dialogue key to racial harmony. *Richmond Review,* p. 10.

McCullough, D. (1996, January 27). No use slamming victims of changing times. *Richmond* Review, p. 10.

Miki, R., & Kobayashi, C. (1991). *Justice in our time.* Vancouver: Talonbooks.

Oikawa, M. (2002). Cartographies of violence: Women, memory, and the subject(s) of the "internment." In S. Razack (Ed.), *Race, space, and the law: Unmapping a white settler society* (pp. 71–98). Toronto: Between the Lines.

Pynn, L. (1995, September 18). Aquarium director criticized for "racist" comments on food fishing. *Vancouver Sun,* p. A1.

Ray, B., Halseth, G., & Johnson, B. (1997). The changing "face" of the suburbs: Issues of ethnicity and residential change in suburban Vancouver. *International Journal of Urban and Regional Research,* 21(1), 75–99.

Razack, S. (2002). Gendered racial violence and spatialized justice: The murder of Pamela George. In S. Razack (Ed.), *Race, space, and the law: Unmapping a white settler society.* (pp. 121–57). Toronto: Between the Lines.

Rose, J. (2001). Contexts of interpretation: Assessing immigrant reception in Richmond, Canada. *Canadian Geographer,* 45(4), 474–93.

Ross, L.J. (1979). *Richmond: Child of the Fraser.*

Richmond, BC: Richmond '79 Centennial Society and the Corporation of the Township of Richmond.

Said, E. (1978). *Orientalism.* New York: Vintage.

Tator, C., Henry, F., & Mattis, W. (1998). *Challenging racism in the arts: Case studies of controversy and conflict.* Toronto: University of Toronto Press. A tide in the affairs of Stanley Park. (1995, September 25). [Letters to the Editor.] *Vancouver Sun,* p. A13.

Ward, P. (1978). *White Canada forever: Popular attitudes and public policy toward orientals in British Columbia.* Montreal and Kingston, ON: McGill-Queen's University Press.

11

Others in Their Own Land: Second Generation South Asian Canadian Women, Racism, and the Persistence of Colonial Discourse

ANGELA AUJLA

Cet article montre que la tradition coloniale a fortement marqué la construction de la sexualité et de l'ethnie chez la femme contemporaine issue de l'Asie du Sud. Les Canadiennes de la deuxième generation de l'Asie du Sud se trouvent donc dans un perpétual état d'aliénation.

"Go back to where you came from!"
"Where are you really from?"
"Paki!"

Though born and raised in Canada, the national identity of multigenerational South Asian Canadian women is subject to incessant scrutiny and doubt, as reflected in the phrases above. They are othered by a dominant culture which categorizes them as "visible minorities," "ethnics," immigrants, and foreigners—categories considered incommensurable with being a "real" Canadian, despite the promises of multiculturalism. Never quite Canadian enough, never quite white enough, these women remain "others" in their own land. Not only are they excluded from national belonging, they are haunted by a discourse which has historically constructed non-white women as a threat to the nation-state. Contemporary constructions of South Asian Canadian women are situated in a larger racist, sexist, and colonial discourse which cannot be buried under cries of "unity in diversity."

In this article, I focus on how the gendered racialization of multigenerational South Asian Canadian women excludes them from national belonging and pressures them to assimilate. The literary production of these women reflects the deep repercussions of this exclusion, and provides a location where issues of identity, otherness, and racism may be articulated and resisted. I will look at poetry and personal narratives by multigenerational South Asian Canadian women as points of intervention into these issues. Beginning with a brief overview of racism against South Asians in Canada, I will discuss how racist and colonial discourses of the past continue to influence dominant discourses and perceptions of South Asian Canadian women today.

UNITY AGAINST DIVERSITY

Despite the many differences among multigenerational South Asian Canadian women, similar experiences can be identified. These include experiences of racism, feelings of being "other" and not belonging, colonialism, patriarchy, sexism, and living in a diasporic culture. I use the term "South Asian" because it challenges the geographical locatedness of cultures and identities through its wide scope of reference. Generally, the category "South Asian" refers to those who trace their ancestry to places including India, Pakistan, Sri Lanka, Bangladesh, Bhutan, Tanzania, Uganda, South Africa, and the Caribbean (Henry *et al.*; Agnew). Terms such as "East Indian" and "Indo-Canadian" are problematic because of their narrow reference. Both refer directly to the Indian subcontinent, excluding other South Asian regions. They also refer to nation states and nationalities, implying the idea that ethnicity, identity and "race" are neatly confined within the borders of homogenous states.

Much in the same spirit as colonial cartography, South Asians have been "mapped" and inscribed by the dominant culture through racialized discourse and state practices since they began immigrating to Canada in the late nineteenth century (Buchnigani and Indra). Surrounded by an imposed mythos of being deviant, threatening, undesirable and inferior to the white "race," South Asians were constructed as "other" to the dominant Canadian culture who could not even bear to sit beside them on trains (Henry *et al.*). This attitude is evident in the contemporary phenomena of "white flight" in certain BC municipalities where some white residents have chosen to move rather than live alongside the South Asians who are "ruining the neighourhood." In the early 1900s, they were not permitted to participate as full citizens, the Canadian state controlled where they could live, where they could work, and even what they could or could not wear. Though they were British subjects, they could not vote federally until 1947 (Henry *et al.*). Though in a less overt form, the traces of this mapping continue to effect South Asian bodies today. Dominant representations of South Asian Canadians are largely stereotypical and impose static notions of culture and identity on them, whether they are immigrants or multigenerational.

The history of media images of South Asians attests to this. In the early nineteenth century, the South Asian presence in British Columbia was referred to as "a Hindu Invasion" by the news media; a proliferation of articles in B.C. newspapers stressed the importance of maintaining Anglo-Saxon superiority[1] (Henry *et al.*). Negative media portrayal of South Asians still persists. As Yasmin Jiwani states ". . . even contemporary representations cohere around an "us" versus "them" dichotomy that ideologically sediments a notion of national identity that is clearly exclusionary" (1998: 60).

Canadian Sikhs, for example, have been depicted as over-emotional religious extremists predisposed to violence. Used repeatedly, these images reinforce prejudice against all South Asians, both male and female. The *Vancouver Sun* headlines "Close Watch on City Sikhs" and "Sikh Militancy Grows" have not strayed very far from the cry of "Hindu Invasion" in the early part of the twentieth century. Representations of South Asian Canadian women in the media portray them as the meek and pitiful victims of arranged marriages and abusive husbands or uses them as colourful, orientalized exotica to be fawned over (Jiwani 1998). Such media images subtly exclude South Asian Canadians from national belonging.

---

1 *The Daily Colonist* wrote: "To prepare ourselves for the irrepressible conflict, Canada must remain a White Man's country. On this western frontier of the Empire will be the forefront to the coming struggle. . . . Therefore we ought to maintain this country for the Anglo-Saxon and those races which are able to assimilate themselves to them. If this is done, we believe that history will repeat itself and the supremacy of our race will continue" (Henry *et al.* 71).

Their cultures are represented as barbaric and backwards, as "clashing" and "conflicting" with civilized and modern Canadian society. These portrayals imply that South Asians do not "fit in" here, and that they are certainly not "real" Canadians. Edward Said states,

> [The] imaginative geography of the "our land/barbarian land" variety does not require that the barbarians acknowledge the distinction. It is enough for "us" to set up these boundaries in our own minds; "they" become "they" accordingly, and both their territory and their mentality are designated as different from "ours." (54)

Said describes how the us-them boundary and its accompanying mythos about "others" mentalities has historically been constructed by the dominant culture and imposed onto "others" regardless of their consent. Though Said was referring to relations between colonizer and colonized, his idea remains just as relevant when applied to contemporary relations between South Asian Canadians and the dominant Canadian culture.

Feel-good, multicultural goals of unity in diversity and ending racism are simplistic and certain to fail because they do not acknowledge the deeply rooted racist, sexist, and colonial discourse that has constructed Canada and "Canadian identity." As Ann Laura Stoler argues, "the discourse of race was not on parallel track with the discourse of the nation but part of it" (93). Historically, Canadian identity has not been a First Nations identity, or even a French identity. It has been, and continues to be a white, British, Anglo-Saxon identity. As in other white-settler colonies, and in Britain, the civility and superiority of blood and nation was constructed against the "backwardness" and inferiority of the "darker races" (Stoler; Jiwani, 1998; Dua). For example, the modernity of the Canadian state was juxtaposed to the pre–modern South Asian woman, the blood of the superior Anglo Saxon race was juxtaposed to the degenerate blood of non-white races (Henry *et al.*). White, Anglo-Canadian unity was constructed in opposition to non-white "diversity." But now, with the introduction of multiculturalism, we are suddenly expected to make the very unrealistic leap from unity against "diversity," to unity *in* diversity.

THE PERSISTENCE OF COLONIAL DISCOURSE

South Asian women have been both sexualized and racialized through colonial discourse as oppressed, subservient, tradition-bound, and pre-modern (Dua). They are also constructed as seductive, exotic objects of desire. In another construction they are considered overly-fertile, undesirable, smelly, and oily-haired (Jiwani, 1992; Brah). The legacy of colonial discourse is evident in contemporary racialized and sexualized constructions of South Asian women. In a *Guardian* article published September 5, 1985, a 19-year-old South Asian woman in London recounts the sexualized racist comments she faces walking home from college:

> . . . if I'm on my own with other girls it's, "Here comes the Paki whore, come and fuck us Paki whores, we've heard you're really horny." Or maybe they'll put it the other way around, saying that I am dirty, that no one could possibly want to go to bed with a Paki. . . . (qtd. in Brah 79)

These co-existing sentiments of desire and revulsion can be seen as remnants of British colonial attitudes towards South Asian women. While their colonizers considered non-white women savage, and backwards, they were also thought to possess a "sensual, enticing and indulgent nature" (Smits 61). According to Yasmin Jiwani, in British imperialist fiction by authors including Rudyard Kipling, the Indian woman was characterized by her rampant sexuality and her abundant fertility (1992). As can be inferred from the comments yelled at the 19-year-old South Asian woman walking home from college, contemporary stereotypes of multigenerational South Asian women remain deeply rooted in the colonial tradition.

Race, blood, and nation have historically been deeply interconnected and overlapping concepts in the West. Historically, the immigration and presence of women of colour in Canada, and other western countries was seen as a threat to the nation-state. They brought with them the danger of increasing the non-white population and the possibility of miscegenation—a danger all the more immanent given their "overly fecund" nature. Dua comments that "In Canada, as well as other settler colonies, racial purity was premised on the Asian peril—the danger of Anglo-Saxons being overrun by more fertile races" (252). Non-white women endangered western "civility" and national identity; the proliferation of non-white babies was not just a threat to the racial purity of western societies, but to their dominance and very existence. It was thought that miscegenation and too many non-white births could lead to the demise of the Anglo-Saxon race, and therefore, the demise of the nation state itself. As Dua writes,

> . . . the submissiveness of Hindu women was linked to a decline into pre-modern conditions. While white bourgeois women were racially gendered as mothers of the nation, colonized women were racially gendered as dangerous to the nation-state. (254)

Similarly, in the everyday racist/xenophobic discourse of this country, the "real" Canadians complain that immigrants are invading their neighbourhoods, cities, and the country itself. The *Globe and Mail* warns, "soon there will be more visible minorities than whites in Vancouver and Toronto," and that their number "is the highest in history." Feeding into fears of non-white women's limitless fertility, they also report that the number of visible minorities born in Canada is rising steadily and that they are younger than "the total Canadian population" (Mitchell). Such articles reflect the persistence of colonial discourse; while the white woman's regulated fecundity was supposed to ensure the reproduction of the social body, the non-white woman's "limitless fertility" was seen as endangering the reproduction of the social body. Non-white and "mixed race" bodies signalled a danger to the State.

I AM CANADIAN?

> "Are you Fijian by any chance?" the stranger asked.
> "No," I replied.
> "Are you from India?"
> "No."

During this brief encounter on Vancouver's Robson Street in 1997, various thoughts quickly ran through my head: do I reply with the answer that I know he wants to hear? Or do I explain that I'm Canadian only to be met with the standard reply of "Where are you *really* from?" or "But where are you from *originally*?" I walked away frustrated, glad I didn't give him the answer he expected, but upset that I didn't take the opportunity to challenge his preconceptions further by stating that not all brown people are immigrants, or saying "why do you ask?" taking the spotlight off me and hopefully inciting him to question the motivation behind his intrusive inquiry. Kamala Visweswaran states,

> Certainly the question "Where are you from?" is never an innocent one. Yet not all subjects have equal difficulty in replying. To pose a question of origin to a particular subject is to subtly pose a question of return, to challenge not only temporally, but geographically, one's place in the present. For someone who is neither fully Indian nor wholly American, it is a question that provokes a sudden failure of confidence, the fear of never replying adequately. (115)

Even in "multicultural" Canada, skin colour and ethnicity continue to act as markers of one's place of origin, markers which are used to ascertain traits and behaviours which are associated with certain "races." It is a question that left me with an acute sense of being out of place and being "other"—if I seemed out of place to the man who asked the question, I must appear so to the people around me. Underlying such (frequently asked) questions are racist assumptions about what a "real" Canadian looks like. In that brief encounter, the stranger automatically linked me to a far away land that I have never seen, a place where I would surely be considered an outsider, and certainly not be considered an outsider, and certainly not be considered Indian. His question served as a reminder of my "visible minority" status—that I was not quite Canadian and could never be so.

The "other" does not necessarily have to be "other" in terms of exhibiting strange or "exotic" language and behaviour. Time and time again, the dominant culture reduces identity down to imaginary racial categories. The fact that multigenerational South Asian Canadians are treated as other, as not-quite Canadians, attests to this. At what point do multigenerational South Asian Canadians cease being seen as from somewhere else? As Himani Bannerji comments, "[t]he second generation grows up on cultural languages which are not foreign to them, though they are still designated as foreigners" (1993: 186).

South Asian Canadian women are in a predicament of perpetual foreigness—constantly being asked where they are from and having stereotypical characteristics assigned to them despite their "Canadianness." Though they are in their country of origin, they are not *of* it.

Presentation of self is one way in which we demonstrate our personal identities and recognize those of others. This holds true if we encountered someone who had inscribed her body with tattoos, multiple body-piercings, and blue hair. However, it is quite a different situation when a South Asian Canadian woman tries to ground her personal identity in this way; regardless of whether her hair is covered by a *hijab* or is short and chic, regardless of whether she is wearing a *salwaar-kameez* or jeans, she is still subject to an otherization based on an imaginary "South Asian other" constructed through racist ideology. Her own body inscriptions are ignored, as the only signifier needed for recognition from the dominant culture seems to be phenotypical. These phenotypical characteristics stand, as they have in the past, though perhaps to a lesser extent, as signifiers of difference and inferiority.

In Farzana Doctor's poem "Banu," the narrator traces her changing responses and attitudes towards racism at different stages throughout her life. During childhood and as a young adult, assimilation is her response. Eventually she rejects assimilation in favour of resistance. In "Banu," the racist interpellation, "Paki go home" (218) is directed at the little girl in the poem. According to the Oxford English Dictionary, "Paki" is an abbreviation for Pakistani, and is also described as a slang word. In "Banu," however, the common use of the term does not reflect its literal or etymological meaning. The term has become imbued with racist emotions and signifies detest, hatred and intolerance towards all South Asians, regardless of their geographical place of origin.

A generically used term in places such as Canada, Great Britain, and the United States, "Paki" is a common racist insult directed toward those who appear to be of South Asian ancestry (Bannerji, 1993; Sheth and Handa). Unlike racist insults against South Asians that are based on food or dress such as "curry-eater" or "rag-head," the insult "Paki" is based simply on one's "foreign/other" appearance. The insult "Paki" does not simply express disgust at aspects of South Asian cultures as the previously mentioned insults do. Rather, it expresses disgust or hatred based directly toward one's "race" or ethnic background. For a multigenerational South Asian Canadian to be told "Paki go home" is particularly disturbing because she is told that Canada is not her home, but a far away land which she may have never set foot on. Regardless of being Canadian by citizenship and birth, she remains, under racist eyes, simply a "Paki." When the South Asian *Canadian* girl in the poem is told to "go home," she is not only told that she does not belong in Canadian society, but is also told that she should leave. The man who uttered the slur obviously felt he was a "real" Canadian with the right to tell the "foreigner" what to do. The popularity of this term in racist discourse not only reflects an ignorance about South Asian cultures and their diversity, but also reinforces the opinion that Canada does not have room for non-white "others."

OTHERS IN THEIR OWN LAND

In looking at Canadian multiculturalism and its promotion of diversity and tolerance, one would not find any overt pressures promoting assimilation. If anything, it seems that assimilation is not an issue—they tell us that we can all co-exist harmoniously within our respective tile of the mosaic. Yet, unstated, implied, and subtle pressures to assimilate remain a powerful force. As Michel Foucault stated, "[t]here is no need for arms, physical violence, material constraints, just a gaze" (155). While official Canadian multiculturalism may promote the acceptance of diversity, the lived experience of multiculturalism is quite a different thing. For many South Asian Canadian women the strong desire to "fit in," as a result of being discriminated against, culminates in an internalization of the gendered racism they receive. Frantz Fanon argues that the consequence of racism from the dominant group to the minority group is guilt and inferiority. The inferiorized group attempts to escape these feelings by "proclaiming his [sic] total and unconditional adoption of the new cultural models, and on the other, by pronouncing an irreversible condemnation of his own cultural style"(38–9).

This is a process multigenerational South Asian Canadian women undergo in their attempts to reject South Asian culture and assimilate. Assimilation has often been used as a coping mechanism not only by South Asian Canadians, but by all visible minorities where the majority of the dominant culture is white. Obvious forms of assimilation include speaking English and wearing western-style clothing. A less obvious form is the desire to change one's physical appearance (Bannerji, 1990; Sheth and Handa; James; Karumanchery-Luik). Based on personal experiences and literature by multigenerational South Asian Canadian women, the desire to be white or possess typically western features is, unfortunately, quite common. The impact of this is compounded for multigenerational South Asian Canadian women who have been socialized into the western beauty ideal.

Internalized racism is a theme common to much of the literature by multigenerational South Asian Canadian women. One manifestation of this is illustrated by the proliferation of ads for "Fair and Lovely" skin cream and skin bleaches aimed at South Asian women, and the desire expressed in matrimonial ads for light-skinned wives. Sheth comments that light skin is so desirable in India that "the cosmetics industry [is] continually pitching skin-lightening products to women" (Sheth and Handa 86). Various cosmetic products promising to do this are also found in Vancouver and Surrey's South Asian shops.

The desire for whiteness is demonstrated in second generation South Asian Canadian activist and theatre artist Sheila James' personal narrative about how she unnaturally became a blond because "All the sex objects on TV, film and magazines were blond-haired and blue-eyed. I figured I could adjust the colour in my head to fit the role" (137). Underlying the desire for "whiteness" is a racist ideology which interprets the world associated with the dark skin of Indian and African people with danger, savagery, primitiveness, intellectual inferiority, and the inability to progress beyond a childlike mentality. Meanwhile whiteness is equated with purity, virginity, beauty, and civility (Ashcroft *et al.*; Arora).

Assimilation pressures and internalized racism experienced by the second generation are captured quite forcefully in Himani Bannerji's short story "The Other Family" (1990: 140–145) in which the second-generation South Asian protagonist of the story draws what is supposed to be a picture of her family for a school project. The picture, however, bears very little resemblance to her own family. She draws her family as white with blond hair and blue eyes, and herself as having a button nose and freckles. The drawing can be interpreted as an illustration of the little girl's desire to belong and to be like the other children—to fit in at the cost of the negation of her own body, of her own physical appearance. An essay by a multigenerational South Asian Canadian woman, Nisha Karumanchery-Luik, reflects a similar theme:

> When I was younger, I hated my brown skin. I had wished that I was not so dark, that my skin would somehow magically lighten. When I was younger, I was ashamed and embarrassed of my Indian heritage and the "foreigness" that my skin betrayed. I developed creative strategies of denial and pretense to cope with and survive in a racist environment. (54)

Her choice of phrase that her skin "betrayed" her "foreigness" and Indian heritage is a significant one. It speaks to the circumstance that many multigenerational South Asian Canadian women and other multigenerational visible minorities are in—though they may act "Canadian" in the mainstream-white-Anglo-Saxon-Protestant sense of the word (language, clothes, behaviour), their skin colour and phenotypical characteristics, signifying them as "other," never fail to give them away. Being different from the mainstream is, of course, not a problem in and of itself. It becomes one as the resulting of the othering, gendered racism, and exclusion that multigenerational South Asian Canadian women are subject to. In the following excerpt of a poem by Reshmi J. Bisessar, she reveals the shame she felt over being Guyanese:

> I was there last in '86
> At age fourteen
> Eleven years ago
> When I would say
> Thank you
> If someone told me
> that I didn't look
> *Guyanese.*
> My, how loyalties change. (22)

Often, multigenerational South Asian Canadian women try to hide and mask what it is that singles them out for racist taunts and prying gazes. For example, in another poem, the parent of a young South Asian Canadian woman asks the daughter "why do you cringe when seen by white folks in your sari?/ why are you embarrassed when speaking Gujurati in public?" (Shah 119). Thus the pressures to assimilate and "belong" result in denying aspects of South Asian culture—even to the point of internalizing the dominant ideology and seeing themselves as inferior. Thinking that their food "stinks," that their physical characteristics are less beautiful and undesirable according to western standards, embarrassment over being seen in Indian clothing, or by the accents of their parents, are all aspects of their inferiorization.

AT THE BORDERS OF NATIONAL BELONGING

Multigenerational South Asian Canadian women's efforts at masking their ethnicity are, of course, in vain. The closest they come is to be mistaken for a less marginalized ethnic group or to be bestowed with the status of "honorary white," through comments to the effect of "you're different. . . . you're not like the *rest* of them." I was given this status when deciding where to go for dinner with a group of people. One white woman asked me if I ate meat, implying that I must have "strange" eating habits as a South Asian. Before I could answer, another white woman exclaimed, "Oh of course she does, she's *just like us*!" But despite the "acceptance" of being just like them, I was still othered by the initial curiosity of "do you eat meat?" If I was "just like them" why was I the only one to whom that question was posed? Thus, even the "honorary white" status given to some South Asians fails to appease a sense of not belonging. Suparana Bhaskaran outlines the limiting typology of the "assimilated South Asian" and the "authentic South Asian" which can be applied to the phenomena of the "honorary white" discussed above:

> The logic of purity allows South Asians to be conceptually defined in only two ways: as authentic South Asians or assimilated South Asians. The "authentic South Asian" may range from being conservative, lazy and poor to being spiritual, brilliant, non-materialistic and religious. By this definition, the assimilated South Asian . . . pursues the promise of the "postcultural" full citizenship of Anglo life. (198)

Though some multigenerational South Asian Canadian women may, by the above typology, be considered "assimilated South Asians" and therefore subject to the discrimination faced by the "authentic South Asian," we see in the literature by South Asian Canadian women that seeking this identification and inclusion into "Anglo-life" is, for the most part, unattainable and continues to be fraught with othering and a sense of exclusion.

Being singled out as "other" and the consequent pressures to assimilate has a particularly strong effect on multigenerational South Asian Canadian women. They have been socialized in Canadian society from birth and have thus, unlike their parents, lived their entire lives as "ethnic/other," and different from the dominant culture. For the second generation, the assimilation process begins much earlier and in the more formative years. Therefore, racism and being othered by the dominant culture has a deeper, more detrimental impact on multigenerational South Asian Canadians than it does on their parents who did not grow up in Canada. Though the parents of second-generation South Asian Canadians may be more "othered" due to their accents, the fact that they wear Indian clothing, and from having been socialized in a non-western culture, they have come to Canada with some pre-established sense of identity (though it changes through their experiences in their new country), which is not the case for their children.

It is likely that many Canadians would be quite content if South Asian Canadians and other "visible minorities" simply integrated into Anglo-Canadian society instead of making a fuss about racist immigration policies, or their right to wear *hijabs*. Of course, assimilation can no longer be overtly legislated, although it continues to be suggested in more subtle ways, as reflected in the literature by South Asian women. Because of "subtle" pressures to assimilate, many South Asian Canadian women have interiorized the inspecting gaze of the dominant culture to the point that they are exercising surveillance over themselves. Foucault argues that physical violence and constraints are no longer needed to control a population once they have interiorized the inspecting gaze—"a gaze which each individual under its weight will end by interiorizing to the point that he is his own overseer, each individual thus exercising this surveillance over, and against himself" (155).

The inspecting gaze in this context, is the judgmental eyes of the dominant culture—state officials, journalists, neighbours, teachers, and peers. The pressure to assimilate is no longer over, it is embedded in everyday language and stereotypes used to describe and "other" South Asian Canadian women, in popular culture and media depictions, and in structures such as institutional racism. The content of the literature by multigenerational South Asians discussed earlier reveals that they have interiorized the inspecting gaze of the dominant culture, though it is a gaze which many of them have come to reject. Over and over again, these writers express the desire they have or once had to belong, to be accepted, and to "fit" into the dominant culture.

CONCLUSION

Though I have concentrated on how multigenerational South Asian Canadians have been "raced" and gendered through the dominant ideology, it is important to note that those constructed as other are not merely the passive recipients of power. In many cases, they are remapping themselves by challenging dominant representations of "their kind" through subversive forms of literary production. I would argue that in the tension between imposed identities and those asserted by multigenerational South Asian Canadian women, spaces of resistance have formed in the anthologies and other venues in which they publish, and in the act of writing itself. These venues provide a forum for South Asian Canadian women to creatively express their insights, anger, pain, and reflections. It is a textual space created by and for multigenerational South Asian Canadian women in which their marginalization and repression is both articulated and resisted.

Multigenerational South Asian Canadian women's literature is considered a new, diasporic form of cultural production. It is new in that these women are writing as both insiders and outsiders to Canadian society. Their literature demonstrates an ongoing negotiation of two intertwined cultural contexts and influences. The positionality of these women allows for a unique vantage point from which to comment on Canadian racism, sexism, and other repressions. Their writing poses an important challenge to the idea that culture and identity are fixed within certain national borders.

Angela Aujla is a PhD student in the department of Sociology, York University. Her interests include postcolonial theory, studies of diaspora, and "third world" feminisms. She recently completed a Master's thesis in anthropology at Simon Fraser University entitled, "Contesting Identity in Diasporic Spaces: Multigenerational South Asian Canadian Women's Literature."

REFERENCES

Agnew, Vijay. *Resisting Discrimination: Women from Asia, Africa, and the Caribbean and the Women's Movement in Canada*. Toronto: University of Toronto Press, 1996.

Arora, Poonam. "Imperilling the Prestige of the White Woman: Colonial Anxiety and Film Censorship in India." *Visual Anthropology Review* 11 (2) (1995): (36–49)

Ashcroft, Bill, Gareth Griffiths and Helen Tiffin. *Key Concepts in Post-Colonial Studies*. London: Routledge, 1998.

Bannerji, Himani. "The Other family." *Other Solitudes: Canadian Multicultural Fictions*. Eds. Linda Hutcheon and Marion Richmond. Toronto: Oxford University Press, 1990.

Bannerji, Himani. "Popular Images of South Asian Women." *Returning the Gaze*. Ed. Himani Bannerji. Toronto: Sister Vision Press, 1993.

Bhaskaran, Suparna. "Physical Subjectivity and the Risk of Essentialism." *Our Feet Walk the Sky: Women of the South Asian Diaspora*, Eds. Women of South Asian Descent Collective. San Fransisco: Aunt Lute Books, 1993.

Bisessar, Reshmi J. "Struggle" *Shaktee Kee Awaaz: Voices of Strength.* Eds. Shakti Kee Chatree. Toronto: Shakti Kee Chatri, 1997.

Brah, Avtar. *Cartographies of Diaspora: Contesting Identities.* London: Routledge, 1996.

Buchnigani, N. and D. Indra. *Continuous Journey: A Social History of South Asians in Canada.* Toronto: McLelland and Stewart, 1985.

"Close Watch on City Sikhs." *Vancouver Sun* 20 October 1985a.

Doctor, Farzana. "Banu." *Aurat Durbar.* E. Fauzia Rafiq. Toronto: Second Story Press, 1995.

Dua, Enakshi. "Beyond Diversity: Exploring the Ways In Which the Discourse of Race Has Shaped the Institution of the Nuclear Family." *Scratching the Surface: Canadian Anti-Racist Feminist Thought.* Eds. Enakshi Dua and Angela Robertson. Toronto: Women's Press, 1999.

Fanon, Frantz. *Toward the African Revolution.* New York: Grove Press, 1967.

Foucault, Michel. *Power/Knowledge.* New York: Pantheon, 1980.

Henry, Frances, Carol Tator, Winston Mattis and Tim Rees. *The Colour of Democracy.* Toronto: Harcourt, Brace and Co., 1995.

James, Sheila. "From Promiscuity to Celibacy." *Aurat Durbar.* Ed. Fauzia Rafiq. Toronto: Second Story Press, 1995.

Jiwani, Yasmin. "The Exotic, Erotic, and the Dangerous: South Asian Women in Popular Film." *Canadian Woman Studies* 13 (1) (1992): 42–46.

Jiwani, Yasmin. "On the Outskirts of Empire: Race and Gender in Canadian TV News." *Painting the Maple: Essays on Race, Gender and the Construction of Canada.* Eds. Victoria Strong-Boag *et al.* Vancouver: University of British Columbia Press, 1998.

Karumanchery-Luik, Nisha. "The Politics of Brown Skin." *Shaktee Kee Awaaz: Voices of Strength.* Eds. Shakti Kee Chatree. Toronto: Shakti Kee Chatri, 1997.

Mitchell, Alanna. "Face of Big Cities Changing." *Globe and Mail* 18 February 1998: A1, A3.

Said, Edward W. *Orientalism.* New York: Vintage, 1994.

Shah, Susan. "The Interrogation." *Shaktee Kee Awaaz: Voices of Strength.* Eds. Shakti Kee Chatree. Toronto: Shakti Kee Chatri, 1997.

Sheth, Anita and Amita Handa. "A Jewel in the Frown: Striking Accord Between Indian Feminists." *Returning the Gaze.* Ed. Himani Bannerji. Toronto: Sister Vision Press, 1993.

"Sikh Militancy Grows." *Vancouver Sun* 7 November 1985b.

Smits, David. "Abominable Mixture." *The Virginia Magazine of History and Biography* 95 (2) (1987): 227–61.

Stoler, Ann Laura. *Race and The Education of Desire: Foucault's History of Sexuality and the Colonial Order of Things.* Durham: Duke University Press. 1995.

Visweswaran, Kamala. *Fictions of Feminist Ethnography.* Minneapolis: University of Minnesota Press, 1994.

PART 4

Contemporary Issues

Canadian society continues to build on its history of colonization and racialization while simultaneously proclaiming itself as a world leader in the areas of diversity and multiculturalism. The key difference today is that both explicit and implicit systemic racism are occurring in the context of *formal* rights provided by the Canadian Charter of Rights and Freedoms and Human Rights and Employment Equity legislations. Within this context, we continue to hear of human rights abuses in the name of national security, calls for reasonable accommodation, the documentation of racist comments made during hearings on immigration, and heated debates on Black-focused schools, violent schools, and gang-related killings without any analysis of the increasing racialization of the economic and social inequalities in Canada. Today's issues—the consequences of racialized social exclusion manifested in school dropout rates, depression and other health issues, and sub-standard working and living conditions—are still not central areas of inquiry by Canadian academic disciplines and academics, research institutions, governments, and the private sector. The four articles in this section demonstrate just *how* colonization and racialization continue to thrive in a Canada that structurally remains a "white-settler society."

In "Reconsidering the Constitution, Minorities and Politics in Canada," Yasmeen Abu-Laban and Tim Nieguth analyze the relative absence of discussions of race and ethnicity within Canadian political science, specifically in relation to the Canadian Constitution. The authors argue that political scientists who "de-ethnicize" the British and French origin groups, and who interpret the Constitution without a historical perspective, miss the understanding of Canada as a white-settler society that implicitly and explicitly reflects relations between ethnic collectivities, including Canadians of British and French origins, Aboriginal Peoples, and ethnic minorities. Abu-Laban and Nieguth note, "constitutional politics cannot be separated from ethnocultural and race relations. In Canada, these relations were historically marked by the legalized subordination of nonwhite people" (2000:478–479). These legalized subordination practices included the creation of "Indian" status for Indigenous peoples, the segregation of schools and other institutions for Blacks and Jews, and the denial of the franchise to the Chinese, Japanese, and South Asians. As a result of this absence the authors conclude, "there is little literature in the discipline that assesses the impact of public policies relating to anti-racism and human rights for minorities" (2000:469).

The impacts of public policies are exactly the focus of the next two articles by Grace-Edward Galabuzi and Nandita Sharma. Galabuzi highlights social exclusion, intensified exploitation, and racialization of poverty as the result of the current restructuring of the global and national economies. Social stratification along racial lines is creating an "economic apartheid" in Canada. Sharma, on another level, focuses on migrant workers and analyzes racist and nationalist ideological state practices that create cheap labour. By designating people as "migrant workers," there is little to no access to services and protection by the Canadian State for these workers. The issues of increasing racialization of poverty of its own citizens and legalized slavery by indenturing workers are conspicuously absent from the federal and provincial political agendas.

In the last article in this section, Sedef Arat-Koc asks some critical questions of the social sciences' focus on transnationalism. The question most analyses of transnationalism neglects to ask is why are some Canadian citizens "pushed" to create their communities internationally? What does this reflect about Canadian society? Arat-Koc suggests that this latest focus on transnationalism "necessitates that we pose a set of critical questions about *what* gets to be studied as transnationalism and *how*, as well as about what does and *does not* get to be asked and about *whom*" (2006:216). The issues of national security and the "war on terrorism" have created a transnational hostile space for Canadian Arabs and Muslims. This criminalization of communities creates fear, insecurity, and institutional racism. Arat-Koc also touches on the case of Maher Arar, a Canadian citizen of Syrian origin, to demonstrate how the Canadian government engages in "torture by proxy."

The increasing racialization of Canada's institutions, policies, and practices is not occurring in a vacuum. Indigenous peoples and racialized communities in Canada are responding to these injustices by initiating strategies of resistances. We will examine some of these strategies in the final section of this book.

CRITICAL THINKING QUESTIONS

1. According to Abu-Laban and Nieguth, a historical analysis of the creation of the Canadian constitution reveals power relations between groups of people. How do the authors argue this position? Do you agree or disagree? Why?
2. Galabuzi notes, "In the midst of the socio-economic crisis that resulted, the different levels of government have responded by retreating from anti-racism programs and policies that would have removed the barriers to employment equity" (2004:240). How do you think the racialization of poverty will affect social cohesion in Canada? Will it strengthen Canada's democracy?
3. How has globalization and the pressure to create competitive labour markets shaped Canadian immigration policies?
4. Sedef Arat-Koc sees Arab and Muslim communities in Canada as communities under siege. She notes, "[t]he reality of siege has implications for the ability of individuals and groups not only to safely *cross* borders, but also to live in safety and as equal citizens *within* borders" (2006:216). If you had to design Canada's national security policy, how would you address issues of national security and individual rights?

FURTHER RESOURCES

Borrows, John. "Contemporary Traditional Equality: The Effect of the Charter on First Nations Politics." In Schneiderman and Sutherland, eds. *Charting the Consequences: The Impact of Charter Rights on Canadian Law and Politics.* Toronto: University of Toronto Press, 1997.

Henry, F. and C. Tator. "The Theory and Practice of Democratic Racism in Canada." In M.A. Kalbach and W.E. Kalback, eds. *Perspectives on Ethnicity in Canada.* Toronto: Harcourt, 2000.

Panitch, L. "Reflections on Strategy for Labour." In L. Panitch, C. Leys, G. Albo, and D. Coates, eds. *Socialist Register 2001: Working Classes, Global Realities*. New York: Monthly Review Press, 2001.

Satzewich, Vic and Lloyd Wong. *Transnational Identities and Practices in Canada.* Vancouver: The University of British Columbia, 2006.

Sharma, Nandita. *Home Economics: Nationalism and the Making of 'Migrant Workers' in Canada.* Toronto: University of Toronto Press, 2006.

Taylor, Charles. "The Politics of Recognition." In Amy Gutmann, ed. *Multiculturalism and the Politics of Recognition.* Princeton: Princeton University Press, 1992.

FILMS

One Dead Indian: The Story of Dudley George and Ipperwash. Monguel Media, 2006.
Description: One Dead Indian chronicles the controversial 1995 incident in which Native Canadian protester Dudley George was shot dead by OPP officer Kenneth Deane during a protest at Ontario's Ipperwash Provincial Park. For the past ten years, Dudley's brother Sam and his family have struggled to ensure that Dudley's death was not in vain.

Finding Dawn. Directed by: Christine Welsh, Produced by: Svend-Erik Eriksen. National Film Board of Canada. 2006.
Description: *Finding Dawn* illustrates the deep historical, social, and economic factors that contribute to the epidemic of violence against Native women in this country. It goes further to present the ultimate message that stopping this violence is everyone's responsibility.

No End in Sight. Director: Charles Ferguson. Red Envelope Entertainment. 2007.
Description: A comprehensive look at the Bush Administration's conduct of the Iraq war and its occupation of the country.

In the Shadow of Gold Mountain. Director: Karen Cho. Producer: Tamara Lynch. National Film Board of Canada, 2004.
Description: Stories from the last living survivors of the Chinese Head Tax and Exclusion Act.

KEY CONCEPTS/TERMS

Multiculturalism

Social Exclusion

Indentured Workers

Ideological State Practices

Transnationalism

Torture by Proxy

12

Reconsidering the Constitution, Minorities and Politics in Canada[1]

YASMEEN ABU-LABAN
TIM NIEGUTH

The constitution and constitutional discourse have loomed large on the agenda of Canadian political science since the 1960s. Over time, political scientists have approached the constitution and its role in society from a number of angles reflecting perceived primary axes of power and dominant cleavages within Canadian society. Thus, federalism, regionalism and British-French dualism have been prominent in explorations of constitutional politics, while such questions as the relation of the constitution to class or gender have been less central.

Given Canada's multi-ethnic, multiracial and multilingual character, it is surprising that a less central issue in empirical analysis has related to ethnic minorities,[2] especially considering the relevance of immigration, particularly in Canada's major cities of Montreal, Toronto and Vancouver, and the public debate over the vices or virtues of multiculturalism (as policy and ideology). We suggest that the underexamination and undertheorization of ethnic minorities, and therefore ethnicity, in the context of constitutional politics introduces a number of biases into the study of both ethnicity and the constitution. This article aims to bridge the gap between Canadian political science and Canadian ethnic studies by linking what have been traditional concerns of these respective fields: constitutional politics in political science,[3] and multiculturalism and the "other ethnic groups" of non-British/non-French/non-Aboriginal origin in ethnic studies. In doing so, our objective is to focus on the evolving Canadian symbolic order, and the recognition given to ethnic collectivities as expressed in the constitution and constitutional dialogues.

We take a threefold approach in pursuing this objective. First, we examine the relative absence of discussions of race and ethnicity within Canadian political science. We suggest that as a result, to the extent that minority ethnic groups have been addressed in the context of constitutional politics, an important strand of an otherwise heterogeneous political science literature has been informed by four implicit fundamental assumptions. These assumptions have led to what we label the "watershed approach" to ethnic minorities

1 For helpful and constructive comments, we would like to thank Claude Couture, Kent Weaver and the anonymous reviewers for this JOURNAL. Earlier versions of this article were presented at the Conference on Constitutional Reform and Constitutional Jurisprudence, Institute for Canadian Studies, University of Augsburg, Germany, October 1999, and the Conference on Nationalism, Federalism and Identities, organized by Faculté St. Jean, University of Alberta and Association Canadienne-Française de l'Alberta, Edmonton, December 1999.

2 In this article we use the terms ethnic minorities, ethnocultural minorities and minorities to refer to the collectivity of Canadians who are of non-British, non-French and non-Aboriginal origin.

3 See Allan Tupper, "English-Canadian Scholars and the Meech Lake Accord," *International Journal of Canadian Studies* 7 (1993), 351.

and the Canadian constitution. Stated briefly, the assumptions are: the entrenchment in 1982 of the Canadian Charter of Rights and Freedoms has created new constitutional actors in the form of ethnic minorities, and stimulated an ethnic constitutional discourse; consequent on the introduction of the Charter, the political power of racial and ethnic minorities has greatly increased in the constitutional sphere; this increase in power cannot be reversed; and ethnic minorities utilize this power to pursue objectives which are limited to constitutional self-interest, and thus generate conflict.

Second, we question the fundamental assumptions behind the watershed approach by examining the actual demands and concerns of minority ethnic groups as voiced by leaders of ethnocultural organizations.[4] Specifically, we examine the manner and extent to which multiculturalism or other minority concerns entered into constitutional debates during the decade preceding the patriation of the constitution, during the patriation of the constitution, during debates over the Meech Lake and Charlottetown constitutional accords and during the post–Charlottetown period.

Third, by taking the empirical evidence in the second section as a point of departure, we provide an explanation for our findings which critiques the fact that the temporal horizon of many empirical discussions within Canadian political science on the constitution and ethnic minorities is limited by what is portrayed as a decisive turning point—the introduction of the Charter in 1982. This limited temporal horizon has often led to greater weight being accorded to the Charter than to state policy and societal relations of power.[5] What we take to be a complex relationship between the state, ethnic groups and constitutional politics in Canada is best understood by placing a stronger emphasis on time. We argue that by revisiting the Canadian constitutional story with an eye towards not only 1982, but to Canadian history, post–Charter constitutional politics can be seen as an episode in the ongoing conflict as well as the give and take of recognition between dominant and subordinate social forces in which the state is implicated. Viewed from this perspective, the constitution has always either implicitly or explicitly reflected relations between ethnic collectivities (including Canadians of British and French origin, Aboriginal peoples and ethnic minorities).

4 Specifically, we examine both official position papers/documents as well as presentations made by minority leaders in government fora dealing with constitutional issues since 1980 of both national and umbrella associations. The focus on ethno-cultural leaders/spokespersons has its limitations in that the pertinent question of representation/representativeness is difficult to assess; nonetheless this approach can yield important insight into the dynamics of race, ethnicity and the constitution.

5 While we focus on empirical discussions, it should be noted that Canadian political philosophers like Charles Taylor and Will Kymlicka have focused on identity and minorities, although their internationally recognized work has more to do with illuminating the optics of liberalism, than the specific rough-and-tumble of the politics of constitutional change. In contrast, the work of James Tully is a major normative contribution to the study of constitutionalism, diversity and recognition which draws from a range of historical and contemporary examples to address whether modern constitutions can recognize and accommodate cultural diversity. See Charles Taylor, "The Politics of Recognition," in Amy Gutmann, ed., *Multiculturalism and the Politics of Recognition* (Princeton: Princeton University Press, 1992); Will Kymlicka, *Multicultural Citizenship* (Oxford: Oxford University Press, 1995); and James Tully, *Strange Multiplicity: Constitutionalism in an Age of Diversity* (Cambridge: Cambridge University Press, 1995).

ETHNIC STUDIES, POLITICAL SCIENCE AND THE WATERSHED APPROACH

The field of Canadian ethnic studies came out of the Royal Commission on Bilingualism and Biculturalism. As so-called third force (non-British, non-French, non-Aboriginal) Canadians challenged the notion of a bilingual and bicultural Canada, the Commission provided an impetus for greater research in the 1960s on the role of "other ethnic groups."[6] The introduction of multiculturalism as official federal policy in 1971 further contributed to the development of the field of Canadian ethnic studies through funding provided to academics.[7] Despite its multidisciplinary character, early in its evolution, ethnic studies tended to be skewed by the interests of those trained in some disciplines more than others. In fact, in 1977, historian Howard Palmer observed that in Canadian ethnic studies "sociologists have made the greatest contribution," but "political scientists in Canada have been relatively slow to do research on ethnic voting behaviour."[8] Indeed, ethnic studies (in Canada and elsewhere, including Australia) was described as marked by a "shortage of solid research on the political behaviour of ethnic groups."[9]

The descriptions of practitioners of ethnic studies concerning the laggardly pace of political scientists to engage in the field of ethnic studies are echoed by the observations political scientists have made about the place of race and ethnicity within the discipline. V. Seymour Wilson argued in his 1993 Presidential Address to the Canadian Political Science Association that:

> As political scientists it seems we have not been particularly comfortable dealing with cultural and racial pluralism and their effects on political life. In this country we approach the study of societal pluralism almost completely from our perspective on Quebec nationalism, despite the varied nature of the subject matter.[10]

The lack of attention to race and ethnicity is not confined to Canadian political scientists: in 1996, Rupert Taylor, writing on the study of race and ethnicity in American political science, noted that "literature to date on 'race' and 'ethnicity' within political science does not constitute a great body of work."[11]

6 Howard Palmer, "History and Present State of Ethnic Studies in Canada," in Wsevolod Isajiw, ed., *Identities: The Impact of Ethnicity on Canadian Society,* Canadian Ethnic Studies Association Vol. 5 (Toronto: Peter Martin, 1977), 173.

7 Evelyn Kallen, "Academics, Politics and Ethnics: University Opinion on Canadian Ethnic Studies," *Canadian Ethnic Studies* 13 (1981), 121–22.

8 Palmer, "History and Present State of Ethnic Studies in Canada," 174–75.

9 Jerzy Zubrzycki, "Research on Ethnicity in Australia and Canada," in Isajiw, ed., *Identities,* 186.

10 V. Seymour Wilson. "The Tapestry Vision of Canadian Multiculturalism," this JOURNAL 26 (1993), 646.

11 Rupert Taylor, "Political Science Encounters 'Race' and 'Ethnicity,'" *Ethnic and Racial Studies* 19 (1996), 891.

The reasons for the relative inattention within political science to the role of race and ethnicity in politics are numerous.[12] Yet there are areas of study relating to race and ethnicity and politics that warrant continued and greater exploration. There is little literature in the discipline that assesses the impact of public policies relating to antiracism and human rights for minorities.[13] There is also much work to be done on the political participation and the representation of minorities and ethnocultural associations in national, provincial and urban political processes.[14] Notably, among Canadian politics specialists, the area of race, ethnicity and politics that has attracted relatively greater attention relates to the constitution. Canadian constitutional expert Alan Cairns was among the first to point out the importance of this topic when he argued that political scientists must begin to grapple with the theme of ethnicity, a theme which could not be addressed from the more traditional focus on federal-provincial government relations. Writing at the time of the debate over the Meech Lake constitutional amendment, Cairns lamented:

> . . . there are good reasons to fear that political scientists will lose ground as constitutional analysts in the future. In the absence of a significant intellectual reorientation they will correctly come to be viewed as too wedded to institutional arrangements, such as federalism, of diminished constitutional importance.[15]

While legal scholars have increasingly been addressing both the historical and contemporary relationship of race, racism and the law,[16] to the extent that Canadian political scientists have heeded Cairns's call in analyzing constitutional politics and the role of ethnic minorities, a series of assumptions have emerged which truncate historical analysis. While different authors adhere to these assumptions to varying degrees (even within their own writings) the cumulative effect is the genesis of an overall approach to ethnicity and the constitution which is relatively coherent, and characteristic of much of the discipline as a whole. What we call the watershed approach contains four fundamental assumptions.

12 Wilson argues that the reasons include the sense amongst some political scientists that ethnicity should disappear as a political force; that class is more important than ethnicity; and that Canadian political science's traditional concern over federalism addresses at best only those ethnic/national groups that may be territorially accommodated. (Wilson, "The Tapestry Vision," 646–47).

13 See the discussion in Audrey Kobayashi, "Advocacy from the Margins: The Role of Minority Ethnocultural Associations in Affecting Public Policy in Canada," in Keith G. Banting, ed., *The Nonprofit Sector in Canada: Roles and Relationships* (Montreal: McGill Queen's University Press, 2000), 233–35.

14 Daiva Stasiulis, "Participation by Immigrants, Ethnocultural/Visible Minorities in the Canadian Political Process," in Canada, Department of Canadian Heritage, *Immigrant and Civic Participation: Contemporary Policy and Research Issues* (November 1997) 12–29.

15 Alan C. Cairns, "Political Science, Ethnicity and the Canadian Constitution," in David P. Shugarman and Reg Whitaker, eds., *Federalism and Political Community: Essays in Honour of Donald Smiley* (Peterborough: Broadview, 1989), 117.

16 See Carol A. Aylward *Canadian Critical Race Theory: Racism and the Law* (Halifax: Fernwood, 1999).

One underlying assumption is that the entrenchment of the Charter created a new set of constitutional actors and, at the very least, invigorated the political activities of ethnic minorities and other subordinate groups (including women and Aboriginal peoples). Thus, for example, Cairns has argued that the Charter created new actors "defined inter alia, by gender, language and ethnicity" with explicit constitutional concerns.[17] According to Guy Laforest's examination of Prime Minister Pierre Trudeau's legacy, the "Charter created a whole series of new constitutional players: women and their organizations, multicultural groups and visible minorities, native peoples, and official-language minorities."[18] Similar judgments abound in much political science literature which is more specifically concerned with constitutional politics. Jennifer Smith, for example, draws on Alan Cairns to argue that the Charter has given constitutional status to a number of groups. These "Charter minorities" are "groups who consider particular provisions of the Charter of Rights and Freedoms to be relevant primarily to themselves."[19]

Some authors have taken this assumption a step further by suggesting that the Charter did not merely create ethnic constitutional actors, or invigorate their engagement in constitutional politics, it also triggered an "ethnic discourse . . . that debates the relative status to be accorded to the two 'founding' British and French peoples and later arrivals who have made Canadians a multicultural and multiracial people."[20] F.L. Morton broadly suggests that with

> the adoption of the Charter of Rights in 1982, the political issues raised by government policies toward [a number of different] minorities in Canada became inextricably linked with the constitutional issues raised by the equality rights provisions of section 15. What were once essentially policy issues to be resolved through the political accommodation of the parliamentary process have taken on a new constitutional dimension and are now subject to judicial resolution.[21]

Likewise, it is alleged that the Charter was not only instrumental in enhancing the role of ethnic and other minorities in the judicial process; it was also crucial in establishing the

17 Alan C. Cairns, "Citizens (Outsiders) and Governments (Insiders) in Constitution-Making: The Case of Meech Lake," *Canadian Public Policy* 14 Supplement (1988), S138.

18 Guy Laforest, *Trudeau and the End of a Canadian Dream* (Montreal: McGill Queen's University Press, 1995), 137–38.

19 Jennifer Smith, "Representation and Constitutional Reform in Canada," in David Smith, Peter MacKinnon and John Courtney, eds., *After Meech Lake: Lessons for the Future* (Saskatoon: Fifth House, 1991), 75.

20 Alan C. Cairns, *Charter versus Federalism: The Dilemmas of Constitutional Reform* (Montreal: McGill-Queen's University Press, 1992), 74.

21 F.L. Morton, "Group Rights Versus Individual Rights in the Charter: The Special Cases of Natives and Quebecois," in Neil Nevitte and Allan Kornberg, eds., *Minorities and the Canadian State* (Oakville: Mosaic, 1985), 71.

very discourse which called for and legitimized their involvement in the process of constitutional politics. According to Cairns, the

> inclusion of section 27 in the Charter, with its reference to the multicultural heritage of Canada, inevitably generates a specific debate on the relevance of ethnicity for how we treat each other in the public domain and how we view ourselves as people. . . .[There is now] an ethnic constitutional discourse stimulated by section 27. . . .[22]

The second assumption guiding the watershed approach is that the Charter has greatly increased the political power of a variety of subordinate social groups by granting them formal constitutional recognition. Their constitutional standing provided these groups with a stake in constitutional politics and thus contributed to a politicization of the social conflicts they represent. A comment by Neil Nevitte and Roger Gibbins is indicative of the extent of this consensus, despite (or precisely because of) the fact that it was made in the context of discussing political dynamics surrounding group rights and their implications for political culture, rather than in the context of discussing constitutional politics per se. Nevitte and Gibbins argue that "the constitutional entrenchment of the Canadian Charter of Rights and Freedoms in 1982 has given a variety of minority groups—the new 'Charter Canadians'—greater constitutional leverage, and has given their claims greater political saliency, than was the case in Canada before 1982. . . ."[23] More dramatically, Guy Laforest has suggested that the "interest groups that represent [these new constitutional] players have in a sense colonized the constitution."[24]

This development is treated as a subversion of the "rationality" of traditional constitutional discussions between governments. Cairns has argued that the "constitutional language of ethnicity wielded by ethnic elites is emotional and passionate—a Mediterranean language—rather than calculating and instrumental. Its affinities are with such concepts as shame, envy, resentment, honour, and pride."[25] In addition, the enhanced status of ethnic minorities and other subordinated groups is perceived as a subversion of the democratic process by special interest groups—the "Court Party"—through a judicialization of Canadian politics undermining the power of "elected," "representative" and thus, presumably, more legitimate institutions.[26]

---

22 Alan C. Cairns, *Reconfigurations: Canadian Citizenship and Constitutional Change* (Toronto: McClelland and Stewart, 1995), 120–21. Cairns's position on this is somewhat contradictory. He has also said that "the written constitution has always been sensitive to ethnicity, with a preamble referring to a 'Constitution similar in Principle to that of the United Kingdom,' allocation of legislative authority over 'Indians and Lands reserved for Indians,' to the federal government (Section 9 [24]), and indirectly in the limited French- and English-language requirements of Section 133." (*Charter versus Federalism*, 109). Nonetheless, he does not discuss how social power is differentially reflected in the provisions or absences of the Constitution.

23 Neil Nevitte and Roger Gibbins, *New Elites in Old States: Ideologies in the Anglo-American Democracies* (Toronto: Oxford University Press, 1990), 88.

24 Laforest, *Trudeau*, 138.

25 Cairns, "Political Science," 122; Laforest, *Trudeau,* 138, and Roger Gibbins (*Conflict and Unity* [2nd ed.; Scarborough: Nelson, 1990], 262), among others, incorporate Cairns's description of the nature of ethnic minority constitutional discourse.

26 Rainer Knopff and F. L. Morton, *Charter Politics* (Scarborough: Nelson, 1992).

A third assumption guiding the watershed approach is that the political power of ethnic minorities and other nondominant groups, once recognized on a constitutional level, has consequently been cemented and is thus a permanent feature on the political landscape. Cairns, for example, observes that

> if one looks at the organizations that participated in the Meech Lake fora of Manitoba, New Brunswick, Ontario, and the federal government, one is struck by the fact that the organizations were overwhelmingly drawn from two broad categories: aboriginal organizations and organizations representing Charter communities. These organizations dominated the expression of organized interests. . . . As a result, the constitutional debate in this country has been about issues of community-social issues, ethnic issues, linguistic issues, issues of gender cleavage.[27]

The assumption of solidified minority power is reflected, for instance, in the widely accepted notion of "constitutional minoritarianism" (a term originally coined by Cairns) which describes changes in Canada's political culture fostered by the Charter. Thus, the Charter is said to have "catalyzed a group of 'single-clause particularisms' into political existence. You could refer to it as 'constitutional minoritarianism': a bunch of actors devoted to the protection and enhancement of their niche-identity status in the constitution."[28] In his analysis of Trudeau's politics, Laforest—drawing on Cairns—contends that the Charter promoted

> a political culture founded on constitutional minoritarianism. For various reasons, the power of these [new constitutional] actors, and their ability to influence the later rounds of constitutional negotiation, can only grow. They henceforth possess rights, status within the system, and real constitutional identity. The social movements with which they are associated—new ethnicities, feminism, native resurgence—are passing through a period of growth throughout the world. Moreover, these groups have developed their own bureaucratic infrastructure, which can rely on experts, university curricula, and specialized journals, as well as on sympathetic treatment by the media. Immigration is gradually in the process of transforming Canada into a country that is ever more open to multiculturalism and racial pluralism.[29]

The assumption that the power of ethnic minority groups has been solidified by the Charter once again extends beyond analyses of constitutional politics, and also influences analyses of the impact of the Charter of Canada's institutional framework. In arguing that the Charter

27 In Richard Simeon and Mary Janigan, eds., *Toolkits and Building Blocks: Constructing a New Canada* (Toronto: C.D. Howe Institute, 1991), 53.

28 Alan Cairns, ibid., 53. See also Cairns, *Reconfigurations.*

29 Laforest, *Trudeau*, 138. A similar thesis is presented by some sociologists. See Giles Bourque and Jules Duchastel, "Les identités, la fragmentation de la societé canadienne et la constitutionnalisation des enjeux politiques," *International Journal of Canadian Studies* 14 (1996), 77–94.

has strengthened the role of the judiciary vis-à-vis elected representative institutions, the increased weight of the judiciary in the policy-making process is presented as benefiting minority groups. Thus, Rainer Knopff and F.L. Morton suggest that "Charter Canadians" pursued constitutional recognition "in order to entrench policies they could not easily achieve through the legislative process. . . . Since it is the courts that 'enforce' Charter rights against reluctant and recalcitrant legislatures, the Charter groups have a vested interest in judicial power."[30] Put differently, the judicialization of politics consequent on the entrenchment of the Charter is seen as serving, among other things, minority interests; it provides an additional access point to the policy-making process and allows minority groups to attain goals which they were not able to secure by more traditional means.

The fourth assumption guiding the watershed approach typically views the constitutional concerns of subordinate groups (including ethnic and cultural organizations) as limited by narrow self-interest.[31] As Jennifer Smith puts it, "they pursue a partial interest—their own, at least as they conceive it—on particular issues."[32] Moreover, the involvement of these groups in the constitutional process is allegedly characterized by, and essentially limited to, a protective interest in those sections of the Charter which these "Charter Canadians" perceive as their property (that is those sections which grant them constitutional standing).[33] In other words, there is a tendency to view minority groups in an implicitly negative light as self-interested, parochial and thus only as conflict-producing, and to ignore the possibility of forms of mutual recognition and coalition-building.

In particular, the constitutional recognition of subordinate groups in the Charter is seen as producing conflict along three axes. First, there is a conflict between new actors (groups) and old actors (governments). As Cairns has observed of the Meech Lake discussions, one is "struck by the vehemence, and bitterness with which various groups challenged the legitimacy of a closed door elite bargaining process restricted to governments."[34] Second, there is an inherent conflict between multiculturalism and dualism.[35] Third, there is conflict among the new actors themselves as they vie with each other for attention and status.[36] In the words of Knopff and Morton, "those who are already members of the [constitutional] club will also fight among themselves for relative advantage."[37]

30 Rainer Knopff and F. L. Morton, "Canada's Court Party," in Anthony Peacock, ed., *Rethinking the Constitution: Perspectives on Canadian Constitutional Reform, Interpretation, and Theory* (Toronto: Oxford University Press, 1996), 66.

31 See Knopff and Morton, *Charter Politics,* 90.

32 Smith, "Representation," 77.

33 Cairns, "Political Science" and *Reconfigurations*; a similar point is made by Smith, "Representation."

34 Cairns, "Citizens (Outsider) and Governments (Insiders)," S125.

35 Pierre Fournier, *A Meech Lake Post-Mortem: Is Quebec Sovereignty Inevitable* (Montreal: McGill-Queen's University Press, 1991) ix–31; Laforest, *Trudeeau,* 138; Cairns, "Citizens (Outsiders) and Governments (Insiders), S.129. Outside the discipline of political science, this thesis is also well-entrenched. See for example, Gilles Bourque and Jules Duchastel, "Pour une identité canadienne post-nationale, la souveraineté partagée et la pluralité des cultures politiques," *Cahiers de recherche sociologique* 25 (1995), 17–58; and Gérard Bouchard, *La nation québécoise au future et au passé* (Montreal: VLB Editeur, 1999).

36 Cairns, "Citizens (Outsiders) and Governments (Insiders), S138.

37 Knopff and Morton, *Charter Politics,* 82.

Most of the assumptions underlying the watershed approach to constitutional politics and ethnicity are closely paralleled by similar assumptions in much of the constitutional literature on other social groups (such as women, First Nations or gays and lesbians) and their relation to the constitutional order. These assumptions have, in turn, been subject to fundamental critique as a result of empirical analysis. For example, dealing with the assumption that the introduction of the Canadian Charter of Rights and Freedoms in 1982 represents a turning point in political activity, Miriam Smith's thorough analysis of the gay liberation movement in the 1970s convincingly demonstrates that long before the introduction of the Charter, rights-based claims and court challenges were a key element of this movement's strategies. As a result, Smith contends that a frequent argument in the literature—that the Charter itself fostered a rights-claiming political culture—is exaggerated.[38]

Similarly, the work of Alexandra Dobrowolsky demonstrates that the Canadian women's movement was active in the constitutional sphere during the late 1970s and early 1980s, long before the introduction of the Charter. As she puts it, "clearly, the citizenry's constitutional involvement predates the Charter."[39] In addition, Dobrowolsky indicates that the constitutional standing granted to women in the Charter did not produce and irreversibly entrench a significantly enhanced position of power for the women's movement in constitutional politics. Instead, women's organizations were excluded from much of the constitutional politics surrounding for example, Meech Lake and, particularly, the Charleottetown Accord.[40] Examining the gay and lesbian movements, Didi Herman has sharply criticized the idea that lesbians and gays gained control over political institutions as a result of the judicialization of politics perceived to stem from the Charter by analysts Knopff and Morton.[41] Similarly, John Borrows has observed that the effects of the Charter, with its emphasis on the language of rights, not only have generated internal divisions within First Nations communities, but may not inevitably produce the desired outcome (that is self-government).[42]

The accuracy of the idea that the women's movement pursued a narrow constitutional agenda in the post–Charter period has been empirically tested by Linda Trimble. Trimble uncovers the diverging positions taken by francophone and anglophone women on the Meech Lake Accord, and demonstrates that the overall constitutional agenda of women was not limited to the Charter, but encompassed a whole range of other issues such as Senate reform or the division of powers, which took into account interests of other "constitutional players" or "Charter Canadians."[43]

38 See Miriam Smith, "Social Movements and Equality Seeking: The Case of Gay Liberation in Canada," this JOURNAL 31 (1998), 285–309.

39 Alexandra Dobrowolsky, *The Politics of Pragmatism: Women, Representation and Constitutionalism in Canada* (Don Mills: Oxford University Press, 2000), 40.

40 Ibid., 159.

41 Didi Herman, "The Good, the Bad and the Smugly: Sexual Orientation and Perspectives on the Charter," in David Schneiderman and Kate Sutherland, eds., *Charting the Consequences: The Impact of Charter Rights on Canadian Law and Politics* (Toronto: University of Toronto Press, 1997), 212.

42 John Borrows, "Contemporary Traditional Equality: The Effect of the Charter on First Nations Politics," in Schneiderman and Sutherland, eds., *Charting the Consequences,* 169–99.

43 Linda Trimble, "'Good Enough Citizens': Canadian Women and Representation in Constitutional Deliberations," *International Journal of Canadian Studies* 17 (1998), 131–56.

Given these important findings in relation to a number of social groups, it might be fruitful as well to re-evaluate empirically the watershed characterization of the objectives of other subordinate groups, including ethnic and racial ones. Indeed, while the Charter (particularly Section 27 that asserts that "this charter shall be interpreted in a manner consistent with the preservation and enhancement of the multicultural heritage of Canadians") undoubtedly provided a new legal framework for the pursuit of individual and group claims by ethnic minorities,[44] whether race/ethnicity mattered constitutionally before the Charter, or the ways and implications of how ethnic and racialized minorities have pursued claims after the Charter deserves re-examination.

REVISITING THE WATERSHED APPROACH: ETHNIC MINORITIES AND THE CONSTITUTION IN CANADIAN HISTORY

From the 1850s to the 1960s

The claim that the Charter created new ethnic constitutional actors and stimulated an ethnic constitutional discourse (that it fostered, in other words, an arena for ethnic politics within constitutional and judicial politics) is clearly deficient. It fails to take into account that the meaning of constitutional politics depends largely on the power relations between different social groups, including ethnic ones, in which the state is embedded. For instance, historically, Canadian law has not been applied without ethnic and racial bias; in fact, Canadian jurisprudence has often upheld practices of ethnic and racial discrimination. Examining a number of landmark Supreme Court cases during 1914–1955 involving race, James Walker observes that it

> has become common in recent years to regard "race" as a social construct. Over the half century represented in these case studies, in Canada as elsewhere, "race" was also a legal artifact. And in the process of its formulation, the Supreme Court of Canada was a significant participant, legitimating racial categories and maintaining barriers among them.[45]

As well, in terms specifically of constitutional politics, the Canadian constitution has always incorporated ethnic and racial groups differentially into the socio-political order. The clearest example for this can be found in the constitutional status accorded

44 See Canadian Human Rights Foundation, ed., *Multiculturalism and the Charter: A Legal Perspective* (Toronto: Carswell, 1987). The evidence on early Charter rulings suggests that Section 27 has mainly been used to modify rights claims from other sections—in particular Section 2 (religious freedom) and Section 15 (equality rights). See G.L. Gall, "Multiculturalism and the Canadian Charter of Rights and Freedoms: The Jurisprudence to Date under Section 27," in Canadian Ethnocultural Council, *Taking Stock: The Jurisprudence on the Charter and Minority Rights* (Ottawa: 1991), 73–186.

45 James Walker, *"Race," Rights and the Law in the Supreme Court of Canada* (Toronto: Osgoode Society for Canadian Legal History, 1997), 302; see also Elizabeth Comack, ed., *Locating Law: Race/Class/Gender Connections* (Halifax: Fernwood, 1999).

to French Canadians and "status Indians" in the *Constitution Act, 1867*. As Evelyn Kallen notes,

> Canada, from Confederation, has been constitutionally predicated on the inegalitarian notion of special group status. Under the Confederation pact and the subsequent *Constitution Act of 1867*, Canada's "founding peoples"—English/Protestant and French/Catholic groups—acquired a special and superordinate status as the majority or dominant ethnic collectivities, each with a claim for nationhood within clearly delineated, territorial boundaries. . . . By way of contrast, under the terms of s.94(24) of the 1867 Constitution, aboriginal nations . . . became Canada's first ethnic minorities. The provisions of s.24(24) gave the Parliament of Canada constitutional jurisdiction to enact laws concerning Indians and lands reserved for Indians. Under ensuing legislation, notably the various *Indian Acts*, once proud and independent aboriginal nations . . . acquired a special and inferior status as virtual wards of the state.[46]

In this context, it should be noted that regardless of their constitutional status, the political and social status of the two majority groups has historically been less than equal, in favour of those of British origin. Indeed, the French were seen as a cultural "other" by many Canadians of British-origin and socially constructed as a separate and inferior race (or racialized).[47] The constitutional recognition of French Canadians was not the product of any normative considerations, but was won as a result of persistent efforts by French Canadians to achieve recognition within the constitutional order. Constitutional recognition was, in other words, the outcome and expression of conflict and power struggles between different ethnic communities. Thus, Stanley Ryerson has

46 Evelyn Kallen, "The Meech Lake Accord: Entrenching a Pecking Order of Minority Rights," *Canadian Public Policy* 14 Supplement (1988), 110–11. On a similar note, Henry and Tator point out that from "the earliest period of Canadian history, the notion of a hyphenated Canadian was part of the national discourse. There was English Canada and French Canada. In the creation of this cultural duality, the Fathers of Confederation disregarded the cultural/racial plurality that existed even at the time of Confederation. Aboriginal cultures and societies were ignored and excluded from the national discourse. As other cultural groups were rendered invisible, Canada imagined a national culture consisting of a unique blend of English and French values. As a result, three categories of citizens were recognized: English Canadians, French Canadians, and 'others.' Only the first two groups had constitutional rights" (Frances Henry and Carol Tator, "State Policy and Practices as Racialized Discourse: Multiculturalism, the Charter, and Employment Equity," in Peter Li, ed., *Race and Ethnic Relations in Canada* [2nd ed.; Toronto: Oxford University Press, 1999], 92).

47 As one well-known example, in his famous report, Lord Durham uses "nation" and "race" interchangeably. Thus he states "I found two nations warring in the bosom of a single state: I found a struggle, not of principles, but of races" (23). Indicating he also saw French Canadians as inferior, he wrote "The superior political and practical intelligence of the English cannot be, for a moment, disputed" (35). See Lord Durham, "Lord Durham's Report: An Abridgement of *Report on the Affairs of British North America* by Lord Durham," ed. by Gerald Craig (Toronto: McClelland and Stewart, 1963).

argued that the establishment of Canada as a federal state was a result of pressures by French Canadians.[48]

Clearly "ethnic" discourse and considerations were already present at the inception of Confederation—as James Tully has observed, constitutions themselves "cannot eliminate, overcome or transcend [the] cultural dimension of politics."[49] By extension, constitutional politics cannot be separated from ethnocultural and race relations. In Canada, these relations were historically marked by the legalized subordination of nonwhite people, as evidenced by the segregation of schools and other institutions for Blacks in Nova Scotia, New Brunswick, Ontario and Quebec, or the denial of the franchise to the Chinese, Japanese and South Asians in British Columbia.[50]

In addition, to imply that the presence of ethnic communities or the ethnic dimension of constitutional politics are recent phenomena indicates a narrow conception of ethnicity as the exclusive preserve of ethnic minorities. In other words, the British and French origin groups become "de-ethnicized." Given that, historically, French Canadians were, in some instances, racialized, this is problematic at best.

Minority ethnic communities faced considerable obstacles in over-coming their subordination and marginalization from Canadian legal and political institutions. The inability of the Canadian Jewish community to reverse the Canadian government's decision not to take in Jewish refugees during the 1930s and 1940s, despite campaigns and the fact that three Jewish MPs were elected in 1935, provides a graphic example.[51] It is therefore not surprising that by 1960, ethnocultural associations like the Canadian Jewish Congress were active in pursuing the entrenchment of a bill of rights in the Canadian constitution.[52]

EXPLAINING ETHNIC MINORITIES AND CONSTITUTIONAL POLITICS

Political scientists traditionally have ignored questions of race and ethnicity, and have not been heavily represented in the development of ethnic studies in Canada. This has consequences for the discipline. In examining the existing empirical discussions of ethnic minorities and constitutional politics within Canadian political science, we have identified

48 Stanley Ryerson, *Unequal Union: Confederation and the Roots of Conflict in the Canadas, 1815–1873* (Toronto: Progress Books, 1968), 362. For a similar point, see (among others) Samuel LaSelva, *The Moral Foundations of Canadian Federalism: Paradoxes, Achievements, and Tragedies of Nationhood* (Montreal: McGill-Queen's University Press, 1996), 136. Christian Dufour suggests that the establishment in 1867 of Canada as a federal state in general and the establishment of a predominantly French province in particular marked the first modest political victory of French Canadians (*A Canadian Challenge* [Lantzville: Oolichan Books, 1990], 66). The modesty of the gains of French Canadians leads Dufour to call for Quebec separation. See also Christian Dufour, *La rupture tranquille* (Montreal: Boréal, 1992).

49 Tully, *Strange Multiplicity*, 6.

50 Frances Henry, Carol Tator, Winston Mattis and Tim Rees, *The Colour of Democracy: Racism in Canadian Society* (2nd ed.; Toronto: Harcourt Brace, 2000), 69–77.

51 For a full discussion, see Irving Abella and Harold Troper, *None is Too Many: Canada and the Jews of Europe 1933–1948* (3rd ed.; Toronto: Lester, 1991).

52 Cynthia Williams, "The Changing Nature of Citizen Rights," in Alan Cairns and Cynthia Williams, eds., *Constitutionalism, Citizenship and Society in Canada* (Toronto: University of Toronto Press, 1985), 105–06.

key assumptions lying behind what we call the watershed approach to minorities and the constitution that do not stand up to empirical or historical analysis. Contrary to these assumptions, the entrenchment of the Canadian Charter of Rights and Freedoms in 1982 did not ethnicize a previously ethnically neutral constitutional discourse, or create "ethnic constitutional actors." Indeed, before the entrenchment of the Charter, minority ethnic collectivities sought to gain recognition for multiculturalism in a constitutionally entrenched charter. While the Charter granted constitutional standing to ethnic minorities through multiculturalism, it is questionable whether this actually produced an increase in political power for ethnic minorities. In fact, as a totality, the discussion of post–Charter constitutional politics suggests that the Charter has certainly not cemented and increased ethnic minority power when it comes to the pursuit of collective recognition.

What explains our findings on ethnic minorities, the constitution and politics in Canada? Primarily, we suggest that many of the inaccuracies associated with the watershed approach result from a lack of historical depth: the attention paid to pre–Charter relations between ethnic minorities and the constitution is scarce at best. However, as Philip Abrams has suggested, any explanation of social phenomena necessarily needs to proceed historically.[53] A greater accent on history permits the development of a more differentiated view of the relation between ethnicity and the Canadian constitution. Thus we suggest that a greater emphasis on time is central to explaining the shifting position of ethnic minorities in Canadian constitutional politics.

Drawing on history is useful, first, for enabling the incorporation of an examination of Canada as a "white settler society" into one's understanding of ethnicity and the constitution. In fact, the superordinate position of British-origin settlers, in relation to French, Aboriginal and other ethnic groups, was etched into state institutions and symbolism early on.[54] In particular, we would stress the fact that Canadian constitutional politics, by virtue of operating in the context of a white settler society characterized by power differentials among ethnic, racial and other groups, has always been subsumed by ethnicity—sometimes by design, sometimes inadvertently. In other words, any analysis of constitutional politics must take into account the social context which stems from Canada's colonial legacy in order to understand the significance of the presence or absence of specific constitutional stipulations.

A greater sensitivity to the historical dimension of constitutional politics further facilitates a re-evaluation of the relative importance of the state, and of social relations of power, in defining the role of ethnicity in the constitution and in constitutional politics. Assumptions guiding the watershed approach place much emphasis on the Charter as a central factor in this respect. However, greater attention to the temporal dimension of constitutional politics suggests that the importance of the constitution might have been overrated. Rather than being a decisive factor in itself, constitutional politics can be

53 Philip Abrams, *Historical Sociology* (Ithaca: Cornell University Press, 1982).

54 For a historical discussion of the development of Canada as a white settler society, and its implications for social power relations, see Daiva Stasiulis and Radha Jhappan, "The Fractious Politics of a Settler Society: Canada," in Daiva Stasiulis and Nira Yuval-Davis, eds., *Unsettling Settler Societies: Articulations of Gender, Race, Ethnicity and Class* (London: Sage, 1995), 95–131.

interpreted as a reflection or manifestation of the broader ongoing development of conflict and forms of recognition between ethnic minorities and other social groups (including British- and French-origin Canadians). In this development, the state itself plays a role through its policies and nonpolicies.[55]

The importance of the state in defining the role of ethnicity in general and in constitutional politics in particular is readily apparent, for example, in the origins of Canada's multiculturalism policy. The policy of multiculturalism within a bilingual framework marked an important symbolic turning point from state policies which at best minimally accommodated the so-called French fact but typically stressed Anglo-conformity.[56] Indeed, minority leaders have generally accepted the multicultural and bilingual framework. This framework, as articulated initially by Prime Minister Trudeau and manifested subsequently in state policy and funding to minority ethnic groups, structured minority activity prior to the Charter.[57] In fact, the fate of many minority ethnic associations, including the Canadian Ethnocultural Council, is directly tied to multiculturalism policy.[58] As well, and perhaps more importantly, multiculturalism was the product of a particular juncture of Canadian politics, characterized by state intervention in society and high state spending. In many ways the account we give of ethnic minorities and constitutional politics over the 1980s and 1990s suggests that the political space to pursue an expanding agenda based on multiculturalism reached its zenith in the 1980s (especially with the inclusion of Section 27 on multiculturalism in the Charter) and began to wane by the 1990s.[59] Notably, in the 1980s there was secure consensus amongst the main political parties of the time (the Liberals, the New Democrats and the Progressive Conservatives) as well as ethnic minority leaders, that multiculturalism was the appropriate framework to pursue state resources and recognition.[60]

The demise of the postwar Keynesian consensus and the ascendancy of neo-liberalism have fundamentally altered ideas of the proper role of the state in society; the return of the minimalist state challenges the very foundations of multiculturalism as a policy, as well as having implications for how marginalized social groups mobilize and influence the political process.[61] Although many Quebec politicians and academics have been critical of multiculturalism since its inception for its perceived weakening of the claims of French

55 While the importance of situating the constitution in relation to state and society has been recognized by Alan Cairns in some of his writings, our position differs in that we see the state itself as important in shaping ethnic relations through its policies and nonpolicies. See Alan Cairns, *Charter versus Federalism,* 99–100.

56 Abu-Laban and Stasiulis, "Ethnic Pluralism Under Siege," 365.

57 For different accounts of the implications of state funding in relation to ethnic minorities, see Daiva K. Stasiulis, "The Political Structuring of Ethnic Community Action: A Reformulation," *Canadian Ethnic Studies* 21 (1980) and Leslie Pal, *Interests of State.*

58 Kobayashi, "Advocacy from the Margins," 229–61.

59 Abu-Laban, "The Politics of Race and Ethnicity," 242–63; and Kobayashi "Advocacy from the Margins," 229–61.

60 Abu-Laban and Stasiulis, "Ethnic Pluralism Under Siege," 366.

61 For a discussion of how neo-liberalism is challenging the women's movement, see Janine Brodie, *Politics at the Margins: Restructuring and the Canadian Women's Movement* (Halifax: Fernwood, 1995).

Quebeckers for recognition, in the late 1980s and 1990s criticisms against multiculturalism became more vociferous outside Quebec. This vigorous and sustained attack was seen in the writings of some academics, as well as in the critical stance adopted towards multiculturalism for its perceived divisiveness and for requiring state spending on "private" matters of ethnicity by the Reform party, and by some well-known members of minority groups such as novelist Neil Bissoondath.[62] This helps us understand the relative ease with which explicit discussion of multiculturalism was left out of the Charlottetown Accord; it was also in the 1990s that multiculturalism began to receive less funding and deemphasize cultural preservation and the arts. In fact, as noted, by the 1993 federal election both the outgoing Conservatives of Kim Campbell and the incoming Liberals of Jean Chrétien actually opted to disband the Department of Multiculturalism and Citizenship—a move that was in keeping with the shifting positions of the traditional political parties in light of the influence of the Reform party.[63] More broadly, as Audrey Kobayashi observes, the CEC, with its commitment to multiculturalism, has "to a significant degree lost sympathy in Ottawa."[64]

Given these observations, it seems most accurate to describe the constitutional recognition granted multiculturalism in Section 27 of the Canadian Charter of Rights and Freedoms as one outcome of an episode of conflict and recognition in which the strong support of the state and the main political parties shifted the balance of power in favour of ethnic minority demands. By the time of the Charlottetown Accord and in the post–Charlottetown period, however, state support for multiculturalism clearly wavered, and a combination of forces led to attacks on the policy and on its symbolism.

62 See Reginald Bibby, *Mosaic Madness: The Poverty and Potential of Life in Canada* (Toronto: Stoddart, 1990) for an academic critique based on survey data. See also Neil Bissoondath, *Selling Illusions: The Cult of Multiculturalism in Canada* (Toronto: Penguin, 1994); and Reform Party of Canada, "Blue Sheet: Principles and Policies of the Reform Party of Canada." (1996–1997).

63 Abu-Laban and Stasiulis, "Ethnic Pluralism Under Siege," 372–76.

64 Kobayashi, "Advocacy from the Margins," 252.

Social Exclusion

13

GRACE-EDWARD GALABUZI

> Perhaps health is not so much a personal matter but the aftertaste of a society's other activities, the residue of all its policies. (Fernando, 1991, p. 1)

INTRODUCTION

Poor social and economic conditions and inequalities in access to resources and services affect an individual's or group's health and well being. Groups experiencing some form of social exclusion tend to sustain higher health risks and lower health status. According to Health Canada, in Canada such groups include: Aboriginal Peoples, immigrants and refugees, racialized groups, people with disabilities, single parents, children and youth in disadvantaged circumstances, women, the elderly and unpaid caregivers, gays, lesbians, bisexuals, and transgendered people (CIHR, 2002). Poverty is a key cause and product of social exclusion. Its impacts on health status are now well established (Wilkinson, 1996; Wilkinson and Marmot 1998; Kawachi et al., 1999; Ross et al., 2000; Raphael, 1999, 2001). Racial and gender differences in health status tend to reflect differences in social and economic conditions (Wilkinson 1996). The "racialization of poverty" compounds inequalities in material conditions in socially excluded communities.[1] Such documented characteristics of racialized poverty as labour market segregation and low occupation status, high and frequent unemployment status, substandard housing combined with violent or distressed neighbourhoods, homelessness, poor working conditions, extended hours of work or multiple jobs, experience with everyday forms of racism and sexism, lead to unequal health service utilization, and differential health status. Recent research shows that the actual experience of inequality, the impact of relative deprivation and the stress associated with dealing with social exclusion tend to have pronounced psychological effects and to negatively impact health status (Wilkinson, 1996; Kawachi and Kennedy, 2002).

Canadian and international research has begun to confirm the links between the minority status of ethnic, immigrant and racialized groups and low health status (Adams, 1995; Anderson, 1995, 2000; Bolaria and Bolaria, 1994; Shaw et al., 1999; Hyman, 2001; Wilkinson and Marmot, 1998). It is now generally agreed that adverse socio-economic conditions in early life lead to increased health risks in adulthood. According to Health Canada, children whose health status is most at risk tend to live in low-income families, single families, or among racialized group populations, including immigrant and refugee families and Aboriginal families (CIHR, 2002). Racialized community members, recent immigrants and refugee women, men and their children, experience the psycho-social stress of discrimination and racism which contribute to such health problems as hypertension, mental health, and behavioural problems such as substance abuse. Vulnerability to compromised health is now documented even among recent immigrants, who have historically enjoyed higher health status because of the stringent health selection process; research shows a loss of ground over time under conditions of social exclusion (Hyman, 2001; Noh et al., 1999). Recent research also shows that the experience of racism and discrimination puts racialized group members and immigrants at higher risk of mental health problems (Beiser, 1988; OACW, 1990; Dossa, 1999; Noh et al., 1999). Research in women's health suggests similar impacts from gender discrimination (Agnew, 2002; Adams, 1995; Janzen, 1998). Canadian health research and the health system as a whole have not always appreciated the multiple influences on the health status of the affected groups imposed by these various dimensions of social exclusion. At a time when Canada's population growth and stability are increasingly dependent on immigration, with racialized group members now forming 13.5% of the population and growing (Census, 2001) and immigrants now 18.4% and projected to account for 25% of the population by 2015, these issues represent an important area of health policy and research.[2]

SOCIAL EXCLUSION

Social exclusion is used to broadly describe both the structures and the dynamic processes of inequality among groups in society which, over time, structure access to critical resources that determine the quality of membership in society and ultimately produce and reproduce a complex of unequal outcomes (Room, 1995; Byrne, 1999; Guildford, 2000; Duffy, 1997; Shaw et al., 1999; Littlewood, 1999; Madanipour et al., 1998). Social exclusion is both process and outcome. While it has its roots in European social democratic discourse, it has been increasingly embraced by mainstream policy makers concerned about the emergence of marginal subgroups who may pose a threat to social cohesion in industrial societies. In industrialized societies, social exclusion is a by-product of a form of unbridled accumulation whose processes commodify social relations and intensify inequality along racial and gender lines (Byrne, 1999; Madanipour et al. 1998).

White (1998) has referred to four aspects of social exclusion. First, social exclusion from civil society through legal sanction or other institutional mechanisms as often experienced by status and non-status migrants. This conception may include substantive disconnection from civil society and political participation because of material and social isolation, created through systemic forms of discrimination based on race, ethnicity, gender, disability, sexual orientation, and religion. In the post-September 11 era, racial profiling and new notions of national security seem to have exacerbated the experience of this form of social exclusion. Second, social exclusion refers to the failure to provide for the needs of particular groups—the society's denial of (or exclusion from) social goods to particular groups such as accommodation for persons with disability, income security, housing for the homeless, language services, and sanctions to deter discrimination. Third is exclusion from social production, a denial of opportunity to contribute to or participate actively in society's social and cultural activities. And fourth, is economic exclusion from social consumption—unequal access to normal forms of livelihood and economy.

Social exclusion can be experienced by individuals and communities, communities of common bond, and geographical communities. The characteristics of social exclusion tend to occur in multiple dimensions and are often mutually reinforcing. Groups living in low income areas are also likely to experience inequality in access to employment, substandard housing, insecurity, stigmatization, institutional breakdown, social service deficits, spatial isolation, disconnection from civil society, discrimination, and higher health risks. The resulting phenomenon is what some have referred to as one of an underclass culture (Wilson, 1987).

SOCIAL EXCLUSION IN THE LATE TWENTIETH AND EARLY TWENTY-FIRST CENTURY

Processes of social exclusion intensified in the late twentieth century. The intensification of social exclusion can be traced to the restructuring of the global and national economies which emphasized deregulation and re-regulation of markets, the decline of the welfare state, the commodification of public goods, demographic changes owing to increased global migrations, changes in work arrangements towards flexible deployment, and intensification of labour through longer hours, work fragmentation, multiple jobs, and increasing non-standard forms of work. These developments have intensified exploitation in workplaces, but also urban spatial segregation processes including the gendered and racialized spatial concentrations of poverty, among others. The emergence of a neo-liberal globalized political economic order has redefined the nature of the state, and redrawn the boundaries of citizenship as part of a process that seeks to institutionalize market regulation of social relations in societies around the world (Jenson and Papillon, 2000; Gill, 1995). Under this neo-liberal order, the social exclusion concept represents a critique of both the commodification of public goods such as health care services, social services, education and the like, brought about by the dismantling of the welfare state, and the market conditioned response to the resulting marginalization, euphemistically characterized simply as the failure of the individual to utilize their opportunities in the marketplace (Del Castillo, 1994). However, this atomization of social exclusion is contested. For instance, Yepez Del Castillo has noted that "the many varieties of exclusion, the fears of social explosions to which it gives rise, the dangers of social disruption; the complexity of the mechanisms that cause it, the extreme difficulty of finding solutions, have made it the major social issue of our time" (Del Costillo, 1994, p. 614).

In this policy environment, the social exclusion concept seeks to shift the focus back to the structural inequalities that determine the intensity and extent of marginalization in society. It represents a shift of the burden of social inequality from the individual back to society, defining it as a social relation and allowing for the reassertion of welfare state-type social rights based on the concept of social protection as the responsibility of society and not the individual. It is in that respect that the policy discourse of social inclusion in response to marginalization has begun to emerge. Yet, a caution is warranted because there is not necessarily a linear relationship between social exclusion—which seeks to unravel the structures and processes of marginalization—and the more liberal conception of social inclusion, as presently constituted in policy discourse, which promises equal opportunity for all without a commitment to dismantling the historical structures of exclusion (Saloojee, 2002).

SOCIAL EXCLUSION IN THE CANADIAN CONTEXT

In the Canadian context, social exclusion defines the inability of certain subgroups to participate fully in Canadian life due to structural inequalities in access to social, economic, political, and cultural resources arising out of the often intersecting experiences of oppression as it relates to race, class, gender, disability, sexual orientation, immigrant status, and the like. Along with the socio-economic and political inequalities, social exclusion is also characterized by processes of group or individual isolation within and from such key Canadian societal institutions as the school system, the criminal justice system and the health care system, as well as spatial isolation or neighbourhood segregation. These engender experiences of social and economic vulnerability, powerlessness, voicelessness, a lack of recognition and sense of belonging, limited options, diminished life chances, despair, opting out, suicidal tendencies and, increasingly, community or neighourhood violence. Aside from numerous health implications, the emergence of the institutional breakdown and normlessness characterized by such phenomena and the turn to an informal economy and community violence represents a threat to social cohesion and economic prosperity.

In Canada, the discourse on social exclusion has tended to focus on Canadians living on low incomes. Guildford's (2000) work on social exclusion and health in Canada is ground breaking but limited because of the focus on the generic low income experiences of social exclusion. Here, we suggest the need to interrogate the multiple dimensions of the phenomena as well as identify the subgroup dimension of the victims of social exclusion, precisely because of the extent to which their experiences are differentiated by the nature of the oppressions they suffer. Social exclusion is an expression of unequal relations of power among groups in society which then determine unequal access to economic, social, political, and cultural resources. The assertion of certain forms of economic, political, social, and cultural privilege, or the normalization of certain ethnocultural norms by some groups occurs at the expense of, and ultimately, marginalization of others. This is especially true in a market regulated society where the impetus for state intervention to reduce the reproduction of inequality is minimal as is increasingly the case under the neo-liberal regime. Social exclusion takes on a time and spatial dimension. In different societies, there are particular groups that are at higher risk of experiencing social exclusion depending on the historical social relations in the societies. In Canada, four groups have been identified as being in that category of special risk: women, new immigrants, racialized group members, and Aboriginal peoples (CIHR, 2002). The focus in this chapter is on the experiences of racialized group members and immigrants.

SOCIAL EXCLUSION, RACIALIZED GROUPS, AND RECENT IMMIGRANTS

> "The name Chinatown continues to express the deeply embedded white desire for us to one day return from whence we came." (Pon, 2000)

Canada welcomed an annual average of close to 200,000 new immigrants and refugees over the 1990s. Immigration accounted for more than 50% of the net population growth and 70% of the growth in the labour force over the first half of the 1990s (1991–96), and it is expected to account for virtually all of the net growth in the Canadian labour force by the year 2011 (HRDC, 2002). According to the 2001 census, immigrants made up 18.4% of Canada's population, projected to rise to 25% by 2015. Since the shift from a European centred immigration policy in the 1960s, there has been a significant change in the source countries, with over 75% of new immigrants in the 1980s and 1990s coming from the so-called Third world or Global south—the majority of them falling into the category of racialized immigrants.

Racialized groups and recent immigrants encounter processes of marginalization in many spheres of life. The racialization of poverty, in particular, represents two increasingly prevalent intersecting experiences of marginalization faced by Aboriginal peoples, and non-Aboriginal women and men from racialized groups (hereafter referred to as racialized peoples) and immigrant groups. Not only are these groups the subject of these processes of marginalization, they have received very limited treatment in the research on the social determinants of health. The preponderance of the experiences the chapter explores are those of the non-Aboriginal racialized populations whose experiences with racial and gender inequality, disproportionate low income status, unemployment and underemployment, low occupation status, low standard housing, intensified workplace exploitation, disproportionate residence in neighbourhoods with social deficits etc., render them disproportionately vulnerable to low health status. The persistence of the racial and gender discriminatory structures, income and employment inequality, economic and social segregation, and political and cultural marginalization means that increasingly, a disproportionate number of racialized group members exist within a reality of exclusion from mainstream Canadian society (Galabuzi, 2001).

DIMENSIONS OF SOCIAL EXCLUSION: THE RACIALIZATION OF POVERTY

The racialized community is divided into Canadian-born members (roughly 33%) and immigrants (about 67%). During the last census period (1996–2001), the growth rate of racialized groups far outpaced that of other Canadians. While the Canadian population grew by 3.9% between 1996–2001, the corresponding rate for racialized groups was 24.6%. Over the same period, the racialized component of the labour force grew by 28.7% for males and 32.3% for females compared to 5.5% and 9% respectively for the Canadian population. Over much of the 1990s, over 75% of Canada's newcomers were members of the racialized group communities. But with that shift has come a noticeable lag in social economic performance among members of the groups. These patterns seem to be holding both during and after the recession years of the late 1980s and early 1990s.

These developments have had numerous adverse social impacts, leading to differential life chances for racialized group members.

A most significant development is the one described as the racialization of poverty. The racialization of poverty refers to the emergence of structural features that predetermine the disproportionate incidence of poverty among racialized group members. What explains these trends are structural changes in the Canadian economy that conspire with historical forms of racial discrimination in the Canadian labour market to create a process of social and economic marginalization. The result is a disproportionate vulnerability to poverty

BOX 13.1

The Impact of Racialization

- A double digit racialized income gap as high as 30% in 1998
- Two to three times higher than average unemployment rates
- Deepening levels of poverty
- Differential access to housing leading to neighbourhood racial segregation
- Disproportionate contact with the criminal justice system (criminalization of youth)
- Higher health risks

among racialized group communities. Racialized groups are also disproportionately highly immigrant communities and suffer from the impact of the immigration effect. However, current trends indicate that the economic inequality between immigrants and native-born Canadians is becoming greater and more permanent. That was not always the case. In fact, immigrants tended to outperform native born Canadians because of their high educational levels and age advantage.

The racialization of poverty is directly linked to the process of the deepening oppression and social exclusion of racialized and immigrant communities on one hand and the entrenchment of privileged access to economic opportunity for a small but powerful section of the majority population on the other. The concentration of economic, social, and political power that has emerged as the market has become more prominent in social regulation in Canada explains the growing gap between rich and poor as well as the racialization of that gap (Yalnyzian, 1998; Kunz et al., 2001; Galabuzi, 2001; Lee, 2000; Dibbs et al., 1995; Jackson, 2001). Racialized community members and Aboriginal peoples are twice as likely to be poor than other Canadians because of the intensified social and economic exploitation of the racialized and Aboriginal communities whose members have to endure historical racial and gender inequalities accentuated by the restructuring of the Canadian economy and more recently racial profiling. In the midst of the socio-economic crisis that has resulted, the different levels of government have responded by retreating from anti-racism programs and policies that would have removed the barriers to economic equity. The resulting powerlessness and loss of voice has compounded the groups' inability to put issues of social inequality and, particularly, the racialization of poverty on the political agenda.

Racialized Group Members Are Twice as Likely as Other Canadians to Live in Poverty

In 1995, 35.6% members of racialized groups lived under the low income cut off (poverty line) compared with 17.6% in the general Canadian population. The numbers that year were comparable in urban areas—38% for racialized groups and 20% for the rest of the population, a rate twice as high (Lee, 2000). In 1996, while racialized group members accounted for 21.6% of the urban population, they accounted for 33% of the urban poor. That same year, 36.8% of women and 35% of men in racialized communities were low-income earners, compared to 19.2% of other women and 16% of other men. In 1995, the rate for children under six living in low income families is an astounding 45 percent—almost twice the overall figure of 26% for all children living in Canada. In Canada's urban centers, in 1996, while racialized groups members account for 21.6% of the population, they account for 33% of the urban poor. The improvements in the economy have not dented the double digit gap in poverty rates. Family poverty rates were similar—in 1998, the rate for racialized groups was 19% and 10.4% for other Canadian families (Lee, 2000; Jackson, 2001).

Some of the highest increases in low income rates in Canada have occurred among recent immigrants. Low-income rates among successive groups of immigrants almost doubled between 1980 and 1995, peaking at 47% before easing up in the late 1990s. In 1980, 24.6% of immigrants who had arrived during the previous five-year period lived below the poverty line. By 1990, the low-income rate among recent immigrants had increased to 31.3%. It rose further to 47.0% in 1995 but fell back somewhat to 35.8% in 2000. In 1998, the annual wages of racialized immigrants were up to one-third less those of other Canadians, partly explaining why the poverty rate for racialized immigrants arriving after 1986 ranged between 36% and 50% (Jackson, 2001). This was happening at a time when average poverty rates have been generally falling in the Canadian population. Studies show that former waves of immigrants were subject to a short term "immigration factor" which over time—not longer than 10 years for the unskilled and as low as 2 years for the skilled—they were able to overcome and either catch up to their Canadian born counterparts or even surpass them in their performance in the economy. Their employment participation rates were as high or higher than the Canadian-born, and their wages and salaries rose gradually to the level of the Canadian-born.

However, recent research indicates persistent and growing difficulties in the labour market integration of immigrants, especially recent immigrants. Rates of unemployment and underemployment are increasing for individual immigrants, as are rates of poverty for immigrant families (Galabuzi, 2001; Ornstein, 2000; Pendakur, 2000; Reitz, 1998, 2001; Shields, 2002). So the traditional trajectory that saw immigrants catch up with other Canadians over time seems to have been reversed in the case of racialized immigrants. Of course the irony is that over that period of time, the level of education, usually an indicator of economic success, has been growing.

Recent Statistics Canada analysis shows that male recent immigrant full time employment earnings fell 7% between 1980 and 2000 (Kazimapur and Hou, 2003). This compares with a rise of 7% for the Canadian born cohort. Among university educated the drop was deeper (13%). For female recent immigrants full time employment earnings rose but by less than other female full time employees. More alarming are the low income implications of these trends. While low income rates among recent immigrants with less than high school graduation increased by 24% from 1980 to 2000, low income rates increased by 50% among high school graduates and a whopping 66% among university educated immigrants!

TABLE 13.1 **Unequal access to full employment (Unemployment rate in %)**

	1981	1991	2001
Total labour force	5.9	9.6	6.7
Canadian born	6.3	9.4	6.4
All immigrants	4.5	10.4	7.9
Recent immigrants	6.0	15.6	12.1

Source: "Unequal access to full employment (%)," adapted from the Statistics Canada publication *The changing profile of Canada's labour force, 2001 census* (Analysis series, 2001 Census), Catalogue 96F0030, February 11, 2003.

TABLE 13.2 **Labour force participation (Employment rate in %)**

	1981	1991	2001
Total labour force	75.5	78.2	80.3
Canadian born	74.6	78.7	81.8
All immigrants	79.3	77.2	75.6
Recent immigrants	75.7	68.6	65.8

Source: "Labour force participation (%)," adapted from the Statistics Canada publication "The changing profile of Canada's labour force, 2001 Census (Analysis series, 2001 Census)," Catalogue 96F0030, February 11, 2003.

Recent immigrants' rates of employment declined markedly between 1986 and 1996. The result is that Canada's immigrants exhibit a higher incidence of poverty and greater dependence on social assistance than their predecessors, in spite of the fact that the percentage of university graduates is higher in all categories of immigrants including family class and refugees as well as economic immigrants than for the Canadian-born (CIC, 2002).

The Ornstein report (2000) revealed that high rates are concentrated among certain groups such as Latin Americans, African Blacks and Caribbeans, and Arabs and West Asians—with rates at 40% and higher in 1996, or roughly three times the Toronto rate. This research is confirmed by accounts in the popular press, which reveal a dramatic increase in the use of food banks by highly-educated newcomers (Quinn, 2002).

A significant factor in these trends is the under-utilization of immigrant skills within the Canadian labour market. Reitz (2001) has looked at the quantitative significance of this issue using a human-capital earnings analysis which identified immigrant earnings deficits as arising from three possible sources: lower immigrant skill quality, or under-utilization of immigrant skills, or pay inequities for immigrants doing the same work as native-born Canadians. He concluded that in 1996 dollars, the total annual immigrant earnings deficit from all three sources in Canada was $15.0 billion, of which $2.4 billion was related to skill under-utilization, and $12.6 billion was related to pay inequity. He observed as well that employers give little credence to foreign education and none to foreign work experience, that discrimination specific to country of origin or visible minority status is mainly related to pay equity rather than skills utilization, and that the economic impact of visible minority status and immigrant status is very similar for both men and women. In addition Reitz noted that race appears to be a more reliable predictor of how foreign education will be evaluated in Canada than the specific location of the origin of the immigrant from outside Europe.

ECONOMIC EXCLUSION IN THE LABOUR MARKET

> The third category of the relative surplus population, the stagnant, forms a part of the active labour army, but with extremely irregular employment . . . hence it furnishes to capital an inexhaustible reservoir of disposable labour-power. Its conditions of life sink below that of the average normal level of the working class; This makes it at once the broad basis of the special branches of capitalist exploitation . . . it is characterized by a maximum of working hours, and a minimum of wages . . . But it forms at the same time a self-reproducing and self-perpetuating element of the working class, taking a proportionally greater part in the general increase of that class than the other elements (Marx, 1977, p. 602)

The neo-liberal restructuring of Canada's economy and labour market towards flexibility has increasingly stratified labour markets along racial lines, with the disproportionate representation of racialized group members in low income sectors and low end occupations, and under-representation in high income sectors and occupations. These patterns emerge out of a context of racial inequality in access to work and in employment income, and the growing predominance of precarious forms of work in many of the sectors racialized group members are disproportionately represented in. They point to racially unequal incidence of low income and racially defined neighbourhood segregation. It is these broader processes that explain the emergence of the phenomenon of the racialization of poverty whose dimensions can primarily be identified by such indicators as disproportionate levels of low income and racialized spatial concentration of poverty in key neighbourhoods.

Economic social exclusion takes the form of labour market segregation, unequal access to employment, employment discrimination, and disproportionate vulnerability to unemployment and underemployment. These are both characteristics and causes of social exclusion. Attachment to the labour market is essential to both livelihood and to the production of identity in society. It determines both the ability to meet material needs but also a sense of belonging, dignity and self-esteem, all of which have implications for health status. Labour market related social exclusion has direct implications for health status not just because of the impact on income inequality, but also because of the extent to which working conditions, mobility in workplaces, fairness in the distribution of opportunities, and utilization of acquired skills all have a direct bearing on the levels of stress that are generated in workplaces.

The Canadian economy and labour market are increasingly stratified along racial lines, as evidenced by the disproportionate representation of racialized group members in low income sectors and low end occupations, and under-representation in high income sectors and occupations. These patterns emerge out of a context of neo-liberal restructuring of the economy conditioned by global competition and demands for flexible deployment of labour, persistent racial inequality in access to employment, and the growing predominance of precarious forms of work in many of the sectors in which racialized group members are disproportionately represented. Labour market research shows this racial stratification as observable in the disproportionate participation of racialized groups in industries increasingly dominated by non-standard forms of work such as textiles, clothing, hospitality and retailing, and an over-representation of racialized group members in low income jobs and low end occupations. On the other hand, they are under-represented in such high income sectors as the public service, automobile making and metal working, which also happen to be highly unionized.

The fastest growing form of work in Canada is precarious work, also referred to as contingent work or non-standard work—contract, temporary, part-time, and shift work with no job security, poor and often unsafe working conditions, intensive labour, excessive hours, low wages, and no benefits. In the early 1990s, it grew by 58%, compared to 18% for full time employment (Vosko, 2000; de Wolf, 2000). Racialized workers are disproportionately represented in this form of work, as a consequence of their vulnerability to the restructuring in the economy (Galabuzi, 2001).

Most of this work is low-skilled and low paying and the working conditions are often unsafe. Such non-regulated service occupations as newspaper carriers, pizza deliverers, janitors and cleaners, dish washers, and parking lot attendants are dominated by racialized group members and recent immigrants who work in conditions with little or no protection—conditions similar to low end work in the hospitality and health care sectors, light manufacturing assembly plants and textile and home-based garment work. Many employees are "self-employed" or sub-contracted on exploitative contracts by temporary employment agencies, with some assigning work based on racist stereotypes.

Racialized women are particularly over-represented in another form of self-employment—unregulated piecemeal homework. Gendered racism and neo-liberal restructuring have conditioned the emergence of what some have called Canada's sweatshops, especially in the garment and clothing industry (Yanz, et al., 1999; Vosko, 2000). The intensity of the experience of exploitation imposes stressors especially on racialized and immigrant women who continue to carry a disproportionate bulk of house work, to go with the subcontract wage work, and many of whom are single parents.

Because of the intensified exploitation characterized by demands for longer working hours and low pay, and/or multiple part-time jobs, the intensity of work under a deregulated labour market becomes a major source of stress and related health conditions. In the case of immigrants and racialized group members, the failure to convert their educational attainment and experience, whether internationally or domestically acquired, due to the structures of racial and gender inequality or barriers in employment relating to immigrant status, has been identified increasingly as a major stress generator.

SOCIAL EXCLUSION AND RACIALIZED NEIGHBOURHOOD SELECTION

> Space is a key function of social exclusion.
> The fact that people live in certain places can either sustain or intensify social exclusion (Hutchinson, 2000)

The racialization of poverty has also had a major impact on neighbourhood selection and access to adequate housing for new immigrants and racialized groups. In Canada's urban centres, the spatial concentration of poverty or residential segregation is intensifying along racial lines. Immigrants in Toronto and Montreal are more likely than non-immigrants to live in neighbourhoods with high rates of poverty as Table 13.3 below shows. Social exclusion is increasingly manifest in urban centres where racialized groups are concentrated through the emergence of racial enclaves and a growing set of racially segregated neighbourhoods. In what is becoming a segregated housing market, racialized groups are relegated to substandard, marginal and often overpriced housing. These growing neighbourhood inequalities act as social determinants of health and well-being, with limited access to social services, increased contact with the criminal justice system, and social disintegration, and violence engendering higher health risks.

Recent studies by Hou awnd Balakrishnan (1998), Kazemipur and Halli (2000), Eric Fong (2000), and Ley and Smith (1997) suggest that these areas show characteristics of 'ghettoization' or spatial concentration of poverty, concentrating in urban cores, tightly clustered, with limited exposure to majority communities. Increasingly these geographical areas represent racialized enclaves subject to the distresses of low income communities.

A racialized spatial concentration of poverty means that racialized group members live in neighbourhoods that are heavily concentrated and "hypersegregated" from the rest of society and often with disintegrating institutions and increasingly dealing with social deficits such as inadequate access to counseling services, life skills training, child care, recreation, and health care services (Kazemipur and Halli, 1997, 2000; Lo et al., 2000).

TABLE 13.3 **Toronto-area Racialized Enclaves and Experience of High Poverty Rates**

	University degree	Unemployment	Low income	Single parent
Chinese	21.2%	11.2%	28.4%	11.7%
South Asian	11.8%	13.1%	28.3%	17.6%
Black	8.7%	18.3%	48.5%	33.7%

Source: "Toronto Area racialized enclaves and experience of high poverty rates," adapted from the Statistics Canada publication ***Visible minority neighbourhood enclaves and labour market outcomes of immigrants*** (Analytical Studies Branch research paper series), Catalogue 11F0019, July 2003.

Young immigrants living in low income areas often struggle with alienation from their parents and community of origin, and from the broader society. The social services they need to cope with dislocation are lacking, the housing on offer is often sub-standard, or if it is public housing it is largely poorly maintained because of cutbacks. They face crises of unemployment, despair, and violence. They are disproportionate targets of contact with the criminal justice system.

Finally, a word about homelessness and the recent immigrant and racialized groups. Homelessness is said to be proliferating among racialized group members because of the incidence of low income and the housing crises in many urban areas (Lee, 2000; Peel, 2000). Homelessness is an extreme form of social exclusion that suggests a complexity of causes and factors. Increasingly, recent immigrants and racialized people are more likely to be homeless in Canada's urban centres than they were ten years ago. It compounds other sources of stresses in their lives. Homelessness has been associated with early mortality, health factors such as substance abuse, mental illness, infectious diseases, and difficulty accessing health services. The complex interactions among these factors, homelessness and access to health services have not received enough study and represent a key gap in both the anti-racism and social determinants of health discourses.

RACIALIZATION, SOCIAL EXCLUSION AND HEALTH

> The importance of power relations, social identity, social status, and control over life circumstances for health status follows from the evidence upon which the population health perspective is based. This evidence can be usefully grouped into three broad categories: social inequalities in health within and between societies; social support and health; and workplace characteristics and health. (Dunn and Dyck, 1998, p. 2)
>
> If today's immigrants have higher rates of illness than the native-born, the increased risk probably results from an interaction between personal vulnerability and resettlement stress, as well as lack of services, rather than from diseases they bring with them to Canada. (Health Canada, 1998, p. 1)

There is no doubt that universal access to health care is now a core Canadian value, espoused broadly by all segments of the political elite as defining Canadian society. But beyond the policy articulation of universality of coverage, other determinants such as income, gender, race, immigrant status, and geography increasingly define the translation of the concept of universality as unequally differentiated. A review of the limited available literature indicates that the processes of social exclusion we have discussed above affect the health status of racialized and recent immigrant communities. The extent of exclusion is expressed through the gap between the promise of universal access to health care and the reality of unequal access to health service utilization, or inequalities in health status arising out of the inequalities in the social determinants of health. It is the gap between the promise of citizenship and the reality of exclusion that represents the extent of social exclusion and the unequal impact on the well-being of members of racialized groups and immigrants in Canada. While there is limited empirical research to draw on, there is significant anecdotal evidence to make the case.

It follows though that given the landscape of exclusion we have painted above, a perspective based on a synthesis of a diverse public health and social scientific literature, which suggests that the most important antecedents of human health status are not medical care inputs and health behaviours (smoking, diet, exercise, etc.), but rather social and economic characteristics of individuals and populations, would suggest significant convergence between social exclusion and health status (Evans et al., 1994; Frank, 1995; Haye and Dunn, 1998). For racialized groups and recent immigrants, power relations, identity and status issues, and life chances are influential on the processes of immigration and integration (Dunn and Dyck, 1998).

However, one of the more significant studies of immigrants and health by Dunn and Dyck (1998) using the "social determinants of health" approach and based on a review of NPHS data perspective found no obvious, consistent pattern of association between socio-economic characteristics and immigration characteristics on the one hand, and health status on the other (Dunn and Dyck, 1998). Hyman (2001) has observed, neither did it find evidence to the contrary.

RACIALIZATION AND HEALTH STATUS

It is now generally agreed that racism is a primary source of stress and hypertension in racialized group communities. Everyday forms of racism, often compounded by sexism and xenophobia, and the related conditions of underemployment, non-recognition of prior accreditation, low standard housing, residence in low income neighbourhoods with significant social deficits, violence against women and other forms of domestic and neighbourhood violence, and targeted policing and disproportionate criminalization and incarceration define an existence of those on the margins of society, an existence of social exclusion from the full participation in the social, economic, cultural, and political affairs of Canadian society. They are also important socio-economic and psycho-social determinants of health. While empirical research is under-developed, there is significant qualitative evidence, collected from group members, service providers, and some qualitative community based research to suggest that these act as determinants of health status for socially marginalized groups such as racialized women, youth and men, immigrants and Aboriginal peoples (Agnew, 2002; Tharoa and Massaquoi, 2001). These conditions contribute to and mediate the experience of inequality into powerlessness, hopelessness, and despair contributing to the emotional and physical impact on health of the members of the groups. These conditions in turn negatively impact attempts by affected individuals, groups, and communities to achieve full citizenship because of their inability to claim social and political rights enjoyed by other Canadians—including the right to physical and mental well-being of residents (Canada Health Act, 1984).

While there is limited literature in the Canadian context, research done internationally shows the connection between race and health more clearly. Research in the United States shows the connection between racism and health status (Randall, 1993). Wilkinson has investigated the processes of racialization, which result in the social and economic marginalization of certain social groups and shown that "racial" differences in health status can largely be accounted for by differences in individuals' social and economic circumstances (Wilkinson, 1996; Anderson, 1987, 1991).

Institutionalized racism in the health care system characterized by language barriers, lack of cultural sensitivity, absence of cultural competencies, barriers to access to health service utilization and inadequate funding for community health services has been identified as impacting the health status of racialized group members. Mainstream health care institutions are Eurocentric, imposing European and white cultural norms as standard and universal and, by extension, their cultural hegemony imposes a burden on racialized and immigrant communities. Insights from the critical race discourses help us understand that the cumulative burden of the subtle, ordinary, persistent everyday forms of racism, compounded by experiences of marginalization also determine health status. The psychological pressures of daily resisting these and other forms of oppression add up to a complex of factors that undermine the health status of racialized and immigrant group members. They are compounded by low occupation, housing and neighbourhood status, high unemployment, and high levels of poverty. Racist stereotypes by health practitioners also tend to impact health status.

SOCIAL EXCLUSION, STRESS, AND MENTAL HEALTH

Many racialized group members and immigrants with mental health issues and mental illnesses identify racism as a critical issue in their lives. The magnitude of the association between racism and poverty and mental health status was said by low income racialized group community members surveyed to be similar to other commonly studied stressful life events such as death of a loved one, divorce or job loss (Healing Journey, 1999).

Racism is a stress generator as are family separation through immigration, the intensification of work, devaluation of one's worth through decredentialism, and the very experience of inequality and injustice. Stress in turn is a major cause of a variety of health problems. It has been observed that one of the reasons the health status of immigrants declines is because of the experiences of discrimination and racism (Hyman, 2001). State imposed barriers to family reunification through immigration policy that discourages reunification in favour of independent class immigration leads to extended periods of family separation. Family separation, and failure to effect reunification robs family members of their support network but also engenders separation anxiety, thoughts of suicide, lack of sufficient support mechanisms and even death.

Racism and discrimination based on immigrant status intensify processes of marginalization and social exclusion, compounding the experiences of poverty and its impacts on mental health status. The everyday darts that arise from put downs and diminishing self-esteem tend to undermine the mental health of racialized group members.

The stigma of mental illness often bars members from seeking treatment, some being afraid that such a stigma would compound their marginalization. The Canadian Task Force on Mental Health Issues Affecting Immigrants identified a mental health gap between immigrants and the Canadian-born population based on the socio-economic status of immigrants. Concluding that this was a determinant of mental health, it called for increased access to mental health services for immigrants, more appropriate culturally sensitive and language specific services to help close the gap (Beiser, 1988).

The serious gap in the research on the mental health of immigrants can be significantly closed by using a framework that recognizes the impact of racism and immigrant status on the process of social exclusion and social determination of health. Beiser et al. (1993) identify the persistence of the gap in health care utilization between immigrants and native born Canadians, its impact on the mental health status of immigrants, and the need for research to better understand the phenomenon.

SKILL DEPLOYMENT: INTERNATIONALLY TRAINED

Many skilled immigrants are experiencing mounting barriers to making full use of the skills and talents in both the economic sphere and in public life. Increasingly they are dealing with frustration at the barriers they face. Such a strong sense of inequality and injustice has implications for their mental health (Beiser, 1988). Moreover, as the Anderson et al. (1993) study on chronic illness shows, for immigrant women living on meager incomes and sustaining a marginal status in the labour market, the daily struggles of this meagre existence and the desire to hold on to their low paying jobs tend to take precedence over disclosing chronic illness to ensure its active management with the support of health professionals. Along with such livelihood considerations, they often face the daunting prospect of navigating the mainstream health care system, with its barriers to access, lack of culturally appropriate services, and inability of health care professionals to understand the choices that those living in poverty and at the margins have to make to survive.

Research shows that immigrant youth sometimes find the stresses of integration on top of the challenges of adolescence overwhelming. Their feelings of isolation and alienation are linked to perceptions of cultural differences and experiences of discrimination and racism. While these are often complicated by intergenerational issues, support from friends, family and institutions is key to overcoming the challenges—in essence it presents them with a recreated community in response to the exclusion they face in mainstream institutions like the school system (Kilbride, 2000).

CONCLUSION: THE DEARTH OF RESEARCH ON HEALTH AND RACE

I have suggested that social exclusion describes both the structures and the dynamic processes of inequality among groups in society which, over time, structure access to critical resources that determine the quality of membership in society and ultimately produce and reproduce a complex of unequal outcomes. Social exclusion speaks of both the process of becoming and the outcome of being socially excluded. It has received the attention of policy-makers concerned about the impact of marginal subgroups on social cohesion and has provided them with the social inclusion framework for responding to this phenomenon. However, its use in health policy and health research is limited although the social determinant of health approach seems to share its philosophical orientation. Its potential is especially suggestive when dealing with the complexity of issues faced by racialized and immigrant communities and their impact on health status, an area where there is limited research. Racialized groups and immigrant groups are disproportionately impacted by labour market segregation, unemployment and income inequality, poverty, poor neighbourhood selection, to go with experiences of discrimination based on race, gender, and immigrant status. The multidimensional approach of social exclusion as a framework for understanding the multiplicity of non-behavourial influences on health status may provide a more adequate basis for assessing the health status of not just racialized and immigrant communities but other socially excluded groups like women, persons with disability, gays and lesbians, bisexual and transgendered, and even the "generic" poor.

NOTES

1. The racialization of poverty here refers to the disproportionate and persistent incidence of low income among racialized groups in Canada.
2. There has been a significant change in the source countries, with over 75% of new immigrants in the 1980s and 1990s coming from the Global South. Most immigrants end up in urban centres—75% in Toronto, Vancouver, and Montreal.

REFERENCES

Adams, D. (1995). *Health issues of women of colour: A cultural diversity perspective.* Thousand Oaks: Sage.

Anderson, J. M. (2000). "Gender, race, poverty, health and discourses of health reform in the context of globalization: A post-colonial feminist perspective in policy research," *Nursing Inquiry* 7 (4), 220–229.

Anderson, J. and Kirkham, R. (1998). "Constructing nation: The gendering and racializing of the Canada health care system," in Strong-Boag, V. and Grace, S. (Eds.), *Painting the maple: Essays on race, gender, and the construction of Canada.* Vancouver: UBC Press.

Anderson, J., Blue, C., Holbrook, A. and Ng, M. (1993). "On Chronic illness: Immigrant women in Canada's workforce: A feminist perspective," *Canadian journal of nursing research* 25/2.

Agnew, V. (2002). *Gender, migration and citizenship resources project: Part II: A literature review and bibliography on health.* Toronto: Centre for Feminist Research, York University.

Alliance for South Asian AIDS Prevention (ASAP). (1999). *Discrimination & HIV/AIDS in South Asian communities: Legal, ethical and human rights challenges: An ethnocultural perspective.* Toronto: ASAP/Health Canada.

Beiser, M. (1998). *After the door has been opened: Mental health issues affecting immigrants and refugees in Canada: Report of the Canadian Taskforce on mental health issues affecting immigrants and refugees.* Ottawa: Health and Welfare Canada.

Bloom, M. and Grant, M. (2001). *Brain gain: The economic benefits of recognizing learning and learning credentials in Canada.* Ottawa: Conference Board of Canada.

Bolaria, B. and Bolaria, R. (Eds.). (1994). *Immigrant status and health status: Women and racial minority immigrant workers in racial minorities, medicine and health.* Halifax: Fernwood.

Byrne, D. (1999). *Social exclusion.* Buckingham: Open University Press.

Chard, J., Badets, J. and Howatson-Lee, L. (2000). "Immigrant women," in *Women in Canada, 2000: A gender-based statistical report.* Ottawa: Statistics Canada.

Chen, J., Wilkins, R. and Ng, E. (1996). "Life expectancy of Canada's immigrants from 1986 to 1991," *Health reports* 8/3 (1996), 29–38.

Cross Boundaries. (1999). *Healing journey: Mental health of people of colour project.* Toronto: Across Boundaries.

DeWolff, A. (2000). *Breaking the myth of flexible work.* Toronto: Contingent Workers Project.

Dossa, P. (1999). *The narrative representation of mental health: Iranian women in Canada.* Vancouver: RIIM.

Dunn, J. and Dick, I. (1998). *Social determinants of health in Canada's immigrant population: Results from the National Population Health Survey.* Working paper series #98–20. Vancouver: Vancouver Centre of Excellence, Research on Immigration and Integration in the Metropolis.

Galabuzi, G. (2001). *Canada's creeping economic apartheid: The economic segregation and social marginisation of racialised groups.* Toronto: CJS Foundation for Research & Education.

Globerman, S. (1998). *Immigration and health care: Utilization patterns in Canada.* Vancouver: RIIM.

Guidford, J. (2000). *Making the case for economic and social inclusion.* Ottawa: Health Canada.

Health Canada. (1998). *Metropolis health domain seminar: Final report.* Ottawa: Health Canada.

Henry, F. and Tator, C. (2000). "The theory and practice of democratic racism in Canada," in Kalbach, M. A. and Kalbach, W. E. (Eds.), *Perspectives on ethnicity in Canada.* Toronto: Harcourt.

Human Resources Development Canada (HRDC). (2002). *Knowledge matters: Skills and learning for Canadians: Canada's Innovation Strategy.* Ottawa: HRDC. Online at www.hrds-drhc.gc.ca/sp-ps/sl-ca/doc/summary.shtml.

Human Resources Development Canada (HRDC). (2001). "Recent immigrants have experienced unusual economic difficulties," *Applied research bulletin* 7(1), Winter/Spring.

Jackson, A. (2001). "Poverty and immigration," *Perception* 24(4), Spring.

Jenson, J. (2002). *Citizenship: Its relationship to the Canadian diversity model.* Canadian Policy Research Networks (CPRN). Online at www.cprn.org.

Jenson. J. and Papillon, M. (2001). *The changing boundaries of citizenship: A review and a research agenda.* Canadian Policy Research Networks (CPRN). Online at www.cprn.org.

Kawachi, I. R. and Kennedy, B. (2002). *The health of nations: Why inequality is harmful to your health.* New York: New Press.

Kawachi, I., Wilkinson, R. and Kennedy, B. (1999). "Introduction," in Kawachi, I., Kennedy, B. and Wilkinson, R. (Eds.), *The society and population health reader: Volume I: Income inequality and health.* New York: New Press.

Kazemipur, A. and Halli, S. (1997). "Plight of immigrants: The spatial concentration of poverty in Canada," *Canadian journal of regional sciences* (special issue), XX 1/2 Spring–Summer 1997, 11–28.

Kazemipur, A. and Halli, S. (2000). *The new poverty in Canada,* Toronto: Thompson Educational Publishing.

Kilbride, K.M, Anisef, P., Baichman-Anisef, E. and Khattar, R. (2000). *Between two worlds: The experiences and concerns of immigrant youth in Ontario.* Toronto: CERIS and CIC-OASIS. Online at www.settlement.org.

Kunz, J. L., Milan, A. and Schetagne, S. (2001). *Unequal access: A Canadian profile of racial differences in education, employment, and income.* Toronto: Canadian Race Relations Foundation.

Kymlicka, W. and Norman, W. (1995). "Return of the citizen: A survey of recent work on citizenship theory," in Beiner, Ronald (Ed.), *Theorizing citizenship.* Albany: SUNY Press.

Lee, Y. (1999). "Social cohesion in Canada: The role of the immigrant service sector," *OCASI* Newsletter 73, Summer/Autumn 1999.

Lynch, J. (2000). "Income inequality and health: Expanding the debate," *Social science and medicine* 51, 1001–1005.

Madanipour, A. (1998). "Social exclusion and space," in Madanipour, A., Cars, G. and Allen, J. (Eds.), *Social exclusion in European cities.* London: Jessica Kingsley.

Marmot, M. (1993). *Explaining socioeconomic differences in sickness absence: The Whitehall II Study.* Toronto: Canadian Institute for Advanced Research.

Marmot, M. and Wilkinson, R. (Eds.). (1999). *Social determinants of health.* Oxford: Oxford University Press.

Marx, K. (1977). *Capital: A critique of political economy, volume one.* New York: Vintage Press.

Noh,S., Beiser, M., Kaspar, V., Hou, F. and Rummens, J. (1999). "Perceived racial discrimination, discrimination, and coping: A study of Southeast Asian refugees in Canada," *Journal of health and social behaviour* 40, 193–207.

Ontario Advisory Council on Women's Issues. (1990). *Women and mental health in Ontario: Immigrant and visible minority women.* Ottawa: Ministry of Health.

Ornstein, M. (2000). *Ethno-racial inequality in the City of Toronto: An analysis of the 1996 Census.* Toronto: City of Toronto. Online at www.ceris.metropolis.net.

Pendakur, R. (2000). *Immigrants and the labour force: Policy, regulation and impact.* Montreal: McGill-Queen's University Press.

Preston, V. and Man, G. (1999). "Employment experiences of Chinese immigrant women: An exploration of diversity," *Canadian women's studies* 19, 115–122.

Quinn, J. (2002). "Food bank clients often well-educated immigrants," *The Toronto Star*, March 31, 2002, A12.

Raphael, D. (1999). "Health effects on economic inequality," *Canadian review of social policy* 44, 25–40.

Raphael, D. (2001). "From increasing poverty to societal disintegration: How economic inequality affects the health of individuals and communities," in Armstrong, P., Armstrong, H. and Coburn, D. (Eds.), *Unhealthy times: The political economy of health and care in Canada.* Toronto: Oxford University Press.

Reitz, J. G. (2001). "Immigrant skill utilization in the Canadian labour market: Implications of human capital research," *Journal of international migration and integration* 2/3, Summer 2001.

Room, G. (1995). "Conclusions," in Room, G. (Ed.), *Beyond the threshold: The measurement and analysis of social exclusion.* Bristol: Policy.

Saloojee, A. (2002). *Social inclusion, citizenship and diversity: Moving beyond the limits of multiculturalism.* Toronto: Laidlaw Foundation.

Shaw, M., Dorling, D. and Smith, G. D. (1999). "Poverty, social exclusion, and minorities," in Marmot, M. and Wilkinson, R. (Eds), *Social determinants of health.* Oxford: Oxford University Press.

Shields, J. (2002). "No safe haven: Markets, welfare and migrants" (paper presented to the Canadian Sociology and Anthropology Association, Congress of the Social Sciences and Humanities), Toronto ON, June 1, 2002).

Siemiatycki, M. and Isin, E. (1997). "Immigration, ethno-racial diversity and urban citizenship in Toronto," *Canadian journal of regional sciences* (special issue) XX: 1, 2, Spring–Summer 1997, 73–102.

Wilkinson, R. (1996). *Unhealthy societies: The afflictions of inequality.* New York: Routledge.

Wilkinson, R. and Marmot, M. (2003). *Social determinants of health: The solid facts.* Copenhagen: World Health Organization. Online at www.who.dk/document/e81384.pdf.

Wilson, W. J. (1987). *The truly disadvantaged: Inner city, the underclass and public policy.* Chicago: University of Chicago Press.

Yanz, L., Jeffcoat, B., Ladd, D. and Atlin, J. (1999). *Policy options to improve standards for women garment workers in Canada and internationally.* Toronto: Maquila Solidarity Network/Status of Women Canada.

Yepz Del Costello, I. (1994). "A comparative approach to social exclusion: Lessons from France and Belgium," *International labour review* 133/5–6, 613–633.

14

On Being *Not* Canadian: The Social Organization of "Migrant Workers" in Canada*

NANDITA SHARMA

Se fondant sur la méthode d'ethnographie institutionnelle de Dorothy E. Smith, l'auteure étudie l'organisation sociale de notre connaissance des gens catégorisés comme non-immigrants ou « travailleurs migrants ». À la suite de l'étude du Non-Immigrant Employment Authorization Program (NIEAP) du gouvernement canadien (1973), elle montre l'importance de la pratique idéologique raciste et nationaliste des États à l'endroit de l'organisation matérielle du marché du travail compétitif « canadien » dans le cadre d'un capitalisme mondial restructuré de même que la réeorganisation qui en résulte des notions d'espirit national canadien. Elle montre aussi que la pratique discursive des parlementaires qui consiste à considérer certaines personnes comme des « problèmes » pour les « Canadiens » ne provient pas de l'exclusion physique de ces « étrangers » mais plutôt de leur différenciation idéologique et matérielle des Canadiens une fois qu'ils vivent et travaillent dans la société canadienne.

Utilizing Dorothy E. Smith's method of institutional ethnography, I investigate the social organization of our knowledge of people categorized as non-immigrants or "migrant workers." By examining Canada's 1973 Non-immigrant Employment Authorization Program (NIEAP), I show the importance of racist and nationalist ideological state practice to the material organization of the competitive "Canadian" labour market within a restructured global capitalism and the resultant reorganization of notions of Canadian nationhood. I show that the parliamentary discursive practice of producing certain people as "problems" for "Canadians" results not in the physical exclusion of those constructed as "foreigners" but in their ideological and material differentiation from Canadians, once such people are living and working within Canadian society.

Expressions such as . . . "foreigner". . . and so on, denoting certain types of lesser or negative identities are in actuality congealed practices and forms of violence or relations of domination. . . . This violence and its constructive or representative attempts have become so successful or hegemonic that they have become transparent—holding in place the ruler's claimed superior self, named or identified in myriad ways, and the inadequacy and inferiority of those who are ruled.

—Himani Bannerji

* I would like to acknowledge the support of the Social Sciences and Humanities Research Council of Canada, award no. 756–2000–0011. I would also like to thank the anonymous reviewers of the CRSA for their helpful suggestions in helping me to strengthen my arguments. Of course, final responsibility for any errors or oversights remains with me. This manuscript was first submitted in January 2001 and accepted in August 2001.

With the introduction of Bill C-11 in March of 2001, the Canadian state is taking steps to increase both the "security" of Canadian borders against those represented as "unwanted intruders" and the "flexibility" of the labour force through the recruitment of (im)migrants as non-citizen, temporary workers. By expanding the issuance of temporary employment authorizations while ignoring various international human rights obligations, greatly expanding state power of detention over refugees and (im)migrants, eliminating certain appeals processes, reinforcing measures of interdiction and broadening provisions of inadmissibility, Bill C-11 can be seen as part of a long-term move that makes it more difficult for certain groups of people to enter, live and work in Canada as permanent residents and eventually formal citizens (see Canadian Council for Refugees (CCR), 2001).

Calls for greater access to temporary, non-immigrant workers have come largely from employers' organizations while demands for greater "order at the border" have come from a diverse number of sources.[1] A number of recent and highly visible cases involving the entry of people from the Global South or from the former eastern bloc countries is evidence of a growing "moral panic"[2] regarding border controls. The entry of 599 undocumented people from China by boat to British Columbia in the summer of 1999 or the entry of women sex workers from eastern Europe throughout the 1990s are two of the many events that have been represented as breaches of "national security" by much of the mass media.

During this time of re-organized state practices concerning the global flow of people, a number of theorists examining the role of the Canadian national state in relation to processes of globalization have argued that the state has lost its once held "sovereign" power to determine "domestic" policy (see Teeple, 1995; 2000; Panitch, 2001).[3] In this paper, I argue that, rather than viewing national governments as having lost control over the "domestic" or "national" space, a reorganized regulation of the international migration of labour along with a revamped nation building project has been part of how processes of globalization have been organized in Canada.

1 A recent public opinion poll commissioned by Citizenship and Immigration Canada asked participants: "Are there too many visible minorities being allowed into the country?" Twenty-seven percent answered "yes" (*The Globe and Mail*, 2000). Another recent popular opinion poll found that approximately sixty percent of respondents thought that the number one priority for Canadian immigration policy should be to stop "illegal immigrants" (*The Globe and Mail*, 1999).

2 Stuart Hall (1978 as cited in P. Gilroy, '*There Ain't No Black in the Union Jack': The Cultural Politics of Race and Nation*. Chicago: University of Chicago Press, 1987: 3) defines a "moral panic" as existing: "[w]hen the official reaction to a person, groups of persons or series of events is out of all proportion to the actual threat offered, when "experts," in the form of police chiefs, the judiciary, politicians and editors, perceive the threat in all but identical terms, and appear to talk "with one voice" of rates, diagnoses, prognoses and solution, when the media representations universally stress "sudden and dramatic" increases (in numbers involved or events) and "novelty" above and beyond that which a sober, realistic appraisal could sustain, then we believe it is appropriate to speak of the beginnings of a moral panic."

3 A thorough examination of the growing body of literature debating the role of national states within processes of globalization is beyond the scope of this paper (see N. Sharma, "The social organization of 'difference' and capitalist restructuring in Canada: The making of 'migrant workers' through the 1973 Non-immigrant Employment Authorization Program (NIEAP)." Ph.D. dissertation. Ontario Institute for Studies in Education at the University of Toronto, May, 2000: 69–121.

It is thus important to investigate how the Canadian state has been active in both the making of the "problem" of international migration as well as in formulating responses to it. The international migration of people has doubled since the mid-1980s (United Nations Population Fund, 1993). The numbers of people entering Canada with or without legal documents has also increased in this period. Despite this, it is still accurate to say that Canadian state policies on immigration have become increasingly restrictive over the last two decades (see Simmons, 1996). These growing restrictions, however, should not be seen as resulting in a drop in the numbers of (im)migrants entering and staying in Canada. Rather, restriction measures have been taken largely in *how* people are able to cross borders—not whether they are able to cross or not (Sharma, 1995; Michalowski, 1996).

In this paper, I will show that the ideology of border control so prevalent in Canada (and other countries in the Global North) over this latest period of globalization has worked not necessarily to exclude people but to cheapen the labour power of a growing number of people once they are inside the country and to leave them increasingly vulnerable to all forms of market relations.[4] Such concerted state practices have been an important part of how nationalized labour markets have become "globally competitive" over the last several decades (see Sassen, 1988; Gardezi, 1995).

In this paper, I emphasize the importance of notions of Canadian "nation-ness" in maintaining nation-state power in the present era of globalization (see Anderson, 1991 for a discussion on "nation-ness" as an ideology). As I have argued elsewhere, borders are both physical and existential (Sharma, 2000a). They define material as well as ideological ground. In other words, the construction of the borders or boundaries of the nation-state affects people's legal-political "rights" as well as the formation of people's consciousness of who "belongs" and perhaps more importantly, of those that do not.

The view of the Self as insider and the Other as foreigner or outsider that nationalist practices aimed at "protecting our borders" organize, consequently helps to naturalize the nation-state system and profoundly shape both material reality as well as a particular ideological understanding of social relationships amongst people. I argue that continued reference to protecting the "nation," and by extension those seen as "belonging" to it, allows those working within the apparatuses of the Canadian state to reorganize the labour market in Canada by recruiting workers categorized as "non-immigrants" (or in the vernacular, as "migrant workers"[5]).

4 The occurrence of calls to "protect our borders" in Canada is not limited to this latest period of globalization. Indeed, it can be said that such calls are related both to *who* is crossing into Canadian territory as well as in which historical period (see F. Iacovetta, "Making 'new Canadians': Social workers, women, and the reshaping of immigrant families," in F. Iacovetta and M. Valverde (eds.) *Gender Conflicts: New Essays in Women's History*. Toronto: University of Toronto Press, 1992; J. Parr, *The Gender of Breadwinners: Women, Men, and Change in Two Industrial Towns, 1880–1950*. Toronto: University of Toronto Press, 1990; G. Creese, "Exclusion or solidarity? Vancouver workers confront the 'oriental problem'," *B.C. Studies*, No. 80 (Winter), pp. 24–51).

5 I am placing the term migrant worker within quotes to emphasize its socially organized character. I will forego the use of this practice, but I continue to problematize it throughout this paper.

The social organization of those categorized as non-immigrants works to legitimize the differentiation of rights and entitlements in Canada along citizen/non-citizen lines by legalizing the indentureship of people classified as migrant workers. However, instead of taking the notion of legitimacy for granted, so that actual state practices are left unexamined, I investigate *how* it is that the Canadian state works at shaping people's consciousness around the boundaries of "Canadianness" in ways that contribute to the "common-sense" realization of the category migrant workers.[6]

My examination of five years of parliamentary debates in the House of Commons during the years 1969 to 1973 allows me to focus on the creation and organization of the category of migrant worker. I investigate how the social relations and practices that organize this category have normalized the denial of rights and entitlements for people so classified. Necessarily, then, I also investigate the social organization of people's consciousness around notions of Canadianness during this time, for it, perhaps more than any other concept organized by national state practices, helps to shape people's common sense of the "imagined community" of Canada (Anderson, 1991).

I argue that acceptance of the oppositional categories of citizen/migrant worker helps to secure the organization of "difference" within Canada, where difference does not mean diversity but inequality. The notion of citizen, then, needs to be understood as the dominant, oppressive half of a binary code of negative dualities. As such, the notion of citizenship is not a philosophical absolute but the mark of a particular kind of *relationship* that people have with one another (see Arat-Koc, 1992; Brown, 1995).

Thus, while there is a growing body of literature examining the working conditions and (direct) employment relations of various groups of migrant workers in Canada, more attention needs to be paid to how the very categorization of people as migrant workers assists in the restructuring of the labour market in Canada (see Bakan and Stasiulis, 1996; Bolaria, 1992; Wall, 1992 for a discussion of migrant workers in specific occupational sectors). In this paper, then, I place the state-organized category of migrant worker at the centre of my inquiry.

Since Canadian immigration policies increasingly emphasize the recruitment of workers admitted on temporary employment authorizations, uncovering the social practices that organize people as migrant workers sheds some light on how concepts of national citizenship are employed in this period of globalization. By examining how parliamentarians in the period 1969 to 1973 organized legitimacy for the creation of a non-immigrant or migrant worker category that strips people of most of the human, civil and other "rights" of citizenship available to "Canadians", we are also in a better position to examine contemporary state practices concerning the regulation of global flows of labour, as well as capital.

6 I am using the notion of "common sense" in ways similar to that of Antonio Gramsci (1971). In this regard, Roxana Ng points out that the notion of common sense allows us to make "good sense" of the ". . . incoherent and at times contradictory assumptions and beliefs held by the mass of the population" (1993:52). She adds that "treating racism and sexism as "common sense" draws attention to the norms and forms of action that have become ordinary ways of doing things, of which we have little consciousness" (1193: 52). The Gramscian use of the term "common sense," thus, is used to refer to notions that have become naturalized or normalized. Therefore, when something is said to be commonsensical, it is understood to have become a hegemonic world view in a particular place and time.

CITIZENSHIP AND (IM)MIGRATION IN CANADA: THE SOCIAL ORGANIZATION OF MIGRANT WORKERS

The state's active participation in the discourse of "protecting Our borders," especially from women and men from the Global South, has been key in the process of "nation"-building during the beginning stages (late 1960s to early 1970s) of this most recent period of capitalist restructuring. Indeed, the work of continuously reimagining Canada features prominently in the discourse concerning proper state practices during this time. In this regard, discursive practices of formulating immigration policy can be said to refract issues of racialized, gendered and nationalized inclusions/exclusions and their relationship to entitlements and disentitlements within Canadian society.

Such practices are of special consequence for the reorganization of the labour market in Canada (see Ng, 1988; Brand, 1993). This is perhaps most apparent when looking at the experiences of those categorized as migrant workers and who, as a result, are made to work in unfree employment relationships as a condition of entering, residing and working in Canada. People so categorized enter through Canada's Non-immigrant Employment Authorization Program (NIEAP) established in 1973. The Canadian system of migrant-worker recruitment raises a number of points that include different elements of ideological practices that make and legitimize the differential categories of citizen and non-citizen (in this case, the non-immigrant).

It needs to be clearly stated, however, that not all persons admitted to the country under this program can be considered unfree wage workers and that the Non-immigrant Employment Authorization Program should not be considered as *only* a labour-recruitment program (Wong, 1984; Michalowski, 1996). Instead, it is heterogeneous in nature. For instance, a large number of the people admitted stay only for a short period of time to do work that normally crosses many national borders and can not be considered to be working as unfree wage labour in the country. These people fall under the following occupational categories: entrepreneurs; artistic, literary, performing arts and related: and sports and recreation professionals, as well as those with "not stated" occupations on their employment authorizations.

The inclusion of people admitted under these categories in the data on migrant workers in Canada has often been cited as a reason to dismiss claims that the Non-immigrant Employment Authorization Program operates as a system of forced, rotational, unfree contract-labour recruitment (Boyd, 1986). Therefore, for the purposes of the present study, out of the total number of people annually issued temporary employment authorizations, those admitted under the above categories are omitted from any statistics presented. Once removed from the data, it becomes evident that such complete dismissals are ungrounded: approximately 75% of those entering through the NIEAP are engaged in non-professional occupations (Sharma, 1995: 128).[7]

7 In 1973, the top three occupational groupings employing unfree contract labour were service (17.2%), farming (13.3%) and fabricating, assembly and repair (11.4%). Together, these three occupations accounted for approximately 42% of all jobs filled that year (CIC, 1995). Ten years later, in 1983, service (29%); farming (9.5%) and fabricating, assembly and repair (13.2%) accounted for about 52% of all temporary visa workers (CIC, 1995). In 1993, the service industry (11.3%) and the farming sector (14.1%) were in the top three of all occupational groupings (CIC, 1995). Fabricating, assembly and repair remained in the top five of all occupations employing unfree contract workers.

For these workers, stipulations regarding the criteria for entering under the NIEAP include having the name of the employer and the location, type, condition and length of employment pre-arranged and stated on the person's temporary employment authorization form prior to arrival in Canada. The person classified as the worker within this process of creating what Smith (1995: 2) calls a "documentary reality" is then bound to ". . . work at a specific job for a specific period of time for a specific employer" (Citizenship and Immigration Canada (CIC), 1994). Migrant workers are unable to change any of their conditions of entry or employment without receiving written permission from an immigration officer. If s/he leaves the stipulated employer or changes occupations without the approval of the government, for instance, s/he is subject to deportation.

Under the regulations of the NIEAP, workers cannot stay in Canada beyond the length of time stated upon their temporary employment authorization form. They are, however, able to renew what the immigration department (CIC, 1994) calls their "foreign worker" visa if the *employer* agrees. Even for those who are able to successfully do so, however, a migrant worker's status in Canada is permanently temporary. This is because people classified as migrant workers cannot apply for permanent residency (or "landed") status.[8] Different people are brought in to work and subsequently expelled to be replaced by others. In this sense, it can be stated that the NIEAP operates as a "revolving door of exploitation" (Ramirez, 1982: 17).

As Table 14.1 shows, there have been some significant fluctuations in both the numbers and percentages of the total number of people entering as either permanent residents or as temporary, migrant workers over the period 1973 to 1993. One clear trend emerges from the data, however. Through the implementation of the NIEAP, the Canadian government has successfully shifted its immigration policy away from a policy of permanent (im)migrant settlement towards an increasing reliance upon unfree, temporary labour.[9]

---

8 After sustained struggle from domestic workers and their allies, those (mainly women of colour) entering through the various domestic workers recruitment schemes (the latest one being the Live-in Caregiver Program) are able to apply for (but are by no means guaranteed) permanent-residency status if they have met certain criteria, including having worked for two years in continuous employment as an indentured domestic worker in Canada. However, the numbers of migrant workers entering through this program are but a tiny fraction of all migrant workers recruited to work in Canada and even these numbers are now in decline (see N. Sharma, "The true north strong and unfree: Capitalist restructuring and Non-immigrant employment in Canada, 1973–1993. "Master's thesis. Simon Fraser University: Burnaby, B.C. 1995; A. Bakan and D.K. Stasiulis, "Structural adjustment, citizenship, and foreign domestic labour: The Canadian case," in *Rethinking Restructuring: Gender and Change in Canada*. I. Bakker (ed.). Toronto: University of Toronto Press, 1996).

9 Margaret Michalowski (1996: 110) has shown that the figures presented on the Non-immigrant Employment Authorization Program represent the number of people granted temporary work visas, rather than the number of visas issued. The numbers do not include multiple visa issuance to the same person. In other words, the number of people entering on temporary work visas is recorded and not the number of jobs employing unfree wage workers in Canada.

TABLE 14.1 **Total Number of (Im)migrant Workers in the Canadian Labour Market by Calendar Year: Permanent Residents "Destined" to the Labour Market and Temporary Visa Workers, 1973 to 1993**

Year	Destined (Immigrant Workers)	Visa ("Non-immigrant" Workers)	Total (All (Im)migrant Workers)[a]
1973	92,228 (57%)	69,901 (43%)	162,129 (100%)
1974	106,083 (60)	71,773 (40)	177,856 (100)
1975	81,189 (51)	77,149 (49)	158,338 (100)
1976	61,461 (47)	69,368 (53)	130,829 (100)
1977	47,625 (41)	67,130 (59)	114,755 (100)
1978	34,762 (71)	14,459 (29)	49,221 (100)
1979	47,949 (60)	31,996 (40)	79,945 (100)
1980	63,479 (39)	98,681 (61)	162,160 (100)
1981	56,676 (37)	96,750 (63)	153,426 (100)
1982	55,023 (35)	101,509 (65)	156,532 (100)
1983	36,540 (29)	87,700 (71)	124,240 (100)
1984	37,468 (25)	113,297 (75)	150,765 (100)
1985	36,949 (22)	134,167 (78)	171,116 (100)
1986	63,479 (30)	150,467 (70)	213,946 (100)
1987	56,676 (26)	157,492 (74)	214,168 (100)
1988	73,134 (27)	194,454 (73)	267,588 (100)
1989	94,412 (36)	169,004 (64)	263,416 (100)
1990	109,840 (38)	176,377 (62)	286,217 (100)
1991	127,870 (40)	191,392 (60)	319,262 (100)
1992	137,360 (43)	178,280 (57)	315,640(100)
1993	65,130 (30)	153,988 (70)	219,118 (100)

a. This category includes all those entering the country under the above "destined" and "visa" categories.

Source: Sharma, 1995.

In 1973, 57% of all people classified as workers "destined" to enter the "Canadian" workforce came with permanent resident status.[10] By 1993, of the total number of workers

10 "Destined" refers to the number of people admitted to Canada as permanent residents who have indicated that they intend to enter the labour market. This category *includes* people admitted under all classes of immigrants (family, refugees, self-employed, retired, assisted relative and independent). These people have the right to choose their occupation, their employer and their location of residence. In other words, they are able to work as free wage workers within Canada. This category excludes: entrepreneurs who were added to this category from 1978 and investors who were included from 1988.

admitted to Canada in 1993, only 30% received this status while 70% came in as migrant workers on temporary employment authorizations.[11]

Through the use of the NIEAP the Canadian state has, to a great extent, controlled the scale, structure and course of labour migration into Canada and has contributed to the creation of a highly flexible (i.e., precarious) labour force. In other words, the Canadian state, through the regulations of the NIEAP, has produced a category of people in Canada that we have come to know as migrant workers.

What allows migrant workers to be used as a cheap and largely unprotected form of labour power are not any inherent qualities of the people so categorized but state regulations that render them powerless (see Sassen, 1988; also Sharma, 1997). Since they have been categorized as non-immigrants, people admitted as migrant workers do not have many of the either *de facto* or *de jure* social, economic or political rights associated with Canadian citizenship.[12] For instance, because they can not stay in the country other than to work for a prespecified employer, migrant workers do not have the ability to access a wide array of social programs and services associated with the entitlements in the Northern welfare states.

Access to these programs and services would provide migrant workers with an alternative to selling their labour power (as it does to those Canadian citizens and permanent residents who are able to access them). However, this would go against the express purpose of their recruitment—to work as unfree labour in the Canadian labour market. In other words, they would no longer be *migrant workers*. Aside from being denied the ability to make changes through Canada's political system (e.g., voting), these workers are placed in a highly vulnerable situation in regards to speaking out for their rights. This is due to the fact that if either the employer or the state finds the worker unsuitable, s/he is subject to deportation (Wall, 1992).

By introducing the NIEAP in 1973, we see, then, that another dimension of coercive state control over certain peoples has been added. This program represents new restrictive conditions of entry for people filtered through it and connects the concepts of citizenship and nationality to a bureaucratic course of action, as well as to the resources of the state (see Ng, 1995; Smith, 1990). While migrant workers are expressly recruited for their contributions to the Canadian labour market, governmental practices categorize them as being part of a *foreign* labour force. The dual construction of a "domestic" and a "foreign" labour market within the space occupied by Canada is accomplished through the category migrant worker.

---

11 "Visa" refers to the number of people admitted to Canada for periods less than or over one year and working in Canada during the calendar year recorded. The total of visa workers includes workers entering through the NIEAP plus the Foreign Domestic Movement Program (1982–1991) and the Live-in Caregiver Program (1992–1993). For the years 1989–1993, the category "backlog clearance," given to refugees granted temporary employment authorizations while waiting for their status to be determined, is also excluded.

12 Of course, even among those holding Canadian citizenship status, there remain profound differences in material realities as a result of social relations shaped by demarcations of "race," gender, dis/abilities, etc. (see B. Cassidy et al., "Silenced and forgotten women: Race, poverty, and disability," in *Feminist Issues: Race, Class, and Sexuality* (3rd ed.). N. Mandell (ed.). Toronto: Prentice Hall, 2001; also see R. Ng, *The Politics of Community Services*. Toronto: Garamond Press, 1988 for a discussion on how inequalities between citizens and permanent residents are organized by the Canadian state, employers and service agencies).

The operation of the migrant worker category can substantially enhance the ability of the Canadian government to attract and/or retain capital investment in its territory by giving employers in the country (whether they are so-called domestic or foreign capitalists) access to a "cheap labour strategy" of global competition (Swanson, 2001). Indeed, capitalist investment has been a key concern of the government during this period of capitalist restructuring. Days after the introduction of regulations that legalized the NIEAP on January 1, 1973, the Liberal government led by Pierre Elliot Trudeau laid down the following priorities in its annual *Speech From the Throne:*

> . . . The Government will introduce legislation establishing a competition policy to preserve and strengthen the market system upon which our economy is based. The new policy will be in harmony with industrial policies in general and foreign investment policy in particular (Hansard, 4 January 1973a: 5).

The migrant workers recruitment program can be said to be situated within the government's stated desire to "strengthen the market system" in Canada and attract capital investment, for it helps to (re)organize a particular kind of labour market in Canada. The operation of the NIEAP enables those in the Canadian government to produce a group of non-citizens who, because of their classification as "non-immigrants," can legally be exempted from laws on minimum employment standards, collective bargaining and the provision of social services and programs such as unemployment insurance, social assistance, old-age pensions, etc.[13] This, in turn, cheapens and weakens the position of these workers. Citizenship, then, has become an important tool in reorganizing the labour market in Canada to the benefit of capital investors (Sharma, 2000d). This can be seen in the even greater emphasis given to recruiting people as temporary migrant workers in Bill C-11, introduced to revamp existing immigration policies (see Sharma, 2000b).

However, while the statistics I have compiled and the immigration regulations I have outlined show the material effects of employing the NIEAP in Canada and the conditions under which migrant workers are made to live and work, they do not, by themselves, explain *how* it is that the government can create a category of non-citizens, such as migrant workers, with relatively little outcry and, even, tacit support from much of the remaining population living and working in Canada.[14] In order to uncover the social and ideological practices that allow for the existence of a group of people we know

13 It is important to note that while the federal government of Canada regulates the entry of people on temporary employment authorizations, the ten provincial governments regulate migrant workers' access to a wide variety of social programs, services, benefits and protections. Therefore, there exists throughout Canada a patchwork of differential access to welfare state provisions for migrant workers. Generally, though, migrant workers are denied access to those programs that provide an alternative to paid employment, e.g., unemployment insurance, welfare, etc.

14 To understand the scope of this passivity, imagine the outcry if all Canadian citizens who were accountants, university professors or autoworkers were legally indentured to their employers and faced harsh penalties for quitting or changing their jobs. Imagine, too, the outcry if all were denied access to minimum employment standards, ability to organize into unions or social programs and services.

as migrant workers in Canada, there is a need to examine state practices, particularly the work done in Parliamentary debates, that aid in the construction of the co-ordinated ideological frames embedded in the concept of "Canadianness" during this early period of capitalist reorganization. I argue that the construction of a new, "tolerant" Canadian identity during this time, especially in regards to people of colour and particularly in the area of immigration, worked to secure the racist and sexist operation of the capitalist labour market in which the organization of the category of migrant worker is referenced.

IDEOLOGICAL TEXTUAL PRACTICES

At the same time that an indentured labour system was being entrenched within Canadian citizenship and immigration policy, the notion of Canada as a tolerant and even a "just" society was being organized. Indeed, one of the striking features of immigration policy-making at this time was the practically simultaneous liberalization and restriction of access to Canadian citizenship. On the one hand, the Canadian government removed *explicitly* racist restrictions on immigration from the South in 1967 through regulatory changes (see Satzewich, 1989). On the other hand, in 1973 the NIEAP was introduced, which served to deny some people access to Canadian entitlements while recruiting them to work in Canada.

This was accomplished through a threefold process, including the further exploitation of the valuable currency of liberal philosophy already embedded within notions of Canadian parliamentary "democracy," selectively reporting accounts so that the NIEAP was rendered more-or-less invisible while the supposedly tolerant Canadian society was highlighted, as well as reproducing (and reworking) commonsensical notions of the entitlements of Canadianness (and, hence, the disentitlements of non-Canadianness) established over time (see Creese, 1988; Abele and Stasiulis, 1989).

By conducting a textual analysis of transcripts (the Hansard) of parliamentary debates both before, during and after the time the NIEAP was introduced in 1973, it is clear that the discursive practices of parliament helped in organizing an ideological understanding of Canadian identity. Indeed, key to the maintenance of legitimacy for coercive state actions at the time the NIEAP was introduced was the reshaping of an identity that was tolerant but still very much exclusionary. Important to this work were the ways in which certain groups of people continued to be excluded from the definition of "Canadian," yet how their exclusion was concealed through the organization of a false, "virtual reality" (Smith, 1990: 62).

The erasure of the colonial and racist foundations of Canada was a key, initial step in the presentation of Canada as a tolerant society. The construction of Canadianness during this period was one where the Canadian national state was portrayed as having sprung into existence through the *overthrowing of colonialism*. This supposedly anti-colonial struggle was waged by the two "founding" English and French "nations" (earlier referred to as founding "races"). In this account, the reality of Canada being built on the *colonization* of Aboriginal peoples and lands by these same people was nowhere in evidence.

This virtual reality was presented in the House of Commons equally well by members of the different political parties. The following quote from David Lewis, Member of Parliament (MP) from York South and soon to be leader of the federal New Democratic Party was representative of how Canada came to be constituted as a previously *colonized—not colonizing*—national state. In speaking of the importance of establishing good relations with "developing" countries in order to foster "international competitiveness," he stated:

> . . . from all our contacts and all our reading *we know* Canada has a special place of trust among the developing nations. We emerged as an independent nation almost a century before them, but we also emerged out of colonial status. We have never [had] an imperial goal or imperialistic intentions (Hansard, 20 January 1969; emphasis added).

Another episode shaping this virtual reality was found in the following statement by Conservative Party MP, Heath Macquarrie, who, speaking in support for the creation of a new statutory holiday, Canada Day, stated:

> . . . A national day in any country reverts back to a time of great achievement. . . . The great moment in Canadian history, one which reflects its unique character among nations, is surely that it has achieved nationhood by peaceful means, by the *getting together* of different communities. . . . (Hansard, 17 February 1970: 3702; emphasis added).

Again, the continuing violence of colonialism against Aboriginal peoples was erased in favour of a Canadianness represented as "peaceful." Macquarrie's comments organized a particular kind of knowledge whereby the English and French empires did not colonize Aboriginal peoples but "got together" with them.

This allowed for an ideological reading of being Canadian, one that obscured the continued substandard conditions under which most Aboriginal peoples continued to live in Canada and one which removed any responsibility for this reality by the English and French colonists (see Frideres, 1988; Goodleaf, 1995). The work that these texts, in part, did then was to have us, their readers (see-ers, hear-ers) come to know Canada (and, hence, Canadians) as bearing no responsibility for the existence of certain oppressive and exploitative social relations. This is where various readers come to be differentiated according to their experiences of being a colonizer or being colonized, being included in things Canadian or being excluded.

Significantly, part of the creation of the tolerant Canadian nation at this time was the exclusion of references to the process of racialization, hence, nowhere was the term "white" attached to being Canadian. This was key to the process of ruling in this period of Canadian society. As neither the Self nor the Other was identified explicitly, we can say an agreement was struck between the producers and (some of the) readers of these texts that allowed reference to the tolerant Canadian to work ideologically to maintain existing relations of ruling while denying them altogether. Not naming who benefited from the existence of Canada allowed for the continuation of these benefits while working to deny complaints from those who were kept from them.

Yet, liberal rhetoric aside, the racialized meaning(s) of Canadianness *remained* embedded within these concepts. This is evident in the following statement by then Prime Minister, Pierre Trudeau, on a report of a Royal Commission on National Security. In addressing parliament at a time when the immigration of people from the South was increasing steadily and commonsensical ideas of being an immigrant were being conflated with being a person of colour, he stated:

> As the commissioners have stated, and I quote: "Canada remains the target of subversive of potentially subversive activities, attempts at infiltration and penetration, and espionage operations" and they emphasize that: "the duty of the state to protect its secrets from espionage, its information from subversion and its policies from clandestine influence is indisputable; what are matters for dispute are the organizations and procedures established by the State to meet this responsibility in an area which can touch closely upon the fundamental freedoms of the individual."... [This requires] a careful and methodical build-up of modern technical facilities directed toward the detection and prevention of large-scale organized crime, as well as the provision of information which the government requires in order to ensure the security and integrity of the state. . . . For this reason . . . the government . . . has decided to accept the commissioners' recommendation for the establishment of a Security Review Board. . . . It is their opinion that such a system of review might be required in the *three areas of employment, immigration and citizenship* (Hansard, 26 June 1969: 10636–37, emphasis added).

The P.M.'s comments signal that citizenship and immigration had become a "security issue" for Canadians and that it was somehow connected to policies on employment. However, Trudeau was careful to recognize, and emphasize, liberal notions of individual freedoms within the context of selecting entire groups, i.e. immigrants, for special attention. In this regard it is significant that there is an association made between citizenship and immigration and the issue of employment (and, hence, unemployment) in the context of identifying national security issues. Without saying "Canadians are under threat from people from the South who will take away Our jobs", Trudeau was able to racialize the "problems" facing Canadians while maintaining a public image of tolerance. This was done by making immigration a *national* problem, rather than a systemic one, as well as by leaving out positive comments regarding immigrants and immigration.

To understand the way that readers can read such notions from the text, it is crucial to remember that each statement was not made in isolation from previous ones. Images of which racialized and gendered bodies "belong" in Canada and which do not are littered throughout the history of governing Canada (see Ward, 1978; Creese, 1988; Bourgeault, 1989). Parliamentary discourse can "quietly borrow" (to use a term of Dorothy Smith's from this history to frame the topic of the entrance of certain people as a "problem" worthy of being called a national security issue. What was identified as a risk by parliamentarians, then, was the identity of the imagined Canadian *nation* as white. This risk, we were told, came from "immigrants" and since the parliamentary representation of immigrants was that of people entering from the Global South, Our "problem" became people of colour who threatened the Canadian character of the state.

The liberal framework, then, can actually be said to strengthen the racist meanings of concepts of Canadianness for they work to naturalize the very categories of difference that the state participates fully in organizing. Within the liberal framework, legitimacy is secured by enshrining the rights of those who are placed (and have placed themselves) within categories that privilege them in relation to Others who are placed within far more inferior categories, such as migrant worker. Liberal practices allow for the "protection of individual freedoms" but only for those that the state vows to protect. Those falling outside of this category are then seen to be *legitimately* denied the state's protection. They are the ones from whom We are to be protected. It can even come to be accepted that this leaves these Others open as targets for state coercive practices designed to strengthen Us.

In this regard, it is important that even while the Other is *ideologically* differentiated from the "norm," the construction of binary codes is intimately connected to the establishment and reproduction of unequal materialities, so that those who are categorized as "different" (from Canadians) *do* become truly differentiated in relation to resources and power, as is the case with migrant workers. This gives social meaning, and not a small modicum of plausibility, to notions that actual differences exist. Our consciousness of ourselves and Others and our respective places in the world are shaped by the fact that we can see that there are tangible consequences stemming from belonging to differently constructed group categories. This is evident in the following discussion that took place in Canada's parliament under the topic of *Manpower: Use of unemployed and students instead of West Indians to pick fruit.*

> Mr. Gérard Laprise (Abitibi): Mr. Speaker, I have a question for the P.M. A few days ago, the Minister of Manpower and Immigration announced that seasonal workers from the West Indies would be hired this summer to help in picking and canning fruits and vegetables in Ontario. Could the P.M. then consider the possibility of assigning this work to our unemployed or to our students who for the most part will not find jobs this summer (Hansard, 23 March 1971: 4508)?

Through his choice of words, Laprise operationalized the negative dualities of us—Canadians/them—West Indians or migrant workers. As we will see, it was also of great import that Mr. Laprise used the word "assign." The P.M. (Mr. Trudeau) responded by stating:

> Mr. Speaker, this is a perennial problem and it must be recognized that this is a type of work that very often students or unemployed will not do. This is why the Department of Manpower and Immigration is admitting *foreign workers* on *our* labour market. Should students be willing to undertake this work, they would certainly have the preference. I am not cognizant with the specific case the honourable member is referring to, but I know that this is a problem which comes up year after year with respect to certain types of work (Hansard, 23 March 1971: 4508, emphasis added).

The "problem," then, was at least partly identified as one of filling jobs that We Canadians do not want. The P.M.'s statement not only activates notions of less entitlement regarding workers from the West Indies but it, again, operationalizes binary notions of us/them. To this, Mr. Laprise responds with a supplementary question:

> Would the P.M. consider inviting the young people to do that work during the holidays, not only in Ontario where fruit and vegetables are grown, but in every province? This would be much more efficient than having them travel" (Hansard, 23 March 1971: 4508).
>
> Mr. Trudeau: Mr. Speaker, I agree with the hon. member on that score. The purpose of the Manpower Centres is to send the unemployed or the students to take part in this work. But, once again, facts reveal that there are in Canada some types of work which the unemployed and the students refuse to do; this proves, by the way, that the rate of unemployment is at times somewhat artificial" (Hansard, 23 March 1971: 4508).

Once again, the comments by the P.M. help to ideologically reframe the recruitment of migrant workers as coming to work at jobs that Canadians will not (normally) take. Why? Because of low wages, unsafe and unacceptable working conditions, the seasonal character of the work, etc. (see Wall, 1992). However, this reality is eclipsed by a virtual one so that the racialized segregation of the labour market in Canada is naturalized through reference to nationality and citizenship.

Contrary to such claims, migrant workers are not only recruited when and where there is an actual *shortage* of workers in Canada. Instead, much of this "shortage" is qualitative. It is about recruiting (and indenturing) workers from outside of Canada for jobs that "Canadians" consistently refuse because *they can* legally and because they have other options, including unemployment and social assistance.[15] It is also about recruiting workers through a category—non-immigrant—that renders these people's unfreedom legal.

Indenturing workers therefore is one of the paramount meanings attached to the classification of some people as migrant workers. Importantly, the articulation of notions of Canadianness with notions of freedom (versus unfreedom) are organized through the operation of this category. Having the ability to work within free employment relationships is one characteristic of being Canadian: the opposite applies to those categorized as migrant workers. The facticity organized by these binary codes that separate migrant workers from Canadians organizes the expectation that differentiated categories of people will, in reality, be treated quite differently by the Canadian state. This is solidified in the following exchange:

> Mr. Roch LaSalle (Joliette):. . . In view of the statement by the P.M. to the effect that some unemployed people would refuse to perform such

15 It is of great import that the ability for formal citizens to access these options is becoming more restrictive. However, while the content of citizenship may be becoming more hollow, the distinction between being a citizen and being a non-citizen within Canada is still of great significance for experiences in the labour force.

> work, would the P.M. consider *compelling Canadians* to work if they receive any social benefits? Would the government favour legislation requiring any government pension recipient to work? (Hansard, 23 March 1971: 4508, emphasis added).
>
> Right Hon. P.E. Trudeau (P.M.): No,. . . the government will not commandeer the work force. *The whole political philosophy of the government is based on freedom of choice for citizens to work where they want* (Hansard, 23 March 1971: 4508, emphasis added).

This statement highlights what is, ultimately, the crux of the issue. The P.M. acknowledges that the Canadian state, at this time, cannot indenture those that it has categorized as its citizens, as the beneficiaries of the existence of Canada itself, at least not without raising serious questions about the liberal democratic character of governance. In the framework of Canadianness it just does not (yet) make common sense to compel Canadians to work where they do not wish to.

Within this same ideological framework, though, people *can be* exploited as indentured labour in Canada, as evidenced by my previous discussion of the NIEAP. Utilizing liberal democratic ideologies to legitimize the state requires that those made unfree by this same state need to be Othered through a classificatory system that deems them as non-citizens. With the non-immigrants or migrant worker category in place, a system of indentured labour in the self-stated liberal, tolerant Canada can proceed.

In this context it is crucial for us to note that nowhere in the preceding debate within the Canadian parliament were any objections raised as to the indenturing of *West Indian* workers on farms in Canada. The fact of a person working as unfree labour was not, in and of itself, the problem. The problem was *who*, citizen or non-immigrant, was going to do the job. As the following statement by MP H.W. Danforth (Kent-Essex) demonstrates, it is presented as natural that we-Canadians get those-foreigners to do this kind of work because, in part, it is natural for Them to do it.

> The attitude of this government has been that if you do not want to work, you should not have to do so. I raise this matter because the P.M. reaffirmed the position of the government that a Canadian should not have to work if he [*sic*] does not want to. Mr. Chairman, many people do not like to work in agriculture. They do not like the monotony, the conditions and the fact that you work sometimes in heat and sometimes in cold. That is all right; they do not like it and they should not be forced to work at it. We all agree with that. . . . How [then] do they [farm owners] obtain labour? Many of them have encouraged offshore labour over the years which comes from three sources, the Caribbean, Portugal and Mexico. We need this labour. . . and these people are used to working in the heat. They are used to working in agriculture, and they are satisfied with the pay scale . . . . Everybody is satisfied: the workers are satisfied, the primary producers are satisfied and the consumers of Canada are satisfied because

> we are getting the crops harvested. . . . I feel that Canadians should provide work for Canadians wherever possible; Canadians should have the first opportunity to work. But . . . if Canadians do not want to work at this job—many of them do not, and have expressed this feeling in no uncertain terms—then I say that the producers of this nation are entitled to offshore, competent labour from wherever it may come, if these people are willing to work under the conditions prevailing in Canada today and produce crops for Canadian consumers (Hansard, 20 July 1973: 5836.)

This statement presents as natural notions that there are actual differences between people that make certain people, i.e., "offshore labour" from the Caribbean, Portugal and Mexico legitimate bodies for coercive state actions to be imposed upon. It is also naturalized that they work under conditions that "Canadians" would not subject themselves to. The discursive organization of this statement conceals the social factors that bring certain groups of people to work in clearly substandard (in relation to Canadians) working conditions and pay scales. Rather than showing the unequal social relations that organize these differences in the first place, these material practices are mystified through parliamentary practices that reproduce the ideologies of racism and nationalism that help to hold in place commonsensical notions about the "natural" superiority/inferiority of differentiated groups of people.

When examining the construction of binary codes that organize difference, then, we see that categorizing a person a citizen or a migrant worker is an ideological practice, for the exploitation of migrant workers is concealed and reproduced through the notion that citizens can expect certain rights and entitlements that non-citizens can not and that this expectation is perfectly "normal." The notion that some people *just are* citizens and Others *just are not*, even within the same borders, comes to be a normative stance. The fact that these are realized through the social organization of human relations in a particularly exclusionary and exploitative way is concealed. As a result, it appears perfectly ordinary that those categorized as non-citizens (migrant workers, for example) would be denied those rights and protections that Canadian citizens are seen as solely entitled to. Why should migrant workers get the same rights as citizens? They are, after all, migrant workers. The circularity of the argument ensnares migrant workers (and others classified as non-citizens) in a particularly vicious way.

Contrary to government claims, then, those made to live and work as migrant workers in Canada do not consist of a "foreign" work force in Canada, one supposedly working separately from and in opposition to Canadians. Rather, the employment of people as migrant workers is very much a part of the labour force that is offered up to capital investors in Canada. In this sense, citizenship can be thought of as a work process that is located within the social organization of productive relations. In other words non-citizens, because they are socially organized to be more malleable to the will of employers, are the quintessential "flexible" employees. Their vulnerability lies at the heart of the flexible accumulation process of this recent period of globalization.

REFERENCES

Abele, F. and D. Stasiulis. 1989. "Canada as a 'white settler colony': What about indigenous and immigrants?" In *The New Canadian Political Economy*. W. Clement and G. Williams (eds.). Kingston: McGill-Queen's University Press, pp. 240–77.

Anderson, B. 1991. *Imagined Communities*. London: Verso.

Arat-Koc, S. 1992. "Immigration policies, migrant domestic workers and the definition of citizenship in Canada." In *Deconstructing a Nation: Immigration, Multiculturalism and Racism in 90s Canada*. Satzewich (ed.). Halifax: Fernwood Publishing, pp. 229–42.

Ashforth, A. 1990. "Reckoning schemes of legitimation: On commissions of inquiry as power/knowledge forms." *Journal of Historical Sociology*, Vol. 3, No. 1, pp. 1–22.

Bakan, A.B. and D.K. Stasiulis. 1996. "Structural adjustment, citizenship, and foreign domestic labour: The Canadian case." *In Rethinking Restructuring: Gender and Change in Canada*. I. Bakker (ed.). Toronto: University of Toronto Press, pp. 217–42.

Bolaria, B. Singh. 1992. "From immigrant settlers to migrant transients: Foreign professionals in Canada." In *Deconstructing a Nation: Immigration, Multiculturalism and Racism in 90s Canada*. Satzewich (ed.). Halifax: Fernwood Publishing, pp. 211–27.

Bourgeault, R. 1989. "Race, class and gender: Colonial domination of Indian women." In *Race, Class, Gender: Bonds and Barriers*. J. Vorst et al (eds.). Society for Socialist Studies and Garamond Press, pp. 87–115.

Boyd, M. 1986. "Temporary workers in Canada: A multifaceted program." *International Migration Review*, Vol. 20, No. 4, pp. 929–50.

Brand, D. 1993. "A working paper on Black women in Toronto: Gender, race and class." In *Returning the Gaze: Essays on Racism, Feminism and Politics*. H. Bannerji (ed.). Toronto: Sister Vision Press, pp. 220–42.

Brown, W. 1995. *States of Injury: Power and Freedom in Late Modernity*. Princeton: Princeton University Press.

Canadian Council for Refugees (CCR). 2001. "New immigration bill reduces newcomer rights." News release, 14 March. Montreal: CCR.

Cassidy, B., R. Lord and N. Mandell. 2001. "Silenced and forgotten women: Race, poverty, and disability." In *Feminist Issues: Race, Class, and Sexuality* (3rd ed.). N. Mandell (ed.). Toronto: Prentice-Hall, pp. 75–107.

Citizenship and Immigration Canada (CIC). 1994. "Hiring foreign workers: Facts for Canadian employers." Ottawa: Minister of Supply and Services.

Citizenship and Immigration Canada (CIC). 1995. Unclassified information provided on request by Albert Redden, Electronic Information Management Office. Hull, Quebec.

Creese, G. 1988. "Exclusion or solidarity? Vancouver workers confront the 'oriental problem.'" *B.C. Studies*, No. 80, pp. 24–51.

Dehli, K. 1993. "Subject to the new global economy: Power and positioning in the Ontario labour market policy formation." *Studies in Political Economy*, No. 41, pp. 83–110.

Foucault, M. 1980. *Power/Knowledge: Selected Interviews and Other Writings: 1972–1977*. C. Gordon (ed.). New York: Pantheon Books.

Foucault, M. 1991. "Questions of method." In *The Foucault Effect: Studies in Governmentality*. G. Burchell, C. Gordon and P. Miller (eds.). Chicago: University of Chicago Press, pp. 73–86.

Frideres, J.S. 1988. "Institutional structures and economic deprivation: native people in Canada." In *Racial Oppression in Canada* (2nd ed.). B. Singh Bolaria and Peter Li (eds.). Toronto: Garamond Press, pp. 71–100.

Gardezi, H.N. 1995. *The Political Economy of International Labour Migration.* Montreal: Black Rose Books.

Gilroy, P. 1987. *'There Ain't No Black in the Union Jack': The Cultural Politics of Race and Nation.* Chicago: University of Chicago Press.

The Globe and Mail. 1999. "Illegal migrants priority, poll finds." *The Globe and Mail.* 22 November.

The Globe and Mail. 2000. "Immigration: Poll shows opposition to minorities rising." *The Globe and Mail.* 11 March.

Goodleaf, D. 1995. *Entering the War Zone: A Mohawk Perspective on Resisting Invasions.* Penticton, B.C.: Theytus Books.

Gordon, C. 1991. "Government rationality: An introduction." In *The Foucault Effect: Studies in Governmentality.* G. Burchell, C. Gordon and P. Miller (eds.). Chicago: University of Chicago Press, pp. 1–51.

Gramsci, A. 1971. *Selections from the Prison Notebooks.* London: Lawrence and Wishart.

Hansard. 1968–69. *House of Commons Debates Official Report.* First Session-Twenty-eighth Parliament, Vol. IV (3 December 1968 to 20 January 1969). Ottawa: Queen's Printer for Canada.

Hansard, 1969. *House of Commons Debates Official Report.* First Session-Twenty-eighth Parliament, Vol. X (18 June to 22 October). Ottawa: Queen's Printer for Canada.

Hansard, 1970. *House of Commons Debates Official Report.* Second Session-Twenty-eighth Parliament, Vol. IV (9 February to 4 March). Ottawa: Queen's Printer for Canada.

Hansard, 1971. *House of Commons Debates Official Report.* Third Session-Twenty-eighth Parliament, Vol. V (22 March to 5 May). Ottawa: Queen's Printer for Canada.

Hansard. 1973a. *House of Commons Debates Official Report.* First Session-Twenty-ninth Parliament, Vol. I (4 January to 9 February). Ottawa: Queen's Printer for Canada.

Hansard, 1973b. *House of Commons Debates Official Report.* First Session-Twenty-ninth Parliament, Vol. VI (19 July to 21 September). Ottawa: Queen's Printer for Canada.

Iacovetta, F. 1992. "Making 'new Canadians': Social workers, women, and the reshaping of immigrant families." In *Gender Conflicts: New Essays in Women's History.* F.
Iacovetta and M. Valverde (eds.). Toronto: University of Toronto Press, pp. 261–303.

Marx, K. 1977. *Capital. A Critique of Political Economy.* Vol. I. New York: Vintage Books.

Michalowski, M. 1996. "Visitors and visa workers: Old wine in new bottles?" In *International Migration, Refugee Flows and Human Rights in North America: The Impact of Trade and Restructuring.* A.B. Simmons (ed.). New York: Center for Migration Studies, pp. 104–22.

Ng, R. 1988. *The Politics of Community Services.* Toronto: Garamond Press.

Ng, R. 1993. "Sexism, racism and Canadian nationalism." In *Returning the Gaze: Essays on Racism, Feminism and Politics.* H. Bannerji (ed.). Toronto: Sister Vision, pp. 182–96.

Ng, R. 1995. "Multiculturalism as ideology: A textual analysis." In *Knowledge, Experience and Ruling/Relations: Studies in the Social Organization of Knowledge.* M. Campbell and A. Manicom (eds.). Toronto: University of Toronto Press, pp. 35–48.

Panitch, L. 2001. "Reflections on strategy for labour." In *Socialist Register 2001: Working Classes, Global Realities*. L. Panitch, C. Leys, G. Albo and D. Coates (eds.). New York: Monthly Review Press.

Parr, J. 1990. *The Gender of Breadwinners: Women, Men, and Change in Two Industrial Towns, 1880–1950*. Toronto: University of Toronto Press.

Ramirez, J. 1982. "Domestic workers organize!" *Canadian Woman Studies*, Vol. 4, No. 2, pp. 17–19.

Sassen, S. 1988. *The Mobility of Labor and Capital: A Study in International Investment and Labor Flow*. New York: Cambridge University Press.

Satzewich, V. 1989. "Racism and Canadian immigration policy: The government's view of Caribbean migration, 1962–1966." *Canadian Ethnic Studies*, Vol. 21, No. 1, pp. 77–97.

Sharma, N. 1995. "The true north strong and unfree: Capitalist restructuring and non-immigrant employment in Canada, 1973–1933." Master's thesis. Simon Fraser University: Burnaby, B.C.

Sharma, N. 1996. "Cheap myths and bonded lives: Freedom and citizenship in Canadian society." *Beyond Law*, Vol. 5, No. 17, pp. 35–61.

Sharma, N. 1997. "Birds of prey and birds of passage: The movement of capital and migration of labour." *Labour, Capital and Society*, pp. 8–38.

Sharma, N. 2000a. "'Race,' class and gender and the making of 'difference': The social organization of 'migrant workers' in Canada." *Atlantis: A Women's Studies Journal*, Vol. 24. No. 2, pp. 5–15.

Sharma, N. 2000b. "Maintaining the master-servant relationship: Canadian immigration policy." *Kinesis* (May).

Sharma, N. 2000c. "The social organization of 'difference' and capitalist restructuring in Canada: The making of 'migrant workers' through the 1973 Non-immigrant Employment Authorization Program (NIEAP)." Ph.D. dissertation. Ontario Institute for Studies in Education, University of Toronto.

Sharma, N. 2000d. "'Citizenship' and 'difference' as a restructuring device: Canada's Non-immigrant Employment Authorization Program." In *Globalization and its Discontents*. S. McBride and J. Wiseman (eds.). London and New York: Macmillan Press and St. Martin's Press, pp. 129–42.

Simmons, A.B. (ed.). 1996. *International Migration, Refugee Flows and Human Rights in North America: The Impact of Trade and Restructuring*. New York: Center for Migration Studies.

Smith, D.E. 1990. *The Conceptual Practices of Power: A Feminist Sociology of Knowledge*. Toronto: University of Toronto Press.

Smith, D.E. 1995. "About Botanizing." Graduate Seminar Handout. Ontario Institute for Studies in Education, University of Toronto.

Swanson, J. 2001. *Poor-Bashing: The Politics of Exclusion*. Toronto: Between the Lines.

Teeple, G. 1995, *Globalization and the Decline of Social Reform*. Toronto: Garamond Press.

Teeple, G. 2000. "What is globalization?" In *Globalization and Its Discontents*. S. McBride and J. Wiseman (eds.). London and New York: Macmillan Press and St. Martin's Press, pp. 9–23.

United Nations Population Fund. 1993. *The State of World Population.* New York.

Wall, E. 1992. "Personal labour relations and ethnicity in Ontario agriculture." In *Deconstructing a Nation: Immigration, Multiculturalism and Racism in 90s Canada.* Satzewich (ed.). Halifax: Fernwood Publishing, pp. 261–75.

Wang, F. Tsen-Yung. 1998. "Disciplining Taiwanese families: A study of family ideology and home care practices." Ph.D. dissertation. Faculty of Social Work, University of Toronto.

Ward, P. 1978. *White Canada Forever: Popular Attitudes and Public Policy towards Orientals in British Columbia.* Montreal and Kingston: McGill-Queen's University Press.

Wong, L. 1984. "Canada's guestworkers: Some comparisons of temporary workers in Europe and North America." *International Migration Review*, Vol. 18, No. 1, pp. 196–204.

15

Whose Transnationalism? Canada, "Clash of Civilizations" Discourse, and Arab and Muslim Canadians

SEDEF ARAT-KOC

In this chapter, I look at Arab and Muslim communities in Canada as communities under siege. The reality of siege has implications for the ability of individuals and groups not only to safely *cross* borders, but also to live in safety and as equal citizens *within* borders. The specific context in which the chapter is being written necessitates that we pose a set of critical questions about *what* gets to be studied as transnationalism and *how*, as well as about what does and *does not* get to be asked and about *whom*. In the environments of racism, anti-immigration, and anti-multiculturalism that prevail in many Western states today, the transnational identities of many ethnic minorities get discussed as a way to interrogate and question their "loyalties" to the nation-state in which they are living. In the political environment that has prevailed since the terrorist attacks of 11 September 2001 in New York and Washington, DC, identities and loyalties of Arabs and Muslims are especially suspect. What are rarely, if ever, interrogated in this context are the "loyalties" as well as the various material transnational connections—economic, cultural, political, and even military—of dominant groups and of the state.

Since the 1990s in Canada, but more specifically since 11 September 2001, we have been in a period of retreat from multiculturalism and a politics of inclusion. During this period racism has not only intensified, but also, and more important, been legitimated through public discourse and mainstream institutions. Precisely at this historical moment, it may be useful, as Enakshi Dua has suggested, to ask "what it means to recodify immigrants and some ethnic and racialized groups as 'transnationals'" as well as to question "whether the concept itself does not contribute to the *dis*location of immigrants from the nation in new ways."[1] There is a danger that the commonly used term "transnational" will be applied almost exclusively to racialized groups in order once again to question their belonging and even their loyalty. Only since the 1960s in Canada, after long struggles by people of colour, has the concept of "Canadian nation" started to include the histories and present experiences of the many peoples that make up this nation. Only recently have "immigrants"—a term used almost exclusively to refer to racialized minorities—become part of the concept of the nation. However, this inclusion has been of a very fragile and tentative nature. The conceptualization—and isolation—of immigrants and most racialized groups as transnational subjects makes their inclusion even more fragile. Their membership in the nation can be more readily questioned and even their deportation more easily legitimated because it may appear to be ethically unproblematic.[2]

Since 11 September 2001 we have witnessed a renewal of nationalism in Canada. This nationalism is of a transnational kind, a white nationalism confirming *some* Canadians' place in "Western civilization." In Canada, Europe, and the United States this reconfigured notion of the nation, based on a "clash of civilizations" perspective (Huntington 1996), effectively serves to exclude those of Arab and Muslim background from Western nations and Western civilization altogether.

As we contemplate the nature and significance of transnationalism, we must not naively approach or celebrate it as a free movement of people around the world. As Ong (1999, 15) has pointed out, it is essential to remember that the state's continuing role is "to define, discipline, control, and regulate all kinds of populations, whether in movement or in residence." Since 11 September 2001 states obsessed with national security have increased and intensified such disciplinary and regulatory roles. In addition, there has also been a reconfiguration of regulatory regimes at the transnational level through cooperation on "antiterror" legislation, intelligence, and border control.

In Sassen's (1996) frequently cited words, economic globalization is "*denationalize(ing)* national economies," whereas immigration is "*renationalizing* politics" (59, emphasis added). If borders were already important as conceived by "Fortress Europe" and "Fortress North America" when Sassen wrote these words, they have gained more significance since 11 September 2001 in an increasingly securitized world.

For the vast majority of people, borders were already far less permeable than optimistic theoreticians of transnationalism might have assumed. They are even less permeable today. We are living at a time when the transnational ties of Arabs and Muslims—whether social, familial, financial, political, or involved in shaping a general sense of identity—are perceived as suspect, if not directly criminalized. While the transnational ties of some groups may be increasing, those of Arabs and Muslims are subject to intense forms of surveillance.

There are empirical as well as ethical difficulties inherent in doing research in communities under siege. Rather than keeping "the gaze" on the diasporic Arab and Muslim "community" in the period since 11 September 2001, this chapter returns "the gaze" and raises questions about the often unscrutinized and unnamed dominant transnationalisms in Canada, transnationalisms that are not just reconfiguring Canadian national identity and its boundaries, but also changing "national" institutions, such as borders, immigration, justice, and the military. I start the chapter with a description of the siege that has gripped Arabs and Muslim Canadians since the attacks on New York and Washington, DC, and the effects of this siege on their identities and transnational connections. I then focus on the nature of the dominant transnationalisms that create and reproduce the siege. I end with a discussion of "alternative transnationalisms"—that is, trans-ethnic and non-ethnic solidarities that are developing to challenge the current national and transnational order.

The category "Arab and Muslim" is problematic because it often does not recognize the complexity and the heterogeneity of the categories "Arab" and "Muslim" but conflates the two. As Suad Joseph (1999) argues, representations of Arabs, Muslims, and Middle Easterners are underpinned by multiple conflations. Even though the categories sometimes overlap, the conflated category makes the diversity of Arabs invisible. Arabs are not just Muslims, but also Christians and Jews. In popular representations in Western countries, "Arab" may be used to include other ethnic groups, such as Turks, Armenians, Persians, or Roma—groups with different languages and ethnic or national backgrounds who do not identify as Arab. Conflations overlook that the category "Muslim" encompasses many ethno-cultural and linguistic groups, the majority of whom are neither

Arab nor Middle Easterner but Indonesian, Malaysian, Filipino, Sudanese, Indian, or Chinese. Joseph argues that in the American context, the inaccurate conflations *erase* difference in order to "serve the *creation* of another difference: the difference between the free, white, male American citizen and this constructed Arab" (260, emphasis in the original).

However, despite the inaccuracy of the conflations and their racist connotations, I make references to the category "Arabs and Muslims." As the category has become "real" socially and politically, I do so in making references to specific racialization. Discourses on "Arabs and Muslims" have become "real" in the subjection of a category of people to specific forms of racialization and *political* designation in North America. The category, which has been politicized by Palestinian resistance to Israeli occupation since 1967, gained special significance in the 1990s during and after the Gulf War and in the post–Cold War political discourses in search of a new enemy of "the West" or globalizing capitalism. Since 11 September 2001 the category, as a concept of racialization, has been raised to the status of "common sense" in depictions of "the enemy," resulting in attacks on many non-Arab and non-Muslim people, often of South Asian background, who are thought to "look like Muslims."

TRANSNATIONALISM AND IDENTITY FOR ARAB AND MUSLIM CANADIANS

The heterogeneity and complexity of the categories "Arab" and "Muslim"—separately and conflated—suggest that it is impossible to form a simple set of assumptions about a type of identity among Arabs and Muslims. There are further complications in conceptualizing the identity of Arabs and Muslims in Canada because the already existing diversity in people's backgrounds is multiplied by different experiences and adaptations to living in the diaspora. The current climate of intensified racialization and vilification also yields different types of responses from people and different ways of *being* Arab and Muslim in Canada.

Different authors have articulated why it is analytically essential to use a transnational perspective in studying immigrants. Spivak (2000, 354) points out the shortcomings of "focusing on the migrant as an effectively historyless object of intellectual and political activism." She argues that one cannot simply treat the postcolonial migrant as a "blank slate," pretending that he or she can be analyzed simply using a class-gender-race calculus that begins and ends in the First World metropolis. While I agree with Spivak that the migrant is not a blank slate, I also think that it is important to conceptualize the history and connectedness of the migrant in non-essentialist ways. When Ong and Nonini (1997a, 327) argue that identities are "constituted through transnational systems," they also offer an approach to identity "as a politics rather than as an inheritance . . . as fluidity rather than fixity, as based in mobility rather than locality, and as the playing out of these oppositions across the world."

For the majority of people, racialization, rather than a common national or religious identity, provides the basis for identification and organizing. As one Muslim has stated: "For me, my whole life in now demarcated by 9/11. It is not pre-9/11 and post-9/11 . . . The attacks were a break with our past as Muslims. Sept. 11, 2001 is not a defining moment in the history of North American Muslims but *the* defining moment" (quoted in Safieddin 2003, B2, emphasis in the original).

As W.E.B. Du Bois (cited in Winant 1997) brilliantly argues, race operates both to assign and to deny people their identities. Hours after the horrific events of 11 September 2001, Arab and Muslim Canadians found themselves racialized in new ways. No longer considered just "exotic" subjects or bearers of irrational traditional cultures, they were now "the enemy." The words of a New York lawyer summarize how many Arab and Muslim Canadians felt: "Before last week, I had thought of myself as a lawyer, a feminist, a wife, a sister, a friend, a woman on the street. Now I begin to see myself as a brown woman who bears a vague resemblance to the images of terrorists we see on television . . . As I become identified as someone outside the New York community, I feel myself losing the power to define myself" (quoted in Deaux 2001).

In an environment where Islam became loaded with political connotations, people of a Muslim background who had never seen religion as a central part of their identity suddenly found others imposing this identity on them:

> For me, Islam was not a factor in structuring my identity at an early age. My parents were both secularists. I wasn't brought up according to Islamic tradition . . .
>
> Sept. 11, 2001 all of a sudden changed that. People started to identify me as a Muslim. They would ask: You are from the Middle East, are you Muslim?
>
> For a while, it was incredibly confusing. It was not a question that I could answer simply. It needed a lot of qualification that people did not have the patience for . . .
>
> It was as if I woke up one day to find that a special type of a Muslim identity was imposed on me. (Quoted in Safieddin 2003, B3)

Racialization is not a recent phenomenon for Arab and Muslim Canadians. In certain periods, such as during the Gulf War, they experienced an intensification of the white gaze and criminalization of their communities as the "enemy within." There were several prevailing racist discourses in Canada prior to 11 September 2001—some used for Arabs and Muslims and some not. These discourses frame immigrants or people of colour as "taking advantage of Canada," as "bogus refugees," and as "welfare cheats." According to Thobani (2004b), after 11 September 2001 discourse on security became *the* racist discourse. Since security is an issue that concerns everyone, it makes the new racist discourse an especially powerful one.

What has been new for Arab and Muslim Canadians since 11 September 2001 is not the experience of racism but its growing public legitimacy, spread, and mainstreaming in all major institutions, from the media to law and policy. Overt acts of violence and expressions of hatred in civil society in the aftermath of 11 September 2001 were soon followed by government "security" measures that not only justify, but also further fuel, racialization and a suspicion of most Arab and Muslim Canadians. Once considered an illegitimate practice, racial profiling has not only become de facto policy, but also gained significant popular legitimacy. A recent survey by Ekos Research Associates revealed that, at 48 percent, close to half of Canadians find it acceptable that "security officials give special attention to individuals of Arabic origin" (Alghabra 2003, A24).

Although negative and politically loaded portrayals of Arabs and Muslims in the media are by no means new or rare, this racialized group has been defined as the "enemy within" since 11 September 2001. George Jonas of the *National Post*, for example, fanned the flames of hatred that were rapidly spreading when he argued that "not all the terrorist caves are in Afghanistan . . . some are in Quebec and Ontario" (quoted in Kutty and Yousuf 2002, A13). The view that Arabs and Muslims belong to a radically different civilization—or perhaps even to a different species!—with a very different set of values than those of "Canadians" has become the basis for news stories and commentaries. In the media—in contrast to white Canadians' portrayal of themselves as rational, civilized, liberal, democratic, and peaceful—depictions of Muslims and Arabs as "extreme, vengeful, irrational, suicidal and fanatical people pathologically predisposed to violence with an incorrigible mindset" (Alghabra 2003, A24) have become regular fare.

Immediately following 11 September 2001 there were countless cases of harassment, intimidation, and violence directed at Arabs, Muslims, and those who were thought to "look like Muslims." Arson at mosques and Sikh temples was accompanied by physical and verbal attacks or by dirty looks directed at Muslims, Arabs, Sikhs, and Hindus. Darker people and those with visible signs of religion or ethnicity—such as women wearing the hijab—bore the brunt of attacks. The Canadian Islamic Congress advised Muslims to stay home from school and to take measures to avoid harassment. It also urged Muslims to avoid crowded areas "where a mob mentality may develop" (Small and DeMara 2001, A2). Perhaps even more disturbing than the actual harassment, intimidation, and violence that took place was that many people who experienced these assaults were so distressed and so insecure about their place in Canada that they did not report them to the police (Raja Khouri, national president of Canadian Arab Federation, personal interview).

Also, Muslim and Arab Canadians faced a number of pressures that differed from blatant attacks on their physical safety. Some business owners and corporate employees were pressured to change their Arab names to more non-Arab or "Anglo-sounding" names (Jamal 2002, 46; Khouri, personal interview). In one case, a business owner decided not to personally conduct the more public aspects of his business and instead sent a "white Canadian" to meet with his clients (Jamal 2002, 46)

Although some political leaders made political statements appealing to "Canadians"—a category that did not seem to include "Arab and Muslim Canadians"—to stop the violence and harassment, they seemed to give double and conflicting messages, as they themselves were participating in a discourse of clashing civilizations and moving rapidly toward legitimization of racism through institutionalization of racial profiling and new anti-terrorism, immigration, and border policies.

A recent survey conducted by the Canadian Arab Federation (2002) in the Arab Canadian community paints an unsettling picture of where Arabs see themselves in the larger Canadian society. According to the survey, 41.3% of the respondents believe that Canadians "don't like Muslims," and 84% believe that Canadians think Muslims and Arabs are violent (17). Although these responses are subjective, they clearly raise questions about Arabs' and Muslims' sense of reception and belonging in Canada. The respondents to the survey almost unanimously agreed that "in general Canadians know very little about Arab culture" and that "what Canadians know about Arab culture stems from negative stereotypes and myths" (17). Thirty-eight percent of the respondents said that they are made uncomfortable by the way that white Canadians look at them, and almost half said that they encounter racism in daily interactions (18).

For 28.1% of respondents, perceptions of racism in the media and experiences of racism in schools and workplaces were accompanied by identification of institutional racism as a problem.[3] The examples that they gave included detentions following 11 September 2001 and interrogations by the Canadian Security and Intelligence Service (CSIS), the introduction of Bill C-36 (the antiterrorism bill), the Immigration Department, and employment in the public sector (Canadian Arab Federation 2002, 18). Nearly three-quarters of the respondents, at 73.9%, strongly or somewhat disagreed with the statement that "the Canadian government is concerned with what Canadian Arabs want from it." The level of dissatisfaction both with the federal government and with foreign policy was particularly high (21).

These responses are especially ironic when juxtaposed with the responses that Arab immigrants gave about why they chose Canada as their immigration destination: 71.9% of the respondents said that it was for Canada's human rights and freedoms, and 51.5% said that it was for its multiculturalism (Canadian Arab Federation 2002, 13). Given the expectations of Arab and Muslim immigrants in Canada, it is not surprising that the intensification of racialization and demonization following 11 September 2001 was a shock to them: "Anxiety, fear, alienation, marginalization, betrayal and disillusionment: This is how Sept. 11, 2001 and its aftermath have left Arab and Muslim Canadians feeling—indeed reeling" (Khouri 2003).

The implications of the racialization and victimization of Arabs and Muslims are very serious not just for members of the communities, but also for the larger society. Reflecting on these implications for Canadian multiculturalism, the current national president of the Canadian Arab Federation, Raja Khouri, is grim:

> Our country has effectively engaged in an exercise of self-mutilation: stripping away civil liberties it holds dear, trampling on citizens' rights it had foresworn to protect, and tearing away at its multicultural fabric with recklessness.
>
> Arab Canadians are today convinced that there is a bigger threat to our way of life from the security agenda than there is from terrorism itself.
>
> The question that remains is: Given that multiculturalism is premised on the equal treatment and respect of all citizens, will multiculturalism survive the security agenda? (Khouri 2003)

Recent research shows that even though children of Arab Canadians born in Canada were "Americanized . . . their Arab identity has been raised as a result of [the] events" of 11 September 2001 (Jamal 2002, 47). I suggest that for Arabs and Muslims in the diaspora, the environment during this period has created a specific *positionality* but not necessarily an *identity* with specific outcomes. This positionality is created in an environment of intensified racism that, partly due to the commonly used "clash of civilizations" discourse, posits every Arab and Muslim as guilty by association, thereby increasing Arabs' and Muslims' sense of exclusion from mainstream notions of what/who constitutes "Canada." The "clash of civilizations" discourse and logic constructs a monolithic conception of both "the West" and some of the communities that live within it. The implications of such monolithic conceptions and of the notion of a naturalized, inevitable "clash" are catastrophic for some diasporic communities who suddenly find themselves outside the borders of what is conceived of as "the West."

Despite the potent role that intensified racialization and exclusion have played in the development of a sense of "we"—forcefully and/or voluntarily—among Arab and Muslim Canadians, this community's racialized positionality has not automatically translated into an identity. Rather, this positionality has given rise to the formation of different identities.

For some, a process of intense racialization has led to identification with others of the same ethnic or religious background with whom one might not have had much in common before: "Prior to 9/11, I never identified with women who wear the veil . . . But all of a sudden, I take the public transportation in the week following the attacks, and the driver allows me in but closes the door in the face of a veiled women waiting at the bus stop. This happened a number of times. At a certain point I started thinking it could have been my cousin or a member of my family, some of whom cover their hair with a scarf" (quoted in Safieddin 2003, B3).

For others, the connotations of the category "Arab and Muslim" were too painful to lead to identification with their racialized coethnics. Some Arabs interviewed by Jamal (2002, 49) regrettably observed that "misunderstanding of [their] culture and prevalent stereotypes have led to a confused Arab identity. Fear has led to people suppressing their Arab identity." Instead of self-identifying as Arab or Muslim, some—especially those whose social class, appearance, accent, or level of assimilation enables them to "pass" as non-Arabs/non-Muslims—have attempted to distance themselves from ethnic identity and community. Some Christian Arabs have tried to distance themselves from their vilified Muslim coethnics—for example, by wearing large, visible crosses to emphasize non-Muslim identity (Khouri, personal interview). As one of Jamal's (2002, 49) interviewees notes, "some . . . retract and retrench from society while others throw away their Arab culture and become fully 'Canadian' by changing their names, habits, and values to the extreme—and neither is happy."

Although some Arabs and Muslims have continued to nurture multiple identities and multiple connections to different communities, for others intensified racialization and othering has led to a rigidness and defensiveness regarding essential identities. Those who are religious Muslims have found themselves especially isolated. Demonization of their religion has been particularly painful, as religion has long been a central part of their identities.

Many Muslims in the diaspora, especially refugee survivors of fundamentalist regimes, find themselves in a bind. Although they are inclined to continue to articulate a critique of fundamentalism, they also find themselves under a new type of attack based on racism and Islamophobia in Canadian society (Tahmasebi 2004). One response to this dilemma has been to emphasize the universal in condemning all forms of hatred and violence. As one Muslim has explained, "what became clearer to me post-Sept. 11 is that everyone should be against hatred, whatever it is, if it is happening from the pulpit or from the government or from ordinary people on the street" (quoted in Safieddin 2003, B2).

The environment of suspicion in which Arab and Muslim Canadians have found themselves since 11 September 2001 has served to create a climate of intimidation and fear. Several members of Arab and Muslim communities have made references to internment in describing what this environment has felt like: "Ghettoization since 9/11 became clear. In reality we could have been physically interned. Instead we have been interned by fear—psychological internment" (quoted in Jamal 2002, 48).

Intimidation has led to different responses by individuals in Arab and Muslim Canadian communities. For some, intimidation has resulted in depoliticization. They have found security in the image of the "good Muslim" and "good immigrant," which has come to be defined in this context as someone who makes generalized negative statements—approaching racism—about Islam and Arabs, gives unconditional support to everything the Canadian government does, and is forever grateful to Canada.

Despite the environment of vilification and intimidation, many Canadian Muslims and Arabs have felt compelled to express their disagreement with the war in Afghanistan, in which Canada continues to participate. Jehad Aliweiwi, the executive director of the Canadian Arab Federation, has observed an intense scrutiny of the loyalties, actions, and beliefs of Arab and Muslim Canadians since 11 September 2001, which he regards as amounting to "a new form of internment." He describes the costs of dissent for people from communities already suspected of disloyalty: "We're perceived as the enemy and as responsible collectively . . . And now, we're seen as guilty because we don't support the bombings. It's a frightening position to be in" (quoted in Mitchell 2001).

Despite the intimidation, some people consider it their duty as Canadian citizens to express what they see as the truth, even if doing so is risky. As Mohamed Elmasry, the president of the Canadian Islamic Congress, states: "Some people feel it may be time to have a low profile and just support the government, no matter what . . . Others feel it is their duty to be a good citizen and voice their opinion" (quoted in Mitchell 2001).

DOMINANT TRANSNATIONALISMS: TRANSNATIONAL IDENTITY OF THE CANADIAN STATE AND "ORDINARY CANADIANS"

> It is a banal fact of contemporary existence that economic forces, communication systems, military interventions, and ecological disasters continually transcend nation-state boundaries, yet state authorities remain deeply suspicious of all international movements, loyalties, and relationships they cannot regulate.
>
> —Asad 1993, 266

There is an irony in the political discourses that have dominated the mainstream since 11 September 2001. Although there has been an inflation in nationalist discourses that interrogate the belonging and loyalties of suspect "ethnics," mainstream "Canadian identity" is more than ever defined in transnational terms and specifically in relation to the United States. The disproportionate focus in the literature on the transnationalism of—often racialized—minorities is misguided. This focus tends to ignore or make invisible the transnationalism of the dominant racial/ethnic groups as well as the transnationalism of the state. The latter has become particularly important in Canada since 11 September 2001, as Canada's already close ties with the United States—in terms of investment, trade, culture, tourism, and so on—have not only solidified, but also reconfigured and redefined, a number of other areas, such as national identity and the harmonization and integration of military policy, security legislation, and immigration and border controls. In the next section, I start by focusing on the redefinition and reconfiguration of Canadian identity as denoting one's belonging in "the West." These changes in Canadian identity, which are constantly fed and refuelled by mainstream media, play a significant role in making possible the racialization and othering of Arab and Muslim Canadians. More important, they continue to legitimize institutional racism.

"Canadian Identity" as "Western" Identity

Even though both President George W. Bush in the United States and former prime minister Jean Chrétien in Canada declared that Islam was not the enemy, the mainstream media in both countries continued to write as though it were. In specifically targeting Arabs and Muslims not just externally, but also in the diaspora, security policies in both countries also strengthened the idea that this image corresponded to reality. The entire discursive framing of the attacks of 11 September 2001 was based on the notion of a "clash of civilizations." After the attacks on New York and Washington, DC, Samuel Huntington's book with this phrase in its title was praised as visionary, ingenious, and brilliant. Five years after its publication in 1996, it rose to the bestseller list. Its thesis was taken up wholeheartedly by the media and came to be used as the common-sense explanation for what had happened.

With the conclusion of the Cold War at the end of the 1980s, some American intellectuals deeply embedded with the state started actively redrawing maps of international conflict and identifying who the new enemy would be in the new world order being created. The concept of a "clash of civilizations" was first introduced by Bernard Lewis to explain what he saw as the conflict between political Islam and "the West" (Lewis 1990). It was then taken up by Samuel Huntington—a long-time Cold War warrior—who initially used the term with a question mark in the title of a 1993 article. Huntington expanded the concept into a universal thesis on the state of the world after the end of the Cold War. In the title of his 1996 book, Huntington got rid of the question mark and suggested that the world was now divided along *civilizational* lines—as opposed to the ideological ones of the Cold War—based on ethnic, cultural, and religious differences. Defining eight major civilizations in the world today—"Western, Confucian, Japanese, Islamic, Hindu, Slavic-Orthodox, Latin American, and 'possibly African'"—Huntington (1996, 128) highlighted the increased significance of ethnic identifications in the post–Cold War world and expressed his expectation that the most serious threats to "the West" would come from the "Islamic" and Chinese civilizations. Huntington articulated the nature of civilizational conflict in this new era in military and potentially catastrophic terms: "In a world where culture counts, the platoons are tribes and ethnic groups, the regiments are nations, and the armies are civilizations."

Huntington's book has been criticized from a number of angles. He uses a very monolithic conception of culture and "civilization identity" that ignores their diversity, internal contradictions, and historical variability. He conceptualizes cultures and civilizations as sealed-off, isolated entities whose historical relationship to each other consists of wars and conflict rather than exchange, interaction, and cross-fertilization. This perspective on cultures denies the significance of a shared history between cultures, even ignoring the recent interconnections through colonization, imperialism, and globalization. Huntington's simplified and monolithic perspective on cultures leads to an exaggeration, absolutization, and mystification of cultural differences. Exaggeration of cultural differences implies that—at least some—cultural differences implies that—at least some—cultural differences are irreconcilable, leading Huntington to naturalize conflict (Arat-Koc 2002b).

It is interesting to observe how rapidly the notion of civilizational clash became the dominant framework in which political leaders and the media interpreted developments leading up to and following 11 September 2001. Columnists in mainstream media made constant references to how "they"—used in a very slippery way to refer sometimes to the terrorists and sometimes to the culture to which they belonged—hated "our" freedoms, democracy, human rights, and women's rights. The article "Why Do They Hate Us So Much?" by *Globe and Mail* columnist Margaret Wente (2001, A13) represents a discourse that was—and continues to be—very common in the media. Wente rejects any reflection on "root causes" stemming from a history of the relations between countries. Opting instead for the essentialism of the cultural-clash perspective, she writes—with the confidence of those who think they speak common sense—that "we" in "the West" have "a culture of peace," while "theirs" is "a culture of violence." Referring to suicide bombers, she declares: "The poison that runs through the veins of the suicide bombers does not come from America. It comes from their culture, not ours. The root causes are in their history, not ours." By the end of October 2001, some columnists were predicting that the war in Afghanistan would only be an "opening battle" in a long war that would "spread and engulf a number of countries in conflicts of varying intensity." They declared that this war would "resemble the clash of civilizations everyone had hoped to avoid" (Kristol and Kagan, quoted in Alam 2003).

The Canadian media played a very significant role in creating an imaginary conception not just of the nation, but also of transnational alliances (Thobani 2002). As the "clash of civilizations" thesis began to be employed daily to define "us" and "them," the definition of "Canada" and Canadian identity rapidly evolved into a transnational one emphasizing Canada's and its citizens' place in "the Western civilization" (Arat-Koc 2002a, 2003; Thobani 2002, 2004a). Of importance here is that this notion of "the West" excluded not only other civilizations, but also the histories and cultures of "non-Western" diasporas living in "the West." Thus the redefinition of Canadian identity as part of "the West" implied a rewhitening of Canadian identity after decades of multiculturalism. In

Canada, given its geography and its history, belonging to "the West" also came to mean strongly identifying with the United States in particular. The repeated statement "we are all Americans now" represented not just an expression of temporary solidarity with an apparently victimized, terrorized neighbour, but also an identification with a white, imperial power.

The new borders of Canadian identity were so rigidly drawn that in the immediate aftermath of 11 September 2001, there was no tolerance among the new "ordinary Canadians" for dissent against strong identification and unquestioning cooperation with the United States. In this environment, Sunera Thobani, the former president of the National Action Committee on the Status of Women, was demonized for remarking that the United States, through its foreign policy, has blood on its hands and that we need to be able to extend the same sympathy shown for the victims of the terrorist attacks of 11 September 2001 to victims of American aggression in many different parts of the world. Not even some "mainstream Canadians" were spared questions about their degree of belonging given this newly defined national identity. A letter to the editor, for example, called Alexa McDonough, then the leader of the New Democratic Party, "un-Canadian" for her opposition to Canada's participation in the war in Afghanistan (Lafond 2001, A25).

Institutionalization of "Western" Identity

> If some "ethnics" were showing the sort of loyalty to another country as Canadian right-wingers are to the United States, they would have been branded traitors to Canada.
>
> —Siddiqui 2003, F1

A significant number of Canada's business, political, and media elites have been pushing hard since 11 September 2001 for an integration of Canada's military, borders, and policies with those of the United States. These elites, who have played an active role in redefining Canadian identity in transnational terms, are also actively pursuing a materialization of this new identity in the form of concrete institutions.[4] Even though I refer to the constitution and institutionalization of this identity as "transnational," it is important to recognize that this is not a transnationalism of equal partners but one based on an agenda defined specifically by the United States and on a union in which Canada is erased. This is a reality that even the proponents of this integration recognize: "Our condition of virtual sovereignty is like that of all the member states of the European Union. With one difference. The political construct that is developing here isn't a comparable North American union. It's just an American union" (Gwyn 2001b, A13).

The discourses used to justify the desire for integration with the United States often conflate business interests and concerns about "security" (Canadian Council of Chief Executives 2004). State discourses explaining new forms of integration with the United States also accept and naturalize this conflation by speaking about trade and security in one breath. For example, the Smart Border Declaration, signed by Canada and the United States on 12 December 2001, states:

> The terrorist actions of September 11 were an attack on our common commitment to democracy, rule of law and a free and open economy. They highlighted a threat to our public and economic security . . .
>
> Public security and economic security are mutually reinforcing. By working together to develop a zone of confidence against terrorist activity, we create a unique opportunity to build a smart border for the 21st century; a border that securely facilitates the free flow of people and commerce; a border that reflects the largest trading relationship in the world. (Government of Canada 2002)

Transnationalization of Justice or Transnationalization of Torture?

As demonstrated by the well-publicized case of Maher Arar, a Canadian citizen of Syrian origin, it appears that in situations when the Canadian intelligence agencies find the task of preparing a case against suspected terrorists under the antiterrorism legislation cumbersome, they may be downloading the work of detention and questioning to foreign governments. In what some are calling "torture by proxy" (Kutty 2004), Canadian intelligence agencies are accused of passing erroneous information about Arar to American authorities, who in turn are accused of Arar's removal to Syria, where he was tortured. In a more recent case, another Canadian citizen, Dr. Khawaja, has argued that he was detained—but not tortured—by Saudi authorities at the request of Canadians. Dr. Khawaja also asserts that the Canadian authorities asked his interrogators to ensure that he would not be able to contact the Canadian consulate in Riyadh (Kutty 2004).

The Arar and Khawaja cases have received some attention in the media, as both men are Canadian citizens. As Wright argues (2003, 2004), much less known and less controversial are the cases of noncitizens, including those with no legal status in Canada. So far, there are at least five known cases of noncitizens detained indefinitely without charge on security certificates. As they are held on the grounds of secret evidence, their lawyers have no access to the evidence against them and therefore no effective way to prepare a defence. There is also the case of more than twenty young Pakistani and Indian nationals arrested in August 2003 under Project Thread. Even though there has been no case for charges, they were detained until most were deported.[5] The question some ask in these cases is, "If there is evidence then why not charge, arrest and prosecute them in Canada to the full extent of the law?" (Kutty 2004). Macklin (2001, 397) may have the answer: when Bill C-36 was introduced, she predicted that criminal prosecution under the antiterrorism law would be difficult, leading her to conclude that in those cases involving noncitizens, the state would use its powers under Bill C-36 to investigate individuals but then issue security certificates under the Immigration and Refugee Protection Act, as it would be "easier to deport than to imprison" suspects.

These and other similar cases reveal a disturbing new form of transnationalization in the treatment of suspected terrorists by Canada. This may be part of a new global pattern of human-rights violations led by the United States. It includes offshore prisons, cross-border arrests that verge on kidnappings, and the rendition of terror suspects to countries where they face torture.

ALTERNATIVE TRANSNATIONALISMS? TRANSETHNIC AND NON-ETHNIC IDENTIFICATIONS AND SOLIDARITY

> We are not lumps of clay, and what is important is not what people make of us but what we ourselves make of what they have made of us.
> —Jean-Paul Sartre, Saint Genêt

In a world that is increasingly multicultural and transnational, for the state and dominant ethno-cultural groups to approach diversity in terms of a "clash of civilizations"—à la Huntington (1996)—would have catastrophic consequences. Such an approach could, in the least, be used to justify deportations, exclusion, and criminalization and could potentially lead all the way to an all-out holocaust to exterminate those identified as the civilizational "others." In this section, I argue that though such apocalyptic possibilities are not inconceivable, there have been significant developments in Canada and abroad of a positive kind that constitute a resistance and challenge to the dominant transnationalisms mentioned in the previous section. I suggest that the kind of transnational, transethnic, and non-ethnic bonds of solidarity established in response to a climate of war, hatred, and racism promise to subvert the logic of a "clash of civilizations" discourse.

In the past decade, several important concepts have been developed to deal with the complexity of identities produced in the context of transnationality. Two of these concepts are hybridity and diaspora. However, even though these concepts represent important attempts to complicate and de-essentialize identity, they do not go far enough as they continue to privilege "culture" as the central element of identity and community (Anthias 2001, 2002). Specifically, the concept of diaspora privileges a notion of "origin" and thus "fails to pay adequate attention to *transethnic,* rather than transnational, processes" (Anthias 2002, 37, emphasis added). Moreover, I would add, it fails to account for *non-ethnic* processes in making sense of the new identities that emerge.

While an atmosphere of racism, and specifically a discourse of "clash of civilizations," attempts to fix racial, religious, national, and civilizational identities, a politics of solidarity—organized around principles of antiracism, anticolonialism, anti-imperialism, peace, or civil rights—has the potential to counter this fixing of identities, leading to the emergence of new democratic subjects.

In a world that has been shaped by "clash of civilizations" discourse since 11 September 2001, there are enormous pressures not just on minorities, but also on the dominant ethnic group, to adopt a fundamentalism of identities. As Ella Shohat (2002, 469) succinctly puts it, "war is the friend of binarisms, leaving little place for complex identities." Thus there is a danger that the widespread acceptance of "clash of civilizations" discourse by "ordinary Canadians" and its internalization by isolated and alienated Arab and Muslim Canadians can turn the discourse into a self-fulfilling prophecy by creating mutual distrust and mutual isolation. The tendency now of some "ordinary Canadians" to see their neighbours with suspicion and fear has been accompanied by a great sense of disappointment and betrayal among Arab and Muslim Canadians who previously enjoyed a sense of belonging based on the professed Canadian ideals of multiculturalism and democracy:

> Our failure as a society in this regard was in not sending a clear signal to the contrary—society did not come to the aid of this maligned minority . . . by and large, Arab and Muslim Canadians were left on their own, having to explain themselves and prove their loyalty; defend their religion and demonstrate its goodness; and at times hide their ethnicity and deny their heritage in a bid to escape scrutiny.
>
> The effect on our communities is that, like our Japanese-Canadian counterparts during World War II, we, too, have become victims of psychological internment. In the meantime, our mainstream institutions, including governments, simply looked the other way. (Khouri 2003)

However, despite the sense of disappointment and betrayal in the community, there have been unprecedented levels of mobilization among Arab and Muslim Canadians. According to Raja Khouri (personal interview) of the Canadian Arab Federation, marginalization has forced some people in the community to become more politicized and, ironically, has led them to be more integrated politically. Activism among Arab Canadians has increased substantially. Participating in interfaith activities and working in civil-rights, antiwar, and antiracist coalitions with other groups, some Arab Canadians have also started to work toward developing a voice in some of Canada's political parties.

There is an indeterminacy about the transnational politics of ethnic groups and their implications for the larger society. Analyzing conflicts in Britain surrounding both the first Gulf War and the Rushdie affair—that is, the controversy and death threats to Salman Rushdie following the publication of his book *The Satanic Verses*—Pnina Werbner (2000, 309) observes that "far from revealing ambiguous loyalties or unbridgeable cultural chasms, British Muslim transnational loyalties have challenged the national polity . . . to explore new forms of multiculturalism and to work for new global human rights causes."

Already reeling from the demonization and attacks that came in the aftermath of 11 September 2001 and during the bombing of Afghanistan, Arab and Muslim Canadians again braced themselves for hostilities in the period preceding the Iraq War. Some Iraqi Canadians even feared internment if Canada participated in the war. There was great relief in the community when the Canadian government decided against participation. Just as important as the decision of the government was the sense of solidarity that developed as a result of the antiwar demonstrations in Canada and worldwide. During the heavily attended antiwar demonstrations in Canada, Arab and Muslim Canadians felt a sense of belonging for the first time since 11 September 2001 (Khouri, personal interview). Because of enormous diversity among the organizations that took part in demonstrations, teach-ins, and various kinds of meetings—including ethnic as well as church, labour, antipoverty, and women's groups—rather than clashing as essentialized civilizational subjects, as had been expected, participants learned from each other, made connections between the various issues that were concerns to different groups, and developed new political subjectivities. In a sense, the emerging picture looked like a multiculturalism based more on intercommunal coalitions than on monoculturalist identities, a configuration proposed by Ella Shohat (1995, 177): "Rather than ask who can speak . . . we should ask how we can speak together . . . How can diverse communities speak in concert? . . . In this sense, it might be worthwhile to focus less on identity as something one 'has,' than an identification one 'does.'"

The worldwide demonstrations against the Iraq War on 15 February 2003, which involved a total of 12 million people, represented a historic moment of transnational solidarity. Subverting the very logic of the discourse of civilizational conflict, these demonstrations were a show of solidarity against imperial power gone out of control and beyond accountability.[6]

Accompanying the widespread violence and harassment of Muslims and Arabs that followed 11 September 2001 were many expressions of support and solidarity from interfaith groups. Some church groups even offered to protect mosques. Although organizations such as the Canadian Muslim Civil Liberties association received many hate e-mail messages, they received five to six times as many messages of support (Kutty and Yousuf 2002, A13).

Several groups expressed support for notions of a Canadian nation that are different from those based on a supposed "clash of civilizations." As early as October 2001, having observed that Arab and Muslim Canadians were being blamed for terrorism and increasingly victimized, the Canadian Federation of Nurses Unions expressed its solidarity by declaring "we are all Muslim Canadians until this crisis is over" (Connors 2001).

Shared experiences of racial profiling have sometimes brought different ethnic and racialized communities together. In the United States, soon after 11 September 2001, there were solidarity demonstrations against the racial profiling that they predicted would follow. The National Asian Pacific American Legal Consortium invited Arab Americans to join it at the National Japanese American Memorial (Stein 2003). In Canada, on the issue of racial profiling, there have been some attempts to express mutual solidarity and to establish transethnic ties between black communities and Arab Muslim groups.

Writing a few weeks after 11 September 2001, Brah (2002, 41–42) made some observations about what she sees as the emergence of a novel transnational political subject. Listing a diverse range of organizations that participated in rallies for the International Day Against War and Racism in San Francisco and Washington, DC, on 29 September—including Women for Afghan Women; Mexican Support Network; Collective for Lesbian, Gay, Bisexual and Trans-Gender Rights; Pastors for Peace; Action for Community and Ecology; Black Voices for Peace; Healthcare Now Coalition; and AFSCME, of the labour movement—Brah sees "a new collective subject being constituted, as speaker after speaker made connections across many different experiences, forms of differentiation and social divisions." Finding this political subject "simultaneously diasporized and localized," Brah suggests that the current conjuncture creates a space for "imagining and negotiating alternative transnational conceptions of the person as 'holder of rights' that are distinct from the current notions of citizenship."

ACKNOWLEDGMENTS

I would like to thank Enakshi Dua, Mustafa Koc, Mary-Jo Nadeau, and Cynthia Wright for discussing aspects of this chapter with me. I am grateful to Raja Khouri, the national president of the Canadian Arab Federation, for an interview and to Audrey Jamal, the executive director of the Canadian Arab Federation, for sharing some of the federation's documents as well as her MA thesis with me.

NOTES

1. Dr. Enakshi Dua, York University, in personal conversation, April 2004.
2. As demonstrated by calls in April 2004 for the deportations of members of the Khadr family—accused of links to al-Qaeda—and for denial of their citizenship rights. Despite the existing legal status of citizenship in Canada, even the membership of some citizens can be open to questioning.
3. Among parents, 11.8% mentioned that their children were teased and called names by fellow students, and 13.2% indicated differential treatment by teachers or school administrators. Examples given from the survey included one case where a teacher called a child "a little terrorist" (Canadian Arab Federation 2002, 18).
4. At the time of writing, Canadian prime minister Paul Martin was planning a trip to the United States. A few weeks before his visit, the Canadian Council of Chief Executives, a group made up of the leaders of Canada's biggest corporations, released a "discussion paper" and headed to Washington for a discussion of their vision of the "Canada-United States Partnership" with some high-level American politicians and bureaucrats. The council proposes a "partnership" with the United States that includes "a common security agenda, joint military institutions, an integrated energy market and harmonized tariffs and regulations" (Goar 2004, A18; Canadian Council of Chief Executives 2004).
5. See the website of Project Threadbare, a group formed in response to the arrest and detention of the South Asian men: http://threadbare.tyo.ca.
6. There are numerous examples of diverse groups being brought together by this kind of organizing. For instance, the groups that were starting to organize a protest against a planned visit by George W. Bush to Ottawa in May 2003—a visit that was subsequently cancelled—were the Anti-Capitalist Community Action and the Committee for Peace in Iraq, groups that were themselves very diverse internally.

REFERENCES

Alam, M.S. 2003. Is this a clash of civilizations? *Counterpunch*. http://www.counterpunch.org/alam02282003.html.

Alghabra, O. 2003. A chance to lift the veil of ignorance about Arabs. *Toronto Star*, 17 October, A24.

Anthias, F. 2001. New hybridities, old concepts: The limits of culture. *Ethnic and Racial Studies* 24 (4 July): 619–41.

——. 2002. Diasporic hybridity and transcending racisms. In F. Anthias and C. Lloyd, eds., *Rethinking anti-racisim: From theory to practice*, 22–43. London: Routledge.

Arat-Koc, S. 2002a. Clash of civilizations? Anti-racist, multicultural feminism and diasporic Middle Eastern identities post September 11th. Paper presented at the conference "Unsettling Imaginations: Towards Re-configuring Borders," Vancouver, 22–24 March.

——. 2002b. Orientalism and the cages of civilizations: Re-articulations of nation, race and culture in Canada, post September 11th. Paper presented at a conference "Critical Race Scholarship and the University, Toronto, 25–27 April.

——. 2003. Tolerated citizens or imperial subjects? Muslim Canadians and multicultural citizenship in Canada after September 11th. Paper presented at the annual meeting of the Canadian Sociology and Anthropology Association, Halifax, 1–4 June.

Brah, A. 2002. Global mobilities, local predicaments: Globalization and the critical imagination, *Feminist Review* 70 (1): 30–45.

Canadian Arab Federation. 2002. *Arabs in Canada: Proudly Canadian and marginalized.* Report. Toronto: Canadian Arab Federation.

Canadian Council of Chief Executives. 2004. New frontiers: Building a 21st century Canada-United States partnership in North America. Discussion Paper. http://www.ceocouncil.ca.

Connors, K. 2001. Canadian nurses: Until this crisis is over, we are all Muslims. http://www.labornotes.org/archives/2001/1001/1001f.html.

Deaux, K. 2001. Negotiating identity and community after September 11. http://www.ssrc.org/sept11/essays/deaux.htm.

Groar, C. 2004. Odd way to build consensus. *Toronto Star*, 12 April, A18.

Government of Canada. 2002. *The Smart Border Declaration.* http://www.canadianembassy.org/border/index-en.asp.

Government of Portual. 2003. Portuguese no Canada e eleicoes em Portugal. Fax communiqué. Ministerio dos Negocios Estrangeiros. 31 December.

Gwyn, R. 2001b. We must accept the inevitable. *Toronto Star*, 30 September, A13.

Huntington, S. 1996. *The clash of civilizations and the remaking of world order.* New York: Simon and Schuster.

Jamal, Audrey. 2002. Arab-Canadians: The other within. MA thesis, Royal Roads University, Victoria, British Columbia.

Joseph, S. 1999. Against the grain of the nation. The Arab. In Michael W. Suleiman. ed., *Arabs in America*, 257–71. Philadelphia: Temple University Press.

Khouri, R. 2003. Can multiculturalism survive security agenda? *Toronto Star*, 9 March, A9.

Kutty, F. 2004. Globalizing the harassment of Muslims. The dirty work of Canadian intelligence. http://www.counterpunch.org/kutty04282004.html.

——, and B. Yousuf. 2002. Climate of distrust threatens our basic values. *Hamilton Spectator*, 11 September, A13.

Lafond, R. 2001. Alexa's view un-Canadian. *Toronto Star*, 16 October, A25.

Lewis, B. 1990. The roots of Muslim rage. *Atlantic Monthly* 266 (September): 47–60.

Macklin, A. 2001. Borderline security. In R.J. Daniels, P. Macklem, and K. Roach, eds., *The security of freedom: Essays on Canada's anti-terrorism bill.* Toronto: University of Toronto Press.

Mitchell, Alanna. 2001. Canadian Muslims, Arabs anxious. *Globe and Mail*, 12 October, A13. http://www.arts.ualberta.ca/peace/articles/alanna_mitchell_oct_12.htm.

Ong, A. 1999. *Flexible citizenship: The cultural logic of transnationality.* Durham: Duke University Press.

——, and D. Nonini. 1997a. Toward a cultural politics of diaspora and transnationalism. In A. Ong and D. Nonini, eds., *Ungrounded empires: The cultural politics of Chinese transnationalism,* 323–32. New York: Routledge.

Safieddin, Hicham. 2003. One religion, 12 voices. Interviews. *Toronto Star,* 11 September, B01.

Sassen. S. 1996. *Losing control? Sovereignty in an age of globalization.* New York: Columbia University Press.

Shohat, E. 1995. The struggle over representation: Casting, coalitions and the politics of identification. In R. de la Campa, E.A. Kaplan, and M. Sprinker, eds., *Late imperial culture,* 166–78. New York: Verso.

——. 2002. Dislocated identities: Reflections of an Arab Jew. In Inderpal Grewal and Caren Kaplan, eds., *An introduction to women's studies: Gender in a transnational world.* Montreal and Boston: McGraw-Hill.

Small, P. and B. DeMara. 2001. Canadian Muslims feel under siege. *Toronto Star,* 14 September, A2.

Spivak, G. 2000. Thinking cultural questions in "pure" literary terms. In Paul Gilroy, L. Grossberg, and A. McRobbie, eds., *Without guarantees: In honour of Stuart Hall,* 335–57. New York: Verso.

Stein, E. 2003. Construction of an enemy. *Monthly Review* 55 (3): 125–29.

Tahmasebi, V. 2004. Islamic identity and secular, progressive voices of Iranian women in the diaspora: Problems and dilemmas. Paper presented at the symposium "Women's Voices from the Middle East," York University, Toronto, 11 March.

Thobani, S. 2002. Racism and Canadian democracy in times of war. Paper presented at the conference "Racism and National Consciousness," University of Toronto, 26 October.

——. 2004a. Contesting empire: A case for anti-racist feminism. Paper presented at the conference "Race, Racism and Empire: The Local and the Global," York University, Toronto, 29 April to 1 May.

——. 2004b. Imperial longings, multicultural belongings. Paper presented at the conference "Race, Racism and Empire: The Local and the Global," York University, Toronto, 29 April to 1 May.

Wente, M. 2001. Why do they hate us so much? *Globe and Mail,* 22 September, A13.

Werbner, P. 2000. Divided loyalties, empowered citizenship? Muslims in Britain. *Citizenship Studies* 4 (3): 307–24.

Winant, H. 1997. Behind blue eyes: Contemporary white racial politics. *New Left Review* 225 (September–October): 73–88.

Wright, C. 2003. Moments of emergence: Organizing by and with undocumented and non-citizen people in Canada after September 11. *Refuge* 21 (3): 5–15.

——. 2004. The nationalism of Maher Arar. Paper presented at the symposium "Acts of Citizenship," York University, Toronto, 25–27 March.

PART 5

Responses and Resistances

In this final section of the book, we bear witness to courageous resistances—both historical and contemporary struggles within specific historical moments. We reflect on what these resistances reveal about the strengths and weaknesses within Indigenous and racialized communities and the nature of dominant, indeed colonial, social relations within Canada. Our critical interpretations of contemporary Canada will be explored in the conclusion as we create new paths toward a vision of a just society and a life worth living to its full potential.

Agnes Calliste, a veteran anti-racism scholar and pioneer, in "Nurses and Porters—'Racism, Sexism and Resistance in segmented labour markets'" (2000), documents and analyzes the similarities between African-Canadian women's resistance to professional exclusion and marginality in nursing from the 1970s to 1990s and sleeping car porters' anti-racist struggles against the racialized split-labour market on Canadian railways in the 1950s and early 1960s (2000:143). These resistances included class-action legal complaints, a national media campaign, and legal test cases. Calliste reveals the limitations of legal state interventions such as Human Right laws and policies: "[I]n 1982-90 in Montreal, there were sixteen human rights complaints of racial discrimination and harassment in the health care sector. All were dismissed" (2000:161). She also highlights the emotional, psychological, and social costs of resistance. Calliste concludes by calling for various levels of political action.

In the second article, "Indigenous Voice Matters: Claiming Our Space through Decolonizing Research," Lina Sunseri provides another form of resistance in the very process of knowledge production. By illustrating how the term "research" is often linked to European imperialism and colonialism, Sunseri analyzes research methodologies and the power relations they structure and maintain. She then highlights the broad spectrum of methodologies she has incorporated in her own research that recognizes Indigenous ways of transmitting knowledge. Sunseri concludes that decolonizing research methodologies have the potential of recognizing "that Indigenous cultures are still strong and positive, despite centuries of attack by colonialism" and such resistances will leave a legacy for future generations (Sunseri, 2007:104).

This white-settler logic is also the target of Taiaiake Alfred's article "Rebellion of the Truth." He calls for the regeneration of authentic Onkwehanwe lives, indigenous consciousness, and ways of being. The struggle against colonization and racism must also be framed by the question, "*What is* the fight? (emphasis in the original), that is, what do we fight for? Reflecting on both colonial legacy and contemporary social exclusion, Alfred writes: "But history has not ended. There are still Onkwehanwe lands, souls, and minds that have not been conquered. For them, a warrior is what a warrior has always been: one who protects the people, who stands with dignity and courage in the face of danger. When lies rule, a warrior creates new truths for the people to believe" (2005:97).

The challenges to bear witness and to "create new truths" is our focus in the conclusion.

CRITICAL THINKING QUESTIONS

1. What were the similarities between the resistances of the African-Canadian nurses from the 1970s to 1990s and the sleeping car porters in the 1950s and early 1960s? Identify some outcomes of these resistances.

2. How can "research" maintain colonizing social relations? Conversely, how can research have liberating potential?

3. According to Taiaiake Alfred, the warrior is significant. How and Why? What are your thoughts of the idea that everyone might have a warrior within?

FURTHER RESOURCES

Dickenson, H. and T. Wotherspoon. "From Assimilation to Self-Government: Towards a Political Economy of Canada's Aboriginal Policies." In Vic Satzewich, ed., *Deconstructing a Nation: Immigration, Multiculturalism and Racism in '90s Canada*. Halifax, NS: Fernwood Publishing, 1992.

Lavell-Harvard, D.M. and J. Corbierre Lavell, eds., *Until Our Hearts Are On the Ground: Aboriginal Mothering, Oppressions, Resistance and Rebirth*. Toronto, ON: Demeter Press, 2006.

Havel, Vaclav. *Disturbing the Peace*. Translated by Paul Wilson. New York, N.Y.: Vintage, 1991.

James, C.E., ed. *Perspectives on Racism and the Human Services Sector: A Case for Change*. Toronto, ON: University of Toronto Press, 1996.

Smith, L.T. *Decolonizing Methodologies: Research and Indigenous Peoples*. London and New York: Zed Books, 1999.

Waters, Frank. *Brave Are My People: Indian Heroes Not Forgotten*. Athens, OH: Ohio University Press, 1993.

FILMS

Keepers of the Fire. Director, Christine Welsh; Producers: Ian Herring, Christine Welsh. Produced by Omni Film Productions in co-production with Studio D and Prairie Centre of the National Film Board of Canada. 1994.

Description: An Aboriginal proverb says that no people is broken until the hearts of its women are on the ground. Presents the stories of Aboriginal women who have participated in important Aboriginal struggles in Canada. Mohawk women tell of their role in the 1990 crisis at Oka. Haida women reminisce about their stand on the picket lines and their arrests in the action that stopped logging on Lyell Island in the Queen Charlottes.

The Promised Land. (Eyes on the prize 2: American at racial crossroads: 4) Executive producer, Henry Hampton. Alexandria, VA: CPB/Blackside Inc., 1990.

Description: Moved by increasing levels of poverty, Martin Luther King Jr. and his staff begin to organize a poor people's campaign, a march of the poor to Washington, D.C. King is called away to help black sanitation workers on strike in Memphis, Tennessee where he is assassinated.

Freedom's Land: Canada and the Underground Railroad. Written and directed by William Cobban. Producer, Marcy Cutler. Narrator, Anthony Sherwood. Toronto, Ont: Canadian Broadcasting Corporation, 2004.

Description: Told through manuscripts, letters, and dramatic reconstructions, this is the story of how Canada and the Underground Railroad became the focal point of the anti-slavery movement in the decade leading to the American Civil War. Tells the stories of Canadian physician Alexander Ross who risked his life in the American South to help escaping slaves, and Henry Bibb who established the first Black-owned newspaper in Canada.

The Road Chosen: The Story of Lem Wong. Directed by: Keith Lock. Producer, Peter Raymont. White Pine Pictures, 1997.

Description: The story of Lem Wong as told by local documents and by family members who emigrated to Canada from China in 1897, paying a head tax of $50.00. Working in a laundry and experiencing anti-Chinese prejudice of the times, Wong worked hard to improve his life. Moving to London, Ontario, he established Wong's Cafe, returned to China and brought over a wife, and raised eight successful children, four sons and four daughters, including three doctors, a professor of chemistry, and a lawyer.

KEY CONCEPTS/TERMS

Split Labour Market Theory and Race

Decolonizing Research Methodologies

Orientalism

Non-violent Militancy

Warrior

16

Nurses and Porters: Racism, Sexism and Resistance in Segmented Labour Markets

AGNES CALLISTE

INTRODUCTION

There are many similarities between African-Canadian women's resistance to professional exclusion and marginality in nursing from the 1970s to 1990s and sleeping car porters' anti-racist struggles against the submerged split labour market (Bonacich 1972) on Canadian railways in the 1950s and early 1960s (in which white men monopolized the higher-paid positions and "locked" Black men into the most menial, low-paying and low-status jobs). Like the porters, Black nurses tend to be streamed into the least skilled and least desirable areas (such as chronic care) irrespective of their qualifications and experience because of managements' perception that these women do not have the qualifications to work in other units (Calliste 1996b; Head 1986; OHRC 1992b). Moreover, the nurses' and porters' struggles for employment equity were carried on within their workplaces and unions, as well as within the political arena, since both groups of workers had to resist inequality practiced by their employers and unions. They also had to pressure the state to enforce fair employment practices and human rights legislation.

Despite the similarities in African-Canadian nurses' and porters' work experiences and anti-racist struggles, to date there is no comparative research on the two groups of Black workers. Recently, two African-Canadian nurses were dismayed that I would even think of comparing professional nurses'[1] experiences and struggles with those of porters—as if I were declassing nurses. This view overlooks the evidence that although racism and gendered racism are historically specific (Essed 1991; Hall 1978), they have been embedded in labour markets and in other societal institutions, and they manifest themselves in different ways in different sites and contexts. Gendered racism refers to the racial oppression of racial/ethnic minority women and men as structured by historically situated racist perceptions of gender roles and behaviour (Essed 1991).

Blatant notions of racial inferiority such as that Blacks are inherently uncivilized and stupid are being replaced by a much more subtle ideology of cultural inferiority, which is articulated by the choice of carefully coded language—for example, portraying them as aggressive and loud (Gilroy 1987). However, the most persistent racist, sexist and classist stereotypes that have been used to justify the racial division of labour and the exploitation and devaluation of Black workers', especially women's, labour are those that label Black workers as less intelligent, less competent, less skilled and less disciplined than white workers (Calliste 1996b; Essed 1991). For instance, although a porter in charge did the jobs of both a porter and a sleeping car conductor, in the 1950s and early 1960s Canadian railway companies and the Order of Railway Conductors (ORC), the Canadian Pacific Railway (CP) conductors' union, argued/implied that Black sleeping car porters were not suited for the position of sleeping car conductor (Blum 1958; Johnstone 1964; Smith 1964b). Similarly, in a human rights case in the early 1990s, management and some white staff members at the Dunfermline Hospital in Toronto tried to justify the dismissal or suspension of seven Black nurses on the grounds that they were either incompetent or unprofessional, thereby blaming the complainants for being the cause of their own misfortunes (Doris Marshall Institute and Arnold Minors & Associates 1994; OHRC 1992a; Ontario Ministry of Labour 1993a, 1993b; interviews, March 31, 1996). Scapegoating nurses and porters diverted attention from a critical analysis of the structures and practices of the workplace, which treated them inequitably and justified the status quo by blaming the victims.

This study focuses on African-Canadian women's resistance to professional marginality and exclusion in nursing and porters' resistance to the submerged split labour market on the railways from an eclectic theroretical perspective. The study draws on segmented and split labour market, social closure, anti-racism and feminist theoretical frameworks (Bonacich 1972, 1975, 1976; Brandt 1986; Calliste 1996b; Collins 1990; Dei 1996; Edwards 1979; King 1990; Parkin 1979). It examines the conditions under which these anti-racist struggles materialized, the constraints placed upon them and the effects of these struggles. Marginality is a process in which "a sense of otherness or peripherality is perpetuated and encouraged" (Brandt 1986: 104). Racial minorities are denied access to positions of power within institutions, and their experiences and perspectives are considered irrelevant. Although some Southern and Eastern Europeans were employed as porters in western Canada during and after World War II, before the war, "Blacks" and "sleeping car porters" were synonymous, and Blacks led the porters' anti-racist struggles in the 1940s to 1960s. Nurses of colour and a few whites have been involved in the nurses' struggle, but I have chosen to focus on the experiences of African-Canadians in Ontario and Quebec in the 1970s to 1990s. These provinces are chosen because there are large numbers of Black nurses there. For example, Reitz et al. (1981) found that Caribbean women[2] in Toronto's workforce were over-concentrated in nursing (except in supervisory positions), nurses' aide positions, orderly work and in other segregated occupations. The late 1970s demarcates the beginning of nurses' political organizing to combat increased racism during economic recession and cuts in health care budgets. The 1990s are characterized by nurses' and the African-Canadian community's intense anti-racism struggles.

THEORETICAL FRAMEWORK

Racially segmented and split labour markets, racial and gender ideologies in segmented labour markets, gender ideology in family units and the capitalist structure in which Black men and women's specific race, gender and social class positions are embedded structure Black men's and Black women's work (Collins 1990: 46). Thus this study on nurses' and porters' resistance to racism and sexism in segmented labour markets draws on split and segmented labour market, social closure, anti-racism and feminist theoretical frameworks, while simultaneously according race, class and gender separate analytical statuses to develop a stronger explanatory model.

Split labour market theory was formulated by Edna Bonacich (1972, 1975, 1976) to explain economic sources of racial/ethnic conflict. She argues that a split labour market exists when the cost of labour differs substantially along racial lines for the same work, or would differ if they did the same work. In labour markets split along racial lines, conflict develops among employers desiring the cheapest possible labour, workers of the higher-paid, dominant racial group resist being undercut or displaced by cheaper labour from the racialized minority group, while the lower-paid racialized minority workers are struggling to find a niche in the economy. Such competition leads to four forms of racial conflict depending on the relative power of the three classes: (1) displacement, (2) exclusion of racialized minority workers, (3) a caste or submerged system which demarcates a colour line beyond which only white workers can advance and (4) radicalism in which the two labour groups form a coalition against employers. A racially split labour market results from the differential political and economic resources available to dearer or cheaper group of workers in the struggle to improve wages and working conditions. However, split labour market theory overemphasizes the role of economics in the process and at best explains the results, not the origins, of white working-class privilege.

Segmented labour market theorists (Edwards 1979; Gordon, Edwards and Reich 1982; Piore 1971) adopt a more historical and structural account of the formation of labour markets segmented by race and gender. They argue that segmented labour markets emerged in the historical context of capitalist development with the formation of separate monopoly and competitive sectors, differential strategies of control in the workplace, changes in productive technology and the differential effects of working-class struggle across economic sectors. In segmented labour markets, jobs and industries are readily divided into a primary or core sector and a secondary or peripheral sector, and this division is reinforced by barriers that make it difficult for workers to move from one sector to another. In the primary sector, jobs are characterized by higher wages and fringe benefits, greater employment stability with possibilities of promotion, a higher degree of unionization, superior working conditions and due process in negotiating job rights. In contrast, the secondary sector includes work in marginal industries. These jobs are low-paying, often seasonal and sporadic, less likely to be unionized and offer little protection against the vagaries of either the individual employer or the ups and downs of the marketplace. Allocations to these labour markets follow existing divisions of race and gender. The secondary sector uses the groups with little bargaining power such as racialized minorities and women who are pushed into particular race or sex-typed jobs to maximize profits, and partly because of systemic racism and patriarchy. Blacks are hired into the dirtiest, most physically demanding and lowest-skilled occupations, while women are pushed toward "helping" and nurturing occupations (Edwards 1979).

Segmented labour market theory helps to explain the segregation of women workers into female-typed jobs, and Blacks into race-typed jobs. However, it overlooks the relationship between social relations in production and race relations in unions, the connection between white labour's alliance with capital and labour's exclusion and marginality of Black and other racialized minority workers and women, and the role of radical labour leaders in promoting racial equality (Calliste 1996b; Glazer 1991; Hartmann 1976, 1981; Wilson 1996). Segmented labour market theory also overemphasizes the role of economics in the subordination of Black workers, and fails to entertain the possibility that racism and sexism are also ways in which "white workers (and men) have come to look at the world"

(Roediger 1991: 10). Race and gender have always been crucial factors in the history of Canada's class formation (see Omi and Winant 1983; Roediger 1991).

The evidence suggests that racism, gendered racism and immigrant/migrant status interacted with class exploitation in the submerged split labour market on the railways, and in the informal segmented labour market in nursing, and that racism and sexism were not accidental to the process. For instance, the racially submerged split labour market practices on the railways developed initially from the employers' demand for cheap labour and systemic racism (Calliste 1987, 1995a; Foner 1981; Gould 1977; Marshall 1965). Racial and class divisions between Euro-Canadian and American railway workers and Black porters made for a fractured brotherhood. White railway unions helped to institutionalize and maintain a racial caste system in the industrial labour force by excluding Blacks as members, or by relegating them to auxiliary and segregated locals, and agreeing to segregated seniority systems which were written into collective agreements. The exclusion and marginalization of Black men by trade unions, especially industrial unions such as the Canadian Brotherhood of Railway Transport and General Workers (CBRT), cannot be explained simply in terms of economic self-interest. It also emerged out of the union's (white men's) quest for higher status and social privileges in the context of a highly stratified industry and a racist and sexist society (Marshall 1974; Roediger 1991; Wilson 1996). White railway workers wanted to maintain their social distance from Blacks; thus they emphasized the fraternal and social benefits of their organizations, and perceived racially mixed unions and integrated sleeping quarters to mean granting social equality to Blacks (Kamarowsky 1961; Marshall 1965; Northrup 1944).

Similarly, white health care administrators and nurses have historically marginalized/excluded racialized minority women to protect their jobs as well as to strive for an increase in the status of the profession. They rationalized the marginality/exclusion of the latter by using ideological constructions of racially specific femininity and sexuality, representing the opposite models of white, middle-class womanhood (Calliste 1993, 1996b, 2000; McPherson 1996). Social closure theory indicates that historical and contemporary patterns of exclusion involve not only discrimination or market mechanisms of job allocation but also privilege (Murphy 1988; Parkin 1979). Moreover, the theory postulates that the more desirable the job, the more likely subordinate groups (such as Black women) would be excluded, and the more a job is filled by Black women, the more it is devalued and deskilled by employers.

Anti-racism theory helps to explain the racialization of gender and class in nursing and on the railways as well as individual and collective resistance. Unlike segmented and split labour market theories that focus on class and treat racism as an epiphenomenon or manifestation of class, anti-racism theory centres racism while acknowledging its intersection with other forms of social oppression (Brandt 1986; Dei 1996). Anti-racism is a critical discourse of social oppressions (such as racism, sexism and classism) through the lens of race. It is also "an educational and political action-oriented strategy" for institutional systemic change to address racism and the interlocking systems of social oppression (Dei 1996: 25). The emphasis on racism is not intended to prioritize oppressions. Instead, the anti-racism discourse highlights that, though "race" is a "fundamental organizing principle of contemporary social life" (Winant 1994a: 115) in racialized societies, the status quo continually denies there is racial discrimination in employment (for example, in hiring, promotion and work relations) and that the racial composition of jobs has an effect on how work is organized (that is, the labour process). However, the logic of capital is not "race" and "gender" blind. Access to work and justice are constrained by relations of oppression such as racism and gendered racism (Brandt 1986; Collins 1990; Essed 1991). Segmented labour theory and anti-racism can inform each other.

Some Black nurses acknowledge that they experience multiple oppressions in the workplace (Congress of Black Women of Canada [CBWC], Toronto Chapter 1991; Nurses, Woodlands Hospital 1992). For instance, in 1991, the Coalition in Support of Black Nurses stated: "Traditionally, Black women have been singled out to bear the brunt of economic hardships in our society, especially during recessionary periods" (CBWC 1991: 1). Similarly, porters argued that they were infantilized or treated like boys by railway management, white railway workers and unions and some passengers (Robinson 1941; interviews, September 1, 1986). For example, porters recalled that some passengers called them "George" after George Pullman, the manufacturer and operator of railway sleeping cars, as if they were his boys or his property, which is similar to the practice of naming slaves after their masters, but also because adult Black males were considered as less than men (see Bederman 1995; Mergen 1974). Despite the interlocking systems of social oppression, Black nurses and porters emphasized racism because it was more visibly salient for them. Nursing, like portering, is a gendered occupation, and many of the oppressors of Black nurses are white female nurses who collaborate with management in harassing Black nurses (interviews, May 25, 1994, and September 8, 1996). Similarly, most of the oppressors of Black porters were white men (Calliste 1987, 1988, 1995a).

Anti-racism goes beyond acknowledging the material conditions that structure social inequalities to confront and question white power and privilege and its rationale for dominance (Dei 1996; Dyer 1997; Frankenberg 1993a; McIntosh 1990). Anti-racism questions the role that the state (including human rights commissions and labour arbitration systems) and societal institutions such as workplaces play in creating and maintaining inequalities. Anti-racism, like feminism (Armstrong and Armstrong 1990; Smith 1977), acknowledges that the state and societal institutions contain and manage the protests of oppressed groups (for example, by individualizing grievances), but they do not examine the underlying causes of discrimination or challenge systemic racism and patriarchy—especially racism—unless they are pressured into doing so (Bolaria and Li 1988; Calliste 1987, 1996b; Henry et al. 1995; Young 1992; Young and Liao 1992). Thus, the state's response to anti-racist, feminist and other minority struggles (whether it crushes, appeases or co-opts them) varies in accordance with the groups' economic, political and social power. Anti-racism theory questions the devaluation of knowledge, credentials and experience of subordinate groups (see Ontario, Ministry of Citizenship 1989a, 1989b), and the marginalization and silencing of certain voices in the workplace and in society. It recognizes the need to confront the challenge of social diversity and difference in Canadian society and the urgency for more inclusive and equitable workplaces and society (see Dei 1995a).

The eclectic theoretical framework is relevant in explaining Black nurses' and porters' resistance to the interlocking systems of social oppression in the workplace. As mentioned previously, historically, Black nurses and porters in Canada (like their counterparts in the United States) have been simultaneously oppressed by racism, classism and sexism (see Barbee 1993; Calliste 1993, 1995a, 1996b; Head 1986; Hine 1989). The divisions between African-Canadian and Euro-Canadian women were most evident in the racist, sexist and classist images of Black women as "mammies," "Jezebels," and "matriarchs" (see Barbee 1993; Calliste 1991, 1993/94; Davis 1983; Hooks 1981). These images, like the Black male images of "Sambo," "Buck" and "Rastus," emphasized physical attributes over social (Fox-Genovese 1988). Sambo captured the image of the stereotypical childlike and subservient male slave or servant—the antithesis of white male honour, masculinity and intelligence (Bederman 1995; Fox-Genovese 1988). Similarly, the "mammy" image of the faithful and obedient domestic servant was created to justify the economic exploitation of female house slaves. These racially specific notions of femininity and masculinity played an important role in justifying the restriction of Black women and men to menial and backbreaking jobs, such as domestic work, chronic care nursing and portering, as well as conditioning managerial strategies of control, and they represented the "normative yardstick used to evaluate" Black women and men (Collins 1990: 71). Porters recalled that some passengers called them "Sam" (Sambo)[3] and Rastus" (interviews, September 1, 1986, and May 1, 1987; see Beckford 1990; Williams 1978; Williamson 1986). Racism, sexism and classism continue to permeate workplace culture which stereotypes Black nurses as childlike, lazy, aggressive, emotional, uncommunicative and trouble-making (Calliste 2000; CBWC 1995; Doris Marshall Institute and Arnold Minors & Associates 1994; OHRC 1992a; interviews, May 25–27 and July 3, 1996). This image of the Black nurse is the antithesis of the Florence Nightingale image of the soft-spoken, compassionate, nurturing, rational and professional nurse (Ehrenreich and English 1973). An African-Canadian nurse also notes that "the Black nurse is still treated as the servant girl in the kitchen" (interview, July 1, 1994). Thus, the Black woman nurse becomes an "undesirable" entity in this context. Black nurses and nursing assistants have helped to maintain a segmented nursing labour force that was initially based on class (see Doyal et al. 1981; Gamarnikow 1978).

Although segmentation usually refers to distinctions between occupations, it includes differentiation within an occupation between professional and nonprofessional strata, for example, between registered nurses (RNs), graduate nurses and registered practical nurses (RPNs). Black nurses in Canada, like their counterparts in Britain, South Africa and the United States, are concentrated at the lower levels of the nursing hierarchy as staff nurses and RPNs—a reflection of persistent inequalities in nursing and in the wider society (Calliste 1996b; Doyal et al. 1981; Glazer 1991; Head 1986; Marks 1994).

In addition to this formal segmentation to protect the class-based and racial/ethnic privileges of some nurses, there are informal ways of segmenting RNs to protect the racial privileges of white RNs. The Dunfermline Hospital in Toronto serves as an illustration of the racially segmented nursing labour force. In 1992 its management was predominantly white, while 56 percent of its nursing staff were people of colour with 30 percent being Black (Das Gupta 1993: 39; OHRC 1992a: 3). Most of the Black nurses were working in the chronic and acute care units and they were the least represented in high-technology and high-status specialty units (where there are greater opportunities for further training and the nurses are encouraged to take courses) (Caissey 1994; Head 1986; OHRC 1992a; ONA 1996). For instance, 41 percent of the nurses in the chronic and acute care units were Black compared to 15 percent white and 44 percent other nurses of colour (such as Canadians of Chinese and Filipino descent). Conversely, only 13 percent of the nurses in the intensive care unit were Black, compared to 55 percent white and 32 percent other nurses of colour (OHRC 1992a: 9).

The state helped to create and maintain segmented labour markets and the subordination of Black nurses and porters through differential immigration policies. As I have argued elsewhere (Calliste 1993, 1993/94, forthcoming), before 1962 the process of immigration control of Black people entering Canada and the manner in which they were incorporated into the labour market—as free immigrant labour with citizenship rights, or unfree migrant labour—were structured by a dialectic of economic, political and ideological relations (the demand of employers for cheap labour and the state's desire to exclude Blacks as permanent settlers). Perceiving Black women and men as inferior, undesirable and likely to create permanent economic, social and race relations problems, immigration officials sought to avoid this difficulty by restricting entry to those whose services were in urgent demand, and only when sources of white labour were unavailable. In the 1950s and early 1960s, a small number of Caribbean nurses were allowed into Canada under differential immigration policy, which stipulated that one of the conditions for admitting them into the country was that the hospital administration which offered them employment had to be "aware of their racial origin" (Acting Chief, Admissions 1956). Moreover, in order for Black nurses to enter Canada as permanent immigrants, they were required to have nursing qualifications that exceeded those of white nurses. Among Black nurses, only those who were eligible for registration with the provincial Registered Nurses Association (RNA) were admitted as landed immigrants. Others were allowed to enter as migrant workers. This differential immigration policy helped to reinforce Black nurses' subordination within a gendered and racially segmented nursing labour force (Calliste 1993).

Similarly, the Canadian state helped the Canadian Pacific Railway (CP) to create and maintain a double split labour market between African-Canadian and imported African-American labour on the one hand, and between Black and white Canadian labour on the other, by allowing the company to import African American porters, particularly from the southern states, both as seasonal workers and also as regular workers on six-month contracts which could be renewed. In 1949, approximately 50 percent of the CP's 600–700 porters were American born. Many of these migrant porters had been with the CP for several years (Calliste 1987: 6, 10). Migrant status hindered solidarity among some Black nurses and porters, given they had no political and citizenship rights, and made them extremely vulnerable to an employer's exploitation (Burawoy 1976; Calliste 1987, 1993; Miles 1982).

The Canadian government played a more direct role in creating and maintaining the submerged split labour market on the CN since it was a crown corporation. Thus in the CN porters' anti-racist struggle, they were pitted against the state, the very institution which was responsible for enforcing the Canada Fair Employment Practices (FEP) Act.

RESEARCH METHODS

My discussion of porters is based upon data drawn from extensive archival research in Canada and the United States, review of literature, oral history interviews of one hundred porters and former porters, and to some extent, porters' families and union leaders conducted mainly in Winnipeg in 1982–83 and in Truro, Halifax, Montreal, Toronto and Vancouver in 1986–89. The sample was drawn through snowball sampling, referrals and from the seniority lists of porters.

My discussion of nurses is based upon data drawn from forty semi-structured interviews conducted in Toronto and Montreal, mainly in 1994–97. The sample consists of thirty registered nurses, three officers from nurses' unions and seven members of community organizations which assisted Black nurses in their anti-racism struggles. The sample was drawn through snowball sampling and referrals, as well as from the membership list of two nurses' organizations. I was also a participant observer at three conferences in Toronto centred on anti-racism and/or integration and advancement of Black nurses in health care. I wrote one of the conference reports (Calliste 1995c). The primary data on nurses were supplemented with analysis of records of the OHRC, the Quebec Human Rights Commission (QHRC), labour arbitration cases and community organizations and other secondary sources. The secondary data also served as checks for possible biases in the primary data. In order to maintain confidentiality, I use fictitious names to identify nurses and health care institutions except those who are already named in secondary sources.

RACISM ON CANADIAN RAILWAYS

Black sleeping car porters had become a fairly entrenched Canadian tradition after the 1880s. Except for the Grand Trunk Railway (GTR) which employed Black male cooks and waiters in its dining cars, the Canadian railway companies followed the southern American practice of hiring Black men almost exclusively as porters until the amalgamation of the porters' and dining car locals of the CBRT in 1964.

The Canadian railway companies employed Blacks as porters because they were cheaper than white workers for comparable work. Moreover, the assumed social distance that existed between whites and Blacks meant that the presence of Black male porters on sleeping cars was considered as part of the furniture and did not serve as a complicating factor in the intimacies which travel by sleeping cars necessitated. Thus, Black men were de-sexed as porters because of the social distance. The railway companies also knew that Blacks were traditionally assigned physically demanding and service roles, and that it was a sign of status among whites to be waited on by them (Anderson 1986; Brazeal 1946; Harris 1977). Portering was a male form of domestic work which reinforced social relationships of white superiority that had been established by slavery. Each Canadian Pacific Railway (CP) porter had to sign a contract acknowledging that the company had the right to dispense with his service at any time without being required to give any reason for doing so (*The Labour Gazette* 1920: 241). The company claimed that this clause was necessary because sleeping car porters' work was of a domestic nature and involved close personal relations between the porters and sleeping car travellers. Dining car employees were not required to sign such a contract. It was the porter's job to give total personalized human service to the passengers, including even the shining of their shoes. In order to guarantee the porter's efficiency and to maximize profits, the railway companies took care that the porter's welfare should be inextricably tied up with the type of service he gave passengers. The porter was paid a low monthly wage and had to depend upon passengers to augment his income through tips. For example, CP porters' monthly salaries in 1920 ranged from $75 to a maximum of $85 after three years (*The Labour Gazette* 1920). Out of his meagre salary, a CP porter had to buy his meals, uniform, a shoe-shine kit and even the polish for the passengers' shoes. Porters in charge performed the duties of both the porter and the conductor (except the porter was usually in charge of one or two cars) but received substantially less pay than conductors. For example, in 1951 a CN conductor's monthly salary during his first year of service was $268 compared to a porter in charge who made $213, and a porter made $188 (CBRT and CN 1951: 37–38). Thus Black porters experienced interlocking systems of social oppression: racism, sexism, class exploitation and, in some cases, super-exploitation based also on their migrant status.

The process of forming a racially submerged split labour market on Canadian railways began in the 1880s and continued in the 1920s when the CN reclassified work categories in order to limit Blacks, except for those already otherwise employed, to being porters only. In 1926 when the CN took over the GTR, it began replacing Black waiters in the dining car service of former GTR cars with white help. Consequently, an arbitration board heard a complaint made by the Black employees that their displacement was contrary to agreement, a violation of status and of seniority rights, which they argued could be attributed to racism. A compromise arrangement was agreed to, allowing the displaced Black waiters to take employment as porters with full seniority. The Black cooks were gradually eliminated through attrition (*The Labour Gazette* 1927; McGuire 1942).

The racially segregated job classification and seniority system in the sleeping and dining car department of the CN were institutionalized in 1927 when the company and the CBRT agreed to a group classification system for seniority purposes (CBRT and CN 1927). Group 1 was for dining car employees and sleeping car conductors. Group 2 was set up exclusively for porters. These groups coincided with the CBRT's segregated dining car and porters' locals. The collective agreement between the CN and the CBRT stated that seniority and promotion of employees would be confined to the groups in which they were hired. The selection of supervisors, such as platform and road inspectors, took place within Group 1. A strictly logical occupational line of promotion for porters, particularly porters in charge, would include the position of sleeping car conductor since porters in charge performed the duties of conductors. However, some whites with far less railway experience and far less education than some Blacks were promoted as inspectors and sleeping car conductors (Grayson 1953; Tobin 1959). Porters often had to assist inexperienced sleeping car conductors.

PORTERS' STRUGGLE AGAINST THE RACIALLY SEGMENTED LABOUR MARKET

The CN porters began the struggle against the submerged split labour market in 1942, but it only became a legitimate discrimination grievance after the enactment of the Canada Fair Employment Act (FEP) in 1953 when they intensified their pressure for promotion. The FEP Act, which prohibited discrimination in employment by employers, unions and employment agencies on the basis of race, national origin, colour and religion, had several weaknesses. For instance, it focused on intentional and individual cases of alleged discrimination. Thus the law was ineffective in dealing with systemic discrimination because it was difficult to prove intent to discriminate (Vizkelety 1987). Moreover, as the United States Supreme Court (*Griggs v. Duke Power Co.*, cited in *Albermarle Paper Co., v. Moody* 1975) and the Fifth Circuit (*Oatis v. Crown Zellerback* 1968) argued, racial discrimination in employment is by definition class discrimination because it is the result of institutional systemic discrimination (i.e., the adverse effects of employment policy and practices rather than motivations for the act) (see Keene 1992). The 1927 collective agreement between the CN and the CBRT discriminated against porters as a group; thus CN porters successfully took class action to effectuate change.

In 1961, the porters launched a massive national media campaign against the discrimination practiced by the CBRT and the CN ("Porters would Like Probe" 1961; Werier 1961).

Unlike CN porters, who took class action, CP porters used test cases to pressure the CP management to promote Black porters to sleeping car conductors' positions.

The CN porters were more successful than the CP porters in their struggle for promotion because they were strongly organized. It took sustained class action from 1955 to 1964 to eliminate the racially segregated job classification and seniority system. Although the CP porters' method of using test cases provided some redress for the aggrieved individuals, the CP maintained its segregated job classification and seniority system until it got out of passenger service in 1978. In addition to CN porters being better organized, as members of the CBRT, they were able to exert group pressure on the Brotherhood. CP porters and conductors were organized by two different unions, the BSCP and the ORC, respectively. This complicated the problem of promoting porters with full seniority rights. The junior conductors who were allowed to retain their seniority as porters had to be members of both unions. The CN porters had another advantage. The CN management was more susceptible to public pressure than the CP, given that the former was a crown corporation. Given the ineffectiveness of the FEP laws, a member of Parliament argued that the most the federal government could do to eliminate CP porters' grievances against lack of promotion was "to use moral persuasion" (Alexander 1972). Both the CN and CP porters' victory effected a significant reduction of outrageous discrimination in Canada. This section examined porters' resistance to the submerged split labour market on Canadian railways. The following section discusses the professional exclusion and marginality of African-Canadian women in nursing.

MARGINALITY AND EXCLUSION IN NURSING

Black women were allowed into nursing as a floating reserve army of labour during the postwar expansion of industrial capitalism and severe nursing shortages. However, they were then and are today marginalized and treated as the "Other." For instance, even at the present time they are concentrated at the bottom of the gendered and racially segmented nursing labour force as staff nurses; they are also more likely than their white counterparts to be employed through the nursing registry, and to be employed part-time and as casuals. Thus, they are less likely to be entitled to fringe benefits. Moreover, Black nurses who work full-time are more likely than their white counterparts to be underemployed, to be denied access to promotions and to be assigned the least desirable shifts and duties involving mostly menial work (Caissey 1994; CBWC 1995; Das Gupta 1996; Head 1986; OHRC 1992a: interviews, September 1, 1996).

During economic crises Black nurses are the first to be suspended, demoted and summarily dismissed, often for alleged incompetence or for not following standard nursing procedures, or for "unacceptable behaviours," especially if they resist racism. In the 1990s, it seemed as if they were being further marginalized and excluded as economic restructuring and rapid downsizing of the health care system had been combined with new (and very old) forms of racism and racialization that had a disproportionate impact on Black health care workers, particularly those who worked in chronic care units and those who were assertive (Calliste 1993 and 1996b; Franklin 1995; Hardill 1993). Black nurses who work in chronic care units are more vulnerable and powerless, given "peripheral" workers are the first to be laid off or displaced by cheaper labour. ONA notes:

> More so than on other units, chronic [care] unit nurses are being replaced by [low-skilled unregulated assistive personnel or] generic workers, while at the same time these same nurses are told they lack the skills to bump into other areas of the hospital, even if they have the seniority to do so. (1996:5)

Overt racism, sexism and other forms of social oppression in nursing, as in other areas of employment, increased during the economic recession and cutbacks in health care budgets in the late 1970s to early 1980s, and intensified during the 1990s fiscal crisis (Bolaria and Li 1988; Hardill 1993; Henry and Ginzberg 1985). As I have discussed elsewhere (Calliste 1996b, 2000), harassment of racialized minority nurses by hospital management takes several forms, including targeting them for discriminatory treatment, and differential documentation and discipline for minor or non-existent problems for which white nurses are not disciplined. For instance, at a meeting of le Ralliement des Infirmières Haitiennes de Montreal (RIIAH) in October 1997, after some Haitian nurses reported on alleged racial discrimination and harassment at their workplaces, forty-two Haitian nurses decided to form a crisis committee to prepare a petition to send to their employers denouncing the numerous allegations of harassment against Black nurses and demanding a work environment free of harassment (Dominique 1997: 6).

Health care institutions reinforce the marginality of Black nurses by denying their experiences of racism. Instead they focus on interpersonal dynamics such as "communication skills" and "personality problems," and blame Black nurses for "their problems" rather than investigate the issue of racism (Doris Marshall Institution and Arnold Minors & Associates 1994: interviews, March 30, June 24 and July 3, 1996). Donna Jones' case serves as an illustration. In 1992, Donna, a public health nurse in Toronto, received her first unsatisfactory performance appraisal in fifteen years of employment at the Belair

Health Unit, partly because of a "negative perception of her personality." However, she was not given a satisfactory explanation for the performance appraisal. She notes: "When I inquired as to why I was not told about the negative perception of my personality and my work prior to this date, I was told that I would have reacted negatively . . . . Most of the comments lack concrete data" (Jones 1993:1). Donna argues that she began experiencing racial harassment when she began to be more assertive and to resist (gendered) racism and ageism in the workplace, such as infantalization by her peers and directors (interview, June 24, 1996). Donna filed several grievances and a human rights complaint of racial discrimination against her employer. However, she resigned in 1994, citing continuing discrimination. She felt that the stressful work environment was impacting upon her health (Ontario Ministry of Labour 1996; Jones 1994; interview, June 24, 1996). This also indicates some of the hidden injuries of racism and sexism as well as the emotional and psychological costs of resistance.

Black women's assertiveness and resistance to the multiple oppressions they experience are perceived as threatening to white patriarchal definitions of femininity. Donna was also harassed to control her assertive behaviour, to discourage solidarity among nurses and to undermine her ability to influence others. Management's divisive strategy also served as a form of containment; at the same time it was a factor in producing a "poisoned" work environment. Hospitals' denial of racism is also reflected in their lack of support for Black nurses who are subjected to racism, sexism or physical assault by patient's families. Management tends to deal with such incidents as crisis management issues, rather than dealing with the issue of racism (OHRC 1992b). Hospital managers who acknowledge racism tend to attribute it to individual biases or misconduct rather than understanding it as systemic (see Young 1992). For instance, in 1990, a public relations director at the Paradise Hospital in Montreal said that she had "never heard of a problem [of racism at the hospital] . . . . I'm sure that it occurs. We have approximately 2500 employees. I'm just going on the assumption that someone will step out of line" (Brown 1990). Given management's denial of racism, it is not surprising that, in 1990, hospitals in Montreal and Ontario did not have racial harassment policies.

Black nurses' everyday experiences of racism reflect systemic racism in the workplace (for example, subjective recruitment, evaluation and promotion processes, as well as differential work assignments). Management practices and employment systems in these hospitals adversely affect Black nurses. The recruitment and promotion processes are often subjective and arbitrary and lead to discriminatory treatment of applicants. For example, some hospitals recruit through employee referrals or word-of-mouth methods that tend to reproduce the status quo. Pre-screening of applications, checking of references, interviewing and internal transfers are often left to one person, a white manager. Moreover, there are no standard questions that are consistently asked when interviewing for nursing staff. This could result in culturally laden questions being asked as well as arbitrary, biased and subjective decisions being made (Henry and Ginzberg 1985; OHRC 1992b). In sum, systemic and everyday racism produce and reproduce each other. The following section discusses Black nurses' resistance to the interlocking systems of social oppression in the health care system.

RESISTANCE IN NURSING

Black nurses in Quebec and health care workers of colour in Ontario began organizing themselves and making formal as well as informal complaints of racial discrimination and harassment in the late 1970s and the early 1980s, as racism and sexism increased with the downturn in the economy and cuts in health care budgets (Calliste 1996b, 2000; Head 1986; interviews, July 5, 1994, June 20, 1998, and June 19, 1999). For instance, in the 1980s, the Black Nurses Association of Quebec (BNAQ) initiated a series of investigations into long-standing cases of racial discrimination against its members at a Montreal hospital and made representations on their behalf before the QHRC (Robertson 1985a, 1985b; Sam 1986; interviews, November 17, 1997). In Ontario, Eva filed a human rights complaint against a university hospital in 1980, alleging racial harassment and wrongful dismissal. Several organizations (such as the local chapter of the International Committee Against Racism and the Afro-Caribbean Students Association) demanded that the university and the hospital "end racist practices in employment and education, rescind the dismissal" of Eva and "drop all charges before the College of Nurses" (Hathiramani 1980; Quigley 1980; "Racism at McMaster" 1980: 1; interview, July 5, 1994).

Although a few Black nurses won individual cases in the 1980s (Ontario Ministry of Labour 1983; Calliste 1996b), neither their organizations nor their communities had the power to pressure local governments and health care facilities for institutional and systemic changes. The status quo (sometimes including the nurses' unions) was able to block resistance because economic and political power resources are unequally distributed. For instance in 1982–90 in Montreal, there were sixteen human rights complaints of racial discrimination and harassment in the health care sector. All were dismissed (Brown 1990).

There are some similarities between the nurses' and the porters' resistance to social oppression in their workplaces. Nurses and porters had to pressure their employers and unions as well as the state to investigate and settle their complaints.

Another similarity between the nurses' and porters' resistance strategies were the effective media campaigns to gain public support and to exert pressure on their employers and unions, as well as the state.

CONCLUSION

This study centred on nurses' and porters' resistance to racism and sexism in segmented labour markets supports an eclectic theoretical perspective: anti-racism, feminism, split and segmented labour markets and social closure by illustrating the relational aspects of social difference. The study also reveals some of the emotional, psychological and social costs of resistance. The settlements of some nurses' and porters' complaints indicate symbolic redress. Black railway workers were not fully compensated for the losses incurred as a result of discrimination (such as years worked with lower salaries and the cyclical effect of discrimination in housing and their children's education). Similarly, the OHRC and the QHRC's maximum compensation ($10,000) for mental anguish is grossly inadequate given some nurses are suffering from severe emotional, physical and psychological effects from their experience with racism, sexism and, in some cases, ageism and ableism

(CBWC 1995; interviews, May 25–26, 1994, and July 6, 1994). In order to minimize reprisals for resisting and to achieve institutional and systemic change, while at the same time addressing the politics of everyday racism and sexism, racialized minority nurses must take political action. They must unite and organize a strong association as well as actively participate and get elected to the executive of their unions, professional associations (Registered Nurses Association) and their licensing bodies (the College of Nurses of Ontario and the Order of Nurses of Quebec) where they could influence policy. Moreover, nurses must form meaningful coalitions and networks at the local, national and international levels with other anti-oppression movements in nursing as well as in other institutions (such as universities and colleges) and workplaces. We cannot eliminate racism, sexism and other interlocking systems of social oppressions in the health care system without simultaneously combating them in education and transforming the social and economic institutions of capitalist society.

NOTES

1. Although Black nurses perceive themselves as educated professional women, white women define the boundaries of professionalism, delineating separate standards for white and Black women, ensuring that the latter are assigned the lowest jobs in nursing, thus maintaining a racially stratified workforce (Calliste 1993, 1996b; Flynn 1998).
2. Although not all Caribbeans are Black, a large proportion are.
3. Some researchers (Beckford 1990; Williams 1978; Williamson 1986) argue that Black slaves never accepted the world view of the white capitalist class—they used the Sambo personality as a guerrilla strategy or role-playing technique to deceive and pacify whites—and that the latter needed the Sambo to feed their egos and mask their terror of slave revolts. Thus the Sambo image functioned to reinforce whites' view of themselves as Christian and civilized and their perspective of the slave as happy and childlike. It is plausible that in the pre-World War II period, some porters "samfied," a behaviour that was encouraged by the tipping system and the absence of due process (for example, a complaint from a passenger could result in demerit marks and even dismissal).

SPECIAL NOTE

I would like to thank the nurses, porters and their families, community nurses' organizations, union officers and others who shared information with me. I would also like to thank Debbie Murphy at the Department of Sociology and Anthropology at St. Francis Xavier University.

REFERENCES

Acting Chief, Admissions. 1956. Letter to the Acting Director, Immigration, May 9. Public Archives of Canada (PAC), Immigration Branch Records, RG 76, Vol. 847, File 553–110.

Albermarle Paper Co. v. *Moody.* 1975. 422 U.S. 405: 16, 417–18.

Alexander, L. 1972. Letter to J. Ewing, August 22.

Anderson, J. 1986. *A. Phillip Randolph.* Berkeley: University of California Press.

Armstrong, P. and H. Armstrong. 1990. *Theorizing Women's Work.* Toronto: Garamond.

Barbee, E. 1993. "Racism in U.S. Nursing." *Medical Anthropology Quarterly* 7(4): 346–62.

Beckford, G. 1990. "Plantation Capitalism and Black Dispossession." In A.W. Bonnett and G.L. Watson (eds.), *Emerging Perspectives on the Black Diaspora.* Lanham: Maryland: University Press of America.

Bederman, G. 1995. *Manliness and Civilization: A Cultural History of Gender and Race in the United States, 1880–1917.* Chicago: University of Chicago Press.

Blum, S. 1958. "Report of the National Committee on Human Rights." January 5, PAC, MG 28, V75, Vol. 14, File 14–13.

Bolaria, S. and P. Li. 1988. *Racial Oppression in Canada.* 2nd edition. Toronto: Garamond Press.

Bonacich. E. 1972. "A Theory of Ethnic Antagonism: The Split Labour Market." *American Sociological Review* 37: 547–59.

——. 1975. "Abolition, the Extension of Slavery, and the Position of Free Blacks: A Study of Split Labour Markets in the United States, 1830–1863." *American Journal of Sociology* 81(3): 601–28.

——. 1976. "Advanced Capitalism and Black/White Relations in the United States: A Split Labour Market Interpretation." *American Sociological Review* 41: 34–51.

Brandt, D. 1986. *The Realization of Anti-Racist Teaching.* London: Falmer.

Brazeal, B. 1946. *The Brotherhood of Sleeping Car Porters.* New York: Harper and Brothers.

Brown, E. 1990. "Hospital Colours: Nurses of Racial Minorities Suffer Harassment in Silence." *Montreal Mirror*, October 18–25.

Burawoy, M. 1976. "The Functions and the Reproduction of Migrant Labour." *American Journal of Sociology* 81(5): 1050–87.

Caissey, I. 1994. "Presentation to the City of North York's Community, Race and Ethnic Relations Committee." Toronto, October 13.

Calliste, A. 1987. "Sleeping Car Porters in Canada: An Ethnically Submerged Split Labour Market." *Canadian Ethnic Studies* 19(1): 1–20.

——. 1988. "Blacks on Canadian Railways." *Canadian Ethnic Studies* 20(2): 37–52.

——. 1991. "Canadian Immigration Policy and Domestics from the Caribbean: The Second Domestic Scheme." In Jesse Vorst et al. (eds.), *Race, Class, Gender: Bonds and Barriers.* Toronto: Garamond.

——. 1993. "Women of Exceptional Merit: Immigration of Caribbean Nurses to Canada." *Canadian Journal of Women and the Law* 6(1): 85–102.

——. 1993/94. "Race, Gender and Canadian Immigration Policy: Blacks from the Caribbean, 1990–1932." *Journal of Canadian Studies* 28(4): 131–48.

——. 1995a. "The Struggle for Employment Equity by Blacks on American and Canadian Railroads." *Journal of Black Studies* 25(3): 297–317.

——. 1995c. *End the Silence on Racism in Health Care. Conference Report.* Written for the CBWC, Toronto Chapter. Toronto: CBWC.

——. 1996b. "Antiracism Organizing and Resistance in Nursing: African Canadian Women." *Canadian Review of Sociology and Anthropology* 33(3): 361–90.

——. 2000. "Resisting Exclusion and Marginality in Nursing: Women of Colour in Ontario." In M. Kalbach and W. Kalbach (eds.), *Race and Ethnicity in Canada*. Toronto: Harcourt Brace.

——. Forthcoming. "Immigration of Caribbean Nurses and Domestic Workers to Canada, 1955–1967."

CBRT and CN. 1927. "Collective Agreement Between the CN and the CBRT for Employees in Sleeping, Dining and Parlour Car Service." June 1: 37–38.

——. 1951. "Collective Agreement Between the CN and the CBRT for Employees in Sleeping, Dining and Parlour Car Service."

CBWC (Congress of Black Women). 1991. Toronto Chapter. Press Conference, 27 March.

——. 1995. "End the Silence on Racism in Health Care: Build a Movement Against Racism, Discrimination and Reprisals." Conference. Toronto, May 25–26.

Collins, P. 1990. *Black Feminist Thought, Knowledge Consciousness and the Politics of Empowerment.* New York: Routledge.

Das Gupta, T. 1993. *Analytical Report on the Human Rights Case involving Northwestern General Hospital.* Toronto, September.

——. 1996. *Racism and Paid Work.* Toronto: Garamond Press.

Davis, A. 1983. *Women, Race and Class.* New York: Vintage Books.

Dei, George J. Sefa. 1995a. "Integrative Anti-Racism: Intersections of Race, Class and Gender." *Race, Gender and Class: Special Edition* 2(3): 11–30.

——. 1996. *Anti-Racism Education: Theory and Practice*. Halifax: Fernwood Publishing.

Dominique, M. 1997. "Racial Harassment in the Workplace and Building a Solidarity to Change the Situation." Address delivered to Professional Training in Health Relations, Program to Help Employers. November. Montreal.

Doris Marshall Institute and Arnold Minors & Associates. 1994. *Ethno-Racial Equality: A Distant Goal? An Interim Report to Northwestern General Hospital.* Toronto: authors.

Doyal, L., G. Hunt and J. Mellor. 1981. "Your Life in Their Hands." *Critical Social Policy* 1(2): 54–71.

Dyer, R. 1997. *White.* London: Routledge.

Edwards, R. 1979. *Contested Terrain.* New York: Basic Books.

Ehrenreich, B. and D. English. 1973. *Witches, Midwives, and Nurses: A History of Women Healers.* Old Westbury, New York: Feminist Press.

Essed, Philomena. 1991. *Understanding Everyday Racism.* Newbury Park, California: Sage Publications.

Flynn, K. 1998. "Proletarianization, Professonalization, and Caribbean Immigrant Nurses." *Canadian Woman Studies* 18(1): 57–60.

Foner, P. 1981. *History of the Labour Movement in the United States. Vol. III: The Politics and Practices of the American Federation of Labour, 1900–1909.* New York: International Publishers.

Fox-Genovese, E. 1988. *Within the Plantation Household: Black and White Women of the Old South.* Chapel Hill: University of North Carolina Press.

Frankenberg, R. 1993a. *White Women, Race Matters: The Social Construction of Whiteness.* Minneapolis: University of Minnesota Press.

Franklin, C. 1995. "Presentation at OHA Conference on Anti-Racism." Toronto, May 29.

Gamarnikow, E. 1978. "Sexual Division of Labour." In A. Kuhn and A. Wolpe (eds.), *Feminism and Materialism*. London: Routledge.

Gilroy, P. 1987. *There Ain't no Black in the Union Jack*. London: Hutchison.

Glazer, N. 1991. "'Between a Rock and a Hard Place': Women's Professional Organizations in Nursing and Class, Racial, and Ethnic Inequalities." *Gender & Society* 5(3): 351–72.

Gordon, D., R. Edwards and M. Reich. 1982. *Segmented Work, Divided Workers*. London: Cambridge University Press.

Gould, W. 1977. *Black Workers in White Unions*. Ithaca: Cornell University Press.

Grayson, M. 1953. Letter to H. Parr, September 15. PAC, MG 28, I 173, Vol. 34.

Hall, S. 1978. *Policing the Crisis*. London: Macmillan.

Hardill, K. 1993. "Discovering Fire where the Smoke Is: Racism in the Health Care System." *Towards Justice in Health* 2(1): 17–21.

Harris, W. 1977. *Keeping the Faith*. Urbana: University of Illinois Press.

Hartmann, H. 1976. "Capitalism, Patriarchy, and Job Segregation by Sex." *Signs* 2: 137–69.

——. 1981. "The Family as the Locus of Gender, Class, and Political Struggle: The Example of Housework." *Signs* 6(3): 366–94.

Hathiramani, H. 1980. "Tribunals Study Nurse Dismissal at Mac-Chedoke." *Silhouette*, September 18.

Head, W. 1986. *An Exploratory Study of Attitudes and Perceptions of Minority and Majority Group Health Care Workers*. Toronto: OHRC.

Henry, F., et al. 1985. *Who Gets the Work? A Test of Racial Discrimination in Employment.* Toronto: Urban Alliance on Race Relations.

Henry, F., C. Tator, W. Mattis and T. Rees. 1995. *The Colour of Democracy: Racism in Canadian Society*. Toronto. Harcourt Brace.

Hine, D. 1989. *Black Women in White*. Bloomington: Indiana University Press.

Hooks, B. 1981. *Ain't I a Woman: Black Women and Feminism*. Boston: South End Press.

Johnstone, T. 1964. Letter to B. Wilson, January 3. PAC, MG 28, I215, Vol. 14.

Jones, D. 1993. Letter to F. White, January 25.

——. 1994. Letter to F. Gallant, September 23.

Kamarowsky, A. 1961. Letter to S. Blum, May 25. PAC, MG 28, V75, Vol. 15, File 8.

Keene, J. 1992. *Human Rights in Ontario*, 2nd edition. Toronto: Carswell.

King, D. 1990. "Multiple Jeopardy, Multiple Consciousness: The Context of Black Feminist Ideology." In M. Malson, E. Mudimbe-Boyi, J. O'Barr and M. Wyer (eds.), *Black Women in America*. Chicago: University of Chicago Press.

Labour Gazette, The. 1920. March (20): 241.

——. 1927. January (27): 17–18.

Marks, S. 1994. *Divided Sisterhood: Race, Class and Gender in the South African Nursing Profession*. New York: St. Martin's Press.

Marshall, R. 1965. *The Negro and Organized Labour*. New York: John Wiley & Sons.

——. 1974. "The Economics of Racial Discrimination: A Survey." *Journal of Economic Literature* 12: 849–71.
McGuire, J. 1942. Letter to A. Mosher, November 6. PAC, MG 28, I215, Vol. 81, File—the Race Issue.
McIntosh, P. 1990. "White Privilege: Unpacking the Invisible Knapsack." *Independent School* 49 (2): 31–36.
McPherson, K. 1996. *Bedside Matters*. Toronto: Oxford University Press.
Mergen, B. 1974. "The Pullman Porter: From 'George' to Brotherhood." South Atlantic Quarterly 73(Spring): 224–35.
Miles, R. 1982. *Racism and Migrant Labour*. London: Routledge and Kegan Paul.
Murphy, R. 1988. *Social Closure: The Theory of Monopolization and Exclusion*. New York: Oxford University Press.
Northrup, H. 1944. *Organized Labour and the Negro*. New York: Harper and Brothers.
Nurses, Woodlands Hospital. 1992. Letter to the OHRC, 5 Dec.
Oatis v. *Crown Zellerback Corp*. 1968. 5th Circuit. 398 F.2d 496.
OHRC (Ontario Human Rights Commission). 1992a. "Case Report." Toronto.
——. 1992b. "Dunfermline Hospital: Systemic Analysis." Toronto.
Omni, M. and H. Winant. 1983. "By the Rivers of Babylon. Part One." *Socialist Review* 13(5): 31–65.
ONA. 1996. "Presentation to the Metro Anti-Racism, Access and Equity Committee." October 16.
Ontario Ministry of Citizenship. 1989a. "Nursing: Background Study Prepared by the TFAPTO. Toronto.
——. 1989b. *Access! Task Force on Access to Professions and Trades in Ontario*. Toronto: Queen's Printer for Ontario.
Ontario Ministry of Labour. 1983. *Sudbury and District Health Unit and ONA*. Toronto.
——. 1993a. *ONA and Dunfermline Hospital*. File 910215, October 13: 199.
——. 1993b. *ONA and Dunfermline Hospital*. File 900727, November 29.
——. 1996. *Belair Health Unit and ONA*. File 940354, July.
Parkin, F. 1979. *Marxism and Class Theory: A Bourgeois Critique*. New York: Columbia University Press.
Piore, M. 1971. *The Dual Labor Market*. Lexington, MA: D.C. Heath.
"Porter would Like Probe." 1961. *Winnipeg Free Press*, May 11.
Quigley, M. 1980. "Racism: A Growing Problem at McMaster." *Silhoutte*, September 18.
"Racism at McMaster." 1980. August 11 (statement).
Reitz, J., L. Calzavara and D. Dasko. 1981. *Ethnic Inequality and Segregation in Jobs*. Research paper, No. 3. Toronto: Centre for Urban and Community Studies, University of Toronto.
Robertson, M. 1985a. "A Victory for Black Nurses." *Afro-Can* 5(5), May: 1, 11.
——. 1985b. "Black Nurse Loses Petition." *Afro-Can* 5(10): 1.
Robinson, J. 1941. Letter to J. McGuire, March 27. PAC, MG 28, I 215, Vol. 81, File—The Race Issue.
Roediger, D. 1991. *The Wages of Whiteness*. London: Verso.

Sam, Y. 1986. "Black Nurses' Association Still Fighting against Discrimination." *Afro-Can* 6(9), September.
Smith, D. 1977. *Feminism and Marxism.* Vancouver, New Star Books.
Smith, W. 1964b. Letter to L. Smelter, November 26. PAC, MG 28, I 215, Vol. 15.
Tobin, L. 1959. Letter to H. Craib, July 15. PAC, MG 28, I 173, Vol. 34.
Vizkelety, B. 1987. *Proving Discrimination in Canada.* Toronto: Carswell.
Werier, V. 1961. "Target of Fight is Equal Rights." *Winnipeg Tribune*, April 22.
Williams, W. 1978. "The 'Sambo' Deception." *Phylon* 39(3): 261–63.
Williamson, J. 1986. *Rage for Order.* Oxford, U.K.: Oxford University Press.
Wilson, T.J. 1996. "Feminism and Institutionalized Racism." *Feminist Review* 52(Spring): 1–26.
Winant, H. 1994a. *Racial Conditions.* Minneapolis: University of Minnesota Press.
Young, D. 1992. *The Donna Young Report.* Toronto: OHRC.
Young, D. and K. Liao. 1992. "The Treatment of Race at Arbitration." *Labour Arbitration Yearbook* 5: 57–79.

Indigenous Voice Matters: Claiming Our Space through Decolonising Research

17

LINA SUNSERI

I begin this paper with a short Kanu helátukslá (thanksgiving), to respect the teachings received as a Ukwehuwé (First Nations person), of the Onyota′ aka (Oneida) nation, a?nowal talót^ (Turtle Clan). I thank Shukwayatisu (Creation) for all that we have been given: all our relations, all human and other animal species, the air, the water, the plants, the medicines, the trees, the fire, the Sun, the Moon, the thunders, and especially our gentle Mother Earth. I don't intend any disrespect for forgetting to name any creation (in our language, in our ceremonies, the address is quite long, so here is just a very small personal version of it). Tane • to niyohtuhake ukwanikula (our minds now stay as one).

During my doctoral studies, I read the work of Linda Tuhiwai Smith and found it both refreshing and validating that a scholar would write that, from an Indigenous colonised position, the "term 'research' is inextricably linked to European imperialism and colonialism. The ways in which scientific research is implicated in the worst excess of colonialism remains a powerful remembered history for many of the world's colonised peoples."[1] This is exactly how many members of my community, the Oneida Nation of The Thames in Canada, had expressed to me that they felt towards "Western" researchers and academics. Karen, a member of my community who participated in my research, explicitly told me, "Lina, you can't forget that we have been researched to death, and has anything good come out of it for our people? I don't think so. I am afraid that the one thing it has done is to make people think that we are screwed up, that our community has only problems."[2] Many experiences and encounters throughout my university education have also taught me that knowledge and power are indeed interconnected, as Foucault conceptualised for us in his works. Knowledge, through discursive formations that become "regimes of truth," has the function to (re)create power.[3] Hence, one has to maintain a critical eye upon existing research approaches practiced on Indigenous communities so as to ensure they do not do more harm to Indigenous communities.

Yet, as an Indigenous woman about to conduct my own doctoral research, I had many questions: "How can I negotiate the contradictions and complexities inherent in research?" "How can I ensure that my own research project is conducted within an Indigenous cultural context?," and "How can my work be part of a broader decolonising movement?" I now realize that asking such questions is part of an Indigenous methodology. To reflect on the political element of social research can only help me to carry out my research in such a way as to be respectful to my community and for my research to be part of the overall decolonisation of our Indigenous nations and of academia.

This paper will address the issues presented by my questions by first examining the historical relationship between research on Indigenous peoples and colonialism. Next, I will review my methodologies, discuss my own position in the research process and set out how I attempted to deal with the power dynamics involved in research. I do so by examining issues of subjectivity, reflexivity and representation. I ultimately argue that a collaborative research methodology can be a part of a wider decolonisation of research methodologies.

INDIGENOUS PEOPLES AS "OTHERS" IN RESEARCH: MISSING OUR OWN VOICES

For the most part, Western research has been part of an imperialist and colonialist agenda towards Indigenous peoples. Historically, social researchers, while achieving the status of authoritative voices of research about Indigenous peoples, have often disrespectfully represented our Indigenous cultural knowledge and disregarded our own established ethical protocols.[4] Such disrespect and/or misrepresentation occurs when researchers, in the name of the pursuit of knowledge, without consent take and dissect parts of dead human bodies, when such an act deviates from some Indigenous cultural norms that consider bodies sacred and not to be so misused. Another example consists in how "historical" accounts of Indigenous peoples have often been written by non-Indigenous researchers after only brief encounters with the communities and most often without integrating the oral histories of Indigenous peoples, discounting them as "biased" or as "fantasies." This results in only a partial "history" being known and in Indigenous voices being dismissed, silenced—although these rich oral histories have survived within Indigenous communities because of the resistance and persistence of our ancestors.

Social research has constituted a major vehicle for representing Indigenous peoples as the "Other" and Western groups as the "Self."[5] In such a representation, the Indigenous "Other" has been portrayed as an exotic figure, a representative of an inferior "dying" civilization. This is directly linked to colonialism because, for colonialism to be achieved and maintained, an active and conscious imagination of a future colonial nation had to be manufactured. This imagination included the necessity to imagine an "Other," a being that was seen in contrast to the colonisers. Within this binary construction of the "Self" and the "Other," the Western European "Self" was attributed with positive and progressive characteristics and the "Other" was constructed as a pre-modern and not totally human subject.[6] By attributing negative characteristics to Indigenous peoples, they have been pathologised and problematised, defining them either as genetically inferior or culturally deviant from the Western "Self."[7] The consequences of producing knowledge in this way have been various. As one example, Indigenous women have been constructed as physically strong, yet also sexually promiscuous, dirty and morally loose. Consequently, this led to strict scrutiny by state institutions of their assumed "unfit" mothering roles, with tragic results such as the experiences of the residential schools in Canada, of the Sixties Scoop in Canada, or of the Stolen Generations case in Australia.[8]

Conversely, Indigenous men have been constructed in other distinctive ways. Colonial discursive formations of Māori men, for example, constructed them as "noble, physically tough, staunch, and emotionless."[9] Such discourses have both homogenised Māori men and restricted them to the "physical" domain of Aotearoa/New Zealand society, a sphere that ultimately does not share the same social and economic status as the "intellectual" one reserved for men of Pákehá descent (from the non-Indigenous European settlers). Similarly, in North America, Indigenous men have been constructed as either noble savages who have a spiritual connection with nature but also a static identity frozen to the pre-contact primitive period, or as ignoble savages who are violent (as seen in popular portrayals of warrior images in the media), emotionally cold, lazy and drunk. The unequal power relationship founded upon colonial constructions demonstrates how knowledge and power are tied together within a colonial context.

Edward Said termed such Western constructions of the Other "Orientalism," where he was specifically referring to Western discourses of the area now known as the Middle East and Asia.[10] Stuart Hall[11] takes Said's analysis a step forward by applying the concept of Orientalism to a more general discourse that the West has constructed about the "Rest," hence making it applicable to an analysis of Indigenous peoples. Within this discourse of the "West" and the "Rest," as I have already discussed, Western societies became defined as developed, industrialised, modern, and progressive, while the "Rest" were defined in opposition to the West. This allowed for the notion of difference to exist, and difference to be here understood as less than the Western norm. Therefore, research about the "Rest" becomes part of a cultural archive, a building of knowledge of those societies constructed as both inferior and different from the West. Within this archive, only certain ways of knowing are viewed as valid epistemologies and normalised as universal truths, including the "truth" of the cultures of the "Rest" as interpreted and written by the West.[12]

Following a Foucauldian perspective, it must be noted that the formation of such a discourse is connected to unequal power relations; the West is the dominant group, with the ability and the resources to represent the world as it sees fit, and is able to write its version of the history between the "West and the Rest" through its own eyes and portray that version as the "truth." Within the social sciences, then, "valid" theories of human existence and development have been based on Eurocentric epistemologies, such as the view that societies are moving in a linear fashion towards a *progressive, modern* continuum[13] and, in terms of epistemology, that there ought to be a separation between body (senses) and mind (reason). The privileging of such an epistemology disallows alternative ways of knowing the same valid space in social research. Instead, only a marginal space—if any at all—is given to Indigenous knowledges. Although there has been an increase in the numbers of Indigenous scholars doing research with/of their own communities, new and/or persistent challenges, contradictions, and complexities exist. For instance, within the academic institutions, we are still placed in marginalised spaces, or not afforded the same deserved credibility as other scholars. In addition, Indigenous knowledges are often pressured to be moulded into forms that mainstream Western thought can better accommodate; in the process they are in danger of becoming appropriated and "translated" into new forms of knowledge.

One example of this is the proliferating use of "cleansing smudges"[14] in universities, performed by individuals who do not have the proper knowledge and/or have not been given the responsibility to do so by their communities. I have been a witness to such events, and when once I humbly asked a non-Indigenous person if she had obtained the proper training and consent to perform such a spiritual act, I was told that I was "too essentialist," that such spirituality is not exclusively owned by Indigenous peoples, and that I should see it as a progressive step that non-Indigenous peoples are "appreciating" Indigenous cultures. I could not perceive her actions as true appreciation, as it misplaced, misappropriated and misrepresented a sacred cultural form of knowledge. A specific challenge that Indigenous women/scholars face is that their research about Indigenous women's experiences, voices and histories is assumed to easily fit into some Western feminist framework, even though many Indigenous women do not feel that feminist theories and movements can always or easily apply when analysing Indigenous women's lives, especially without a deep engagement with colonialism.[15]

The connection between power and knowledge has not escaped some social researchers, some of whom are not Indigenous. Critical research, which often uses qualitative methods of inquiry, rejects, for example, the notion of a value-free science and is invested instead in both critiquing and, more importantly, transforming social relations.[16] Social research methodologies such as action-research, collaborative community-based research, feminist research, and critical ethnography[17] are guided by principles that move us away from the colonial legacy of social research. Researchers from those fields work towards sharing power with research participants, are allies in working towards emancipatory goals, value local (Indigenous in this case) knowledges, have long-term commitment to the communities they work with, and familiarise themselves with the ethics protocols of the communities. With the increasing presence of such researchers, perhaps the justifiable distrust Indigenous peoples have had towards social research will diminish.

Within Canada, something that could encourage a more trusting relationship between social researchers and Indigenous communities is the recent establishment by the Social Sciences and Humanities Research Council of specific guidelines for doing research with Aboriginal peoples, as a response to the criticisms expressed by Indigenous communities of past research practices. Recognising that Aboriginal peoples have specific rights and interests that must be met by researchers, the Council expects researchers to conduct accurate and informed research about Aboriginal peoples; that their research not cause further stigmatisation; that cultural property no longer be expropriated for the sole sake of research; that they respect the cultures and traditions of the Aboriginal groups they work with; that they establish partnerships with the community by involving them as much as possible in the research process; and that they make preliminary and final reports available to the community for review and comments.[18] These are principles that I believe can improve the relationship between social researchers and Indigenous peoples. However, this is a new development and there are still some persistent challenges that some Indigenous scholars have already addressed.

One concern that has been raised is about obtaining consent from the appropriate community representative, when the community and the appropriate body to give consent are not clearly defined. Indigenous communities are quite diverse and dynamic, and this could cause confusion and tension in how to proceed to obtain consent. My community, as with others of the Six Nations League, is made up of those who follow the traditional governing body of the Longhouse and those who are part of the Band Council government. As Martin-Hill[19] points outs, a researcher wanting to do research that is critical of the Band Council or investigating a community issue that the Band

Council is wary about, would encounter some resistance to her/his research, given that the Social Sciences and Humanities Research Council often regards the council to be the representative of an Indigenous nation, just as the Canadian state does. This could involve a very important and necessary piece of research, but if the researcher is restricted to one definition of community and/or community representative/leader, it would most likely be stopped. Hence, one needs to become very familiar with the dynamics of the community, and to be honest about how community is defined. Depending on one's own position and/or the nature and scope of the proposed research, there ought to be fluidity on how community members and leaders are defined and where to go to obtain consent and collaboration. In addition, through the whole process, the researcher needs to be transparent about which communities and community representatives he/she is working with.

ASSERTING OUR OWN INDIGENOUS VOICES IN SOCIAL RESEARCH

For many Indigenous peoples, there is an emotional component of knowledge that cannot be separated from other forms of knowledge. Knowledge is derived from, and connected to, both the external and the inner world.[20] Our knowledge must be acquired through establishing good and right relations. Establishing such relations "requires a balancing of all our capabilities as human beings to know the world around us."[21] Knowing the world requires that we connect to the inner world, to an emotional level of understanding, so as to become consciously aware of our personal connection to the topic and to the participants in our research, and thus clearly present to others our emotions about the ongoing knowledge we acquire. Indigenous methodologies are holistic in nature and include the concept of "relational accountability,"[22] referring to the recognition that we depend on everything and everyone around us and that "all parts of our research are related, from inspiration to expiration, and that the researcher is not just responsible for nurturing and maintaining this relationship but also accountable to 'all your relations'."[23] As Kovach[24] points out, we must speak from the heart, recognise that experience is a valid basis of knowledge, incorporate Indigenous methods such as storytelling into our work, and constantly have the interest of our collective community at heart when doing our research.

Throughout my dissertation research I tried to follow such Indigenous practices. I positioned myself as a member of the community that I was researching, I tried to maintain a healthy connection with all my relations and with the inner world, and I incorporated traditional stories into my writing. This type of research is referred to as a decolonising methodology,[25] meaning that researchers "research back"—a process whereby the researcher firstly acknowledges that Indigenous peoples have been constructed and represented in negative ways and that power and knowledge are interconnected. In doing so, Indigenous researchers provide an analysis of colonialism in their work and, most importantly, their academic work may become part of the larger struggle for self-determination.[26]

I use the term "Indigenous researcher" here to mean an individual of Indigenous descent who is connected with her/his land and community, and is familiarising (I herein stress the fluidity of it) her or himself with the culture, oral stories and teachings. Within the role of an academic researcher, this individual attempts to integrate his/her Indigeneity into the work she/he does and, in the process, is committed to working towards the decolonising embetterment of her/his community; follows Indigenous principles in the research process and in her/his relationships with all relations; aims to increase Indigenous ways of knowing in the academy; and, as an agent of change, struggles to make the academy more responsive and responsible towards Indigenous peoples.

I make clear in my work that, as an Oneida woman scholar, I have a personal and political investment in deconstructing master narratives of colonialism that have portrayed Indigenous nations as less progressive, intellectual and egalitarian than Western ones. As an Indigenous woman, my academic life is never separated from my everyday personal and political lives. My research methods are very much grounded in everyday life experiences and shaped by the connections I have with people, either through clan membership, nationhood, "sisterhood," or broader Indigenous networks.

In my dissertation, I offered an alternative vision of nation and national identity and connected these with gender issues. Ultimately, I hope that my work will offer an opportunity for women's voices in my community to be heard more during our progress on the path towards self-determination. I also want to ensure that these voices will be included in spaces outside of my community, such as in academia and other mainstream spaces. My research reveals that freedom from the destructive forces derived from colonialism entails a movement towards self-determination and a reestablishment of our own Indigenous ways of being and governing. I argue so because of the experiential knowledge gathered from myself and others in my community.

Much of the knowledge that I have obtained about my topic did not derive exclusively from conventional fieldwork, but also from informal, ongoing life-learning experiences acquired from my connections to my community—for example, from attending various traditional ceremonies and healing circles, and from listening to stories about our culture, about our history, and about matters of importance to our nation, told to me by my mothers, my aunties, and elders. When I decided that I wanted to examine the nature of Indigenous nationhood, to learn about our traditional ways of governing, the roles that women held in them and how, through colonialism, those things had changed and how our nation is trying to revitalise them, I talked to my family members and my Clan Mother to get some guidance on how to proceed, whom to contact and how I could learn about appropriate research protocols.

In the end, my "fieldwork" methods included narratives with twenty women, participation in ceremonies, oral histories, incorporation of teachings from traditional creation stories and a review of literature. The narratives were created through conversations I had with the twenty women, wherein together we framed themes for discussion rather than having pre-set questions formulated exclusively by me. It was admitted (both by me and by the participants who expressed their views on this subject to me), that our relationship was not totally equal. The participants and I recognised that, as the academic researcher, I would have responsibility for reviewing the literature, transcribing and doing the initial interpretation of our conversations, and organising and writing the thesis. After the initial interpretation of the content of the narratives, though, I made sure to have a follow-up discussion with them about my interpretations, and at times we had a deeper conversation about a theme that we felt needed further discussion or clarity. I showed them my final draft to ensure that they were in agreement with how the narratives were analysed and integrated within the whole work.

The broad spectrum of the methodologies that I incorporated in my research is in agreement with what is often recognised as Indigenous ways of transmitting knowledge. Indigenous ways of knowing, like dreams or ceremonies, do not necessarily conform to Western academic standards but, as an Indigenous person, I know that often it is exactly through "contextually based, rooted in place and time, spiritual practices"[27] that I can come to a specific understanding of the themes and issues that we are analysing. Being an Indigenous person who is connected with her community has provided me with the privilege to be able to see, hear, feel, and understand through such spiritually enriching experiences that are intrinsically part of most Indigenous cultures. As explained earlier, "Indigenous researcher" is herein defined as someone who is strongly rooted in the land, the people and the culture of her nation and whose one role as a researcher is to integrate that rootedness into her work and to be an agent of change in the various sites she occupies. This status as an Indigenous researcher ultimately leads to a particular analysis of the information that is shared with the participants—one that others who are not so connected may not be able to replicate.

Indigenous methodologies follow culturally-specific guidelines. One major guideline is respect for people.[28] The importance of this was mentioned to me by the women who participated in my own dissertation research, and involves both an individual and a collective level of respect. The first dimension means that I must adhere to a respectful relationship when interacting individually with each participant. For example, I had to respect each woman's wish not to have her name disclosed, and her wish to review my transcript and early drafts of my research. Of course, this is also a practice that is expected by most researchers, as my previous discussion of the principles of conduct established by the Canadian Social Sciences and Humanities Research Council pointed out. However, there are some culturally-specific ways of respecting that are often subtle, unspoken, yet embedded in many of the North American Indigenous traditions, that could be missed by a non-Indigenous researcher—such as the practice of offering tobacco or other medicine (strawberries during specific times, especially by and/or to a woman, as strawberries are considered women's medicine), non-interference when elders speak, offers of food (and being required to accept food offered to you, especially during cultural and spiritual ceremonies), and non-direct eye contact when someone is speaking that could be misinterpreted as disinterest.

The second dimension involves a communal responsibility: I must respectfully follow the community's established code of ethics and protocols, and ensure no harm is inflicted on the community due to my research activities. Doing no further harm to the community is an important principle that any researcher with integrity should follow, especially when working with Indigenous peoples, because of the stigmatisation and other harmful effects that past researchers have inflicted upon them. During my research, the importance of respect for the community's ethics became evident during one of the interactions with my Clan Mother who provided much of the teachings of our clan structures and women's responsibilities to me.

In an earlier draft of my dissertation, I had written the name of the Peacemaker who was mainly responsible for forming the Haudenosaunee (known as the Six Nations) League, of which my nation, Oneida, is a member. During a discussion, my Clan Mother reminded me of our community's tradition not to publicly write or spell out the name of the Peacemaker. I had been told of that tradition from other elders, but had forgotten to follow it while writing my first draft. I am thankful that I was reminded of it, because I feel I would have otherwise unintentionally done harm to my community and my action would have been interpreted as an act of disrespect.

Some of the women shared with me some complaints they had about how many Western academics had not demonstrated proper respect, and they reminded me to:

> make sure you show respect to the women you are interviewing. You have to be careful that you always respect us, respect our ways . . . You should show special respect to the Grandmothers, because you know that they know more about all this stuff you want to find out than you do . . . for example, when there are some traditions that can't be shared with others, because we consider them sacred, you can't disrespect our ways and go ahead and tell everyone all the sacred ways . . . Also have respect for all the Creator's beings . . . when you write about people you disagree with, write with respect, they are children of the Creator too—be gentle. (Lori)

To show respect does not, however, mean that you don't disagree or critique. A few fellow graduate students and others not familiar with Indigenous cultural ways have expressed to me their concerns about how I, as an academic, can keep a critical eye on the "data" if I must "show respect." But Indigenous ways of gathering knowledge contain a coexistence of critique and respect. Indigenous epistemology and methodology demand that the relationship between the researcher and the participants be built on sincere and heartfelt dialogue, so that good and right relations can be nurtured. This means that for a fully honest and respectful relationship to happen, different points of view, positions, experiences and interpretations should be shared. There are culturally appropriate ways through which one has a responsibility to share his/her story with the other, and for each party to be open to hear each other's story. As a researcher, I need to have a "compassionate mind in methodology."[29] I have to develop a method of listening and acquiring knowledge through sharing my own stories with the participants, while listening carefully and with an open mind to their stories. Together we can (re)tell the story, after each of us has had a chance to (re)consider our own position.

Another component of Indigenous methodology is the "seen face,"[30] meaning that the researcher is/becomes familiar with the community he/she is researching. For my research, I chose informal face-to-face collaborative narratives as one method to achieve this. Additionally, I participated in many social, cultural and traditional ceremonies and political activities in my community, some of which became direct sources of knowledge for my research. Although some non-Indigenous researchers could engage in some of these activities through a "participatory ethnography," others are restricted to members of the Oneida and/or Six Nations League. Despite the fact that I had lived a few hours away from the community during my eight years of graduate work, I tried to attend many traditional ceremonies, pow-wows and other gatherings. To a number of the participants in my research, I was a close acquaintance; to others, they "knew [me] by face. I know I have seen you at places" (Debbie). This familiarity was for the most part welcomed by the participants, as they felt that "you are not just using us to get your degree or something. You know what I am talking about when I talk about the way ceremonies are done, because you have been there yourself."(Debbie) Remarks such as these point out the importance of familiarity and connections for Indigenous peoples. I, the researcher, and the participants shared a form of bonding that extended well beyond the duration of the "interview." Through our interactions, some participants and I felt at "home" with each other, shared some common stories about life as Oneida members and had the same passion in our desire to restore a healthy nation.

As an insider to the Indigenous community, I am aware that there are types of knowledge that are often shared among ourselves in specific contexts that some members feel protective of and are reluctant to share with outsiders. Therefore, it is important that, together with the research participants and with the permission of the proper keepers of our culture (in my case, I often checked with my Clan Mother), decisions are made about what it is appropriate to share or not with "others." Of course, this familiarity becomes even more important in some urban spaces where there are few familiar faces with whom an Indigenous person can share this sense of belonging.

Not flaunting one's own knowledge is another principle within an Indigenous methodology. I didn't consider myself an "expert" merely because I was a member of the Oneida Nation or because I had a high level of post-secondary education. I do believe that my experiences and my insider position in the research allowed me to have some tacit knowledge of the topics and issues that "outsiders" may not have. A participant supported my belief by stating that "I know that you understand what I am speaking about. For example, you know what it is like to have experienced those racist looks . . . You have been to a lot of our ceremonies where you know how important women's places are there. I know you can believe me when I say those things, because you have seen it yourself."(Lori) These comments highlight how familiarity and commonality can become characteristics of the relationship between the researcher and the participants when the researcher is perceived to be an insider. Obviously, for Lori, and arguably also for other participants in my research, our common experiences and shared knowledge of cultural traditions gave them a sense that their lived experiences were validated and accepted by the listener. However, I was mostly a learner during my research process.

This learner position was felt by me throughout my conversations with the twenty women participants and during the many traditional ceremonies where teachings about our culture and governance occurred. My learner identity became evident during my discussions with my Clan Mother, who is a lot more knowledgeable than I about the history of our people. Hence, I agree with Hokowhitu[31] that the research done by many Indigenous people might differ from that of others in that we are both the researcher and the researched. This specific position meant, for example, that during the years of "researching" through the historical experiences of Oneida women, I was often left with many deep emotions: I wasn't reading about other peoples' history, but my own, and that of my own ancestors. The experiences which the participants disclosed to me touched my heart very deeply. I had similar experiences to the participants and therefore I relived the pain during our conversations; their anger and hope about their community's ongoing struggles were also felt deep inside of me, since it is also *my* community.

Although a more mainstream research approach would perceive the attachment that I have with the community as a source of bias that could potentially "contaminate" my research, in contrast, an Indigenous research methodology sees this subjectivity as a strength in a more holistic and genuine research process, one that shapes a truly collaborative relationship.

MY COLLABORATIVE NARRATIVE APPROACH

My conversations with the research participants were constructed in such a way that they can be termed "collaborative narratives,"[32] wherein meaning is mutually constructed between all parties and there is a joint reflection on shared experiences between the researcher and the other participants. This type of approach is very suitable for Indigenous research, because it permits Indigenous participants to be equal participants in the initiation, representation, legitimation, accountability and benefits of research.[33] In my particular case, however, I initiated all the research-related processes, although some of the participants did initiate contact with me after they had heard about my ongoing research. The initial contact with most of the participants was done informally during pow-wows, ceremonies or while chatting in somebody's kitchen or family room.

Although I likely benefited from the research mostly because I was able to complete my doctoral degree, I believe that there are some potential benefits my research can bring to my community. My work, for instance, could increase knowledge of my community's issues of colonialism and of the efforts to rebuild a self-determined nation. This knowledge can only help to build a better relationship between Oneida people and other groups in Canada, as the latter can become better informed of the historical context of Oneida's contemporary demands for self-determination. Also, the experiences shared by the women involved in my research highlight the multiplicity and complexity of Indigenous identity politics in North America. Their voices need to be heard and attended to while various Indigenous communities across Turtle Island/Canada move towards forming a decolonised nation. These women reported that a decolonising nationalist movement can bring back women's powerful positions in their nation only if such a movement is inclusive of the different voices and experiences of colonialism and "Native-ness," and issues of gender are fully incorporated into the evolving Oneida's nationalist discourse.

I agreed to give a copy of the finished work to any of the participants who wished to have it. I also intend to write a smaller version that emphasises the narratives and the history of our nation. I feel that these two parts are the ones that are of more interest to most members of the community and could be useful to better understand the complex lives of Oneida women, to teach the young of our history with the use of both mainstream textual sources and oral histories and, finally, to use the historical analysis of colonialism for political strategies by the community within the state and mainstream Canadian society. In addition, as an Indigenous academic who is eager to help in decolonising the academy, I plan to use my research in my courses to provide students with the necessary Indigenous knowledge that they often miss.

Within a collaborative narrative methodology, negotiation occurs between the parties to the research throughout the whole research process. For example, in my case, we agreed that research was to be conducted in a way that was respectful to and reflective of an Indigenous framework, and therefore there had to be consensus about the research process. Construction of meaning through the research, then, was conducted through a joint effort, which allowed for power sharing to occur. I consciously made efforts to devolve power and give opportunities to the participants to be active agents and share control as to how their words were interpreted, written, and what the meaning of their daily life activities were. Together, we constructed meaning, and this meaning-making followed a spiral process whereby we revisited our interpretations until we were satisfied that an agreed-upon interpretation of experiences was achieved.

I want to share with you how illuminating this research method was for me. There was a case where a participant (Lisa), after reading my first draft of the analysis of our first taped conversation, disagreed with how I had interpreted her views on feminism. I had thought that she disliked feminist movements and did not see them as being relevant to an Indigenous worldview and experience. Lisa corrected me and stated that she didn't dislike feminism but only thought that, since most feminists seemed to be exclusively concerned with European women's issues, feminism couldn't as easily apply to Indigenous women's reality. Rather than following a neat, conventional linear progressive direction in the research, my methodology reflected a spiral process. A spiral method is characterised by its continuously revisiting, reexamining, and refining of ideas and theoretical assumptions, due to the ongoing process of collaboration between the participants. This spiral method allowed for us to revisit our conversation, to reexamine our positions and refine our ideas, thereby resulting in a collaborative meaning-making process. During this spiral process, we weaved new stories.[34] Together, the twenty women and I acknowledged our participatory connectedness and denied the distance that some other conventional and/or positivistic research methodologies are characterised by.

This connectedness was also evident in other aspects of my research process. I participated in many activities and traditional ceremonies where I acquired knowledge of traditional ways of governing and of contentious issues within the self-determination movement. Within these activities I witnessed the active roles that women had and continue to have in our nation. This experiential knowledge enriched and complemented the knowledge gained from interacting with the twenty participants and from reviewing the existing literature on my topic. While accompanying some of the women on a trip to a political gathering in our traditional Oneida territory, which is located in what is now known as New York State, I felt connected to a place far away from where we live, but to which all Oneidas are affiliated because of our spiritual connection to that land and all our relations there. No interviews could have shown me how important traditional territory is to Oneida people. The tears in our eyes, our smiles when an eagle flew above, and the warm words spoken by the elders while standing on that land are precious and valid ways of Indigenous knowing. Later on, when some of the women and I spoke of those events, we weaved meaning into our Oneida history by recollecting and interpreting that experience and connecting it to some of the themes that had arisen throughout our conversations.

CONCLUDING REMARKS

Although historically, social research has not always been a friend to Indigenous peoples—often more an imperial gaze distorting views of Indigenous societies—we are now embarking on a new path, wherein Indigenous peoples are reclaiming our own voices in research (as researchers and researched) and demanding that our perspectives of the history between the "West" and the "Rest" be given equal status in the existing literature. We are (re)establishing our ways of doing research and (re)presenting knowledge, with the self-determination of our Indigenous nations as the primary goal. In the process, Indigenous peoples will move beyond being considered as "objects" of study under a Western gaze, to become active participants and producers of our own knowledge. In doing so, Indigenous communities all over the world are increasingly developing spaces where the mind, spirit, body and heart of our peoples can be decolonised. Some non-Indigenous researchers have joined this new path in their own ways by forming more respectful and collaborative relationships with the Indigenous communities that they have been working with and learning and/or using elements of Indigenous epistemologies and methodologies.

At the end of November 2005 I was fortunate to attend and present a paper at the World Indigenous Peoples Conference on Education in Aotearoa/New Zealand, hosted by Te Wa¯nanga o Aotearoa. The event was attended by over ten thousand delegates from all corners of the world, and almost all of these individuals were Indigenous. The theme of the conference was "Te Toi Roa—Indigenous Excellence," and it accurately reflected what I witnessed that week. It was a space where our collective experiences were celebrated, where we shared our values, stories, energies, all in the spirit of Indigenous cultures. From the beginning, we were welcomed in a traditional Ma¯ori ceremony at the Turangawaewae Marae or meeting house at Ngaruawahia and felt at home among the Indigenous people of Aotearoa/New Zealand, truly becoming part of a large Indigenous wha¯nau (family). At the end of this hui (gathering), we learned of interesting and innovative ideas about empowering the various Indigenous communities that exist on Mother Earth: from stories of emerging leadership that are serving to better our communities, to examples of existing research that truly decolonises, to stories by educators of the revitalisation of Indigenous knowledges both inside and outside academia. I left the conference with so much pride in our Indigeneity, realising that Indigenous cultures are still strong and positive, despite centuries of attack by colonialism, and that we are leaving a legacy for our future generation to move forward on the path of decolonisation.

I want to end this paper with some powerful words spoken at that conference by a Hawaiian scholar, Dr Manulani Aluli Meyer. They might not be her exact words, since I am repeating them here as I remember them, as they touched my heart. "It's time. Time to recognise the legitimacy of our own interpretation of the world . . . our Indigenous knowledge is a spiritual act; we are earth and our ways of knowing are embedded in it. Mahalo to our Māori cousins for hosting this event! May you return to your homelands refreshed and uplifted." I did, indeed. I hope that my own words in this paper will also serve to uplift your spirits and help you believe that our Indigenous knowledges are rich and need to be nurtured, so that all of us—Indigenous and non-Indigenous alike—can take care of all our relations on Mother Earth and build decolonised relationships with one another.

ACKNOWLEDGEMENTS

This paper is based on my dissertation research and in particular the part dealing with methodology. An earlier version of this paper was presented at the World Indigenous Peoples Conference on Education in Hamilton, New Zealand/Aotearoa, on 29 November 2005. I want to thank the members of my supervisory and examining committees for their helpful suggestions. I am also grateful to the organisers of the World Indigenous Peoples Conference on Education and those who attended my presentation and provided positive feedback and constructive criticisms. I also want to express my biggest gratitude to the women who participated in my research.

NOTES

1. L.T. Smith, *Decolonizing Methodologies: Research and Indigenous Peoples* (London and New York: Zed Books, 1999), 1.
2. All names of participants used in this paper are pseudonyms as decided and chosen by them.
3. M. Foucault, *Power/Knowledge: Selected Interviews and Other Writings* (New York: Pantheon Books, 1980).
4. R. Bishop, *Collaborative Research Stories: Whakawhanaungatanga* (Palmerston North, NZ: The Dunmore Press, 1996); S. Hall, "The West and the Rest: Discourse and Power," in *Formations of Modernity*, ed. S. Hall and B. Gieben (Cambridge: Polity Press and The Open University, 1992); J.S. Youngblood Henderson, "Postcolonial Ledger Drawing: Legal Reform," in *Reclaiming Indigenous Voice and Vision*, ed. M. Battiste (Vancouver: University of British Columbia Press, 2000); M. Kovach, "Emerging from the Margins: Indigenous Methodologies," in *Research as Resistance: Critical, Indigenous, and Anti-Oppressive Approaches*, ed. L. Brown and S. Strega (Toronto: Canadian Scholars' Press, 2005); R.L. Louis, "Can You Hear Us Now? Voices from the Margins: Using Indigenous Methodologies in Geographic Research," *Geographical Research*, 45:2 (2007), 130–139; E. Said, *Orientalism* (London and Henley: Routledge and Kegan Paul, 1978); Smith, *Decolonizing Methodologies*; K.H. Thaman, "Decolonizing Pacific Studies: Indigenous Perspectives, Knowledges and Wisdom in Higher Education," *The Contemporary Pacific*, 16:2 (2004), 259–84.
5. Ibid.
6. Ibid.
7. Louis, "Can You Hear Us Now?"
8. *Until Our Hearts Are On the Ground: Aboriginal Mothering, Oppressions, Resistance and Rebirth*, ed. D.M. Lavell-Harvard and J. Corbiere Lavell (Toronto: Demeter Press, 2006).
9. B. Hokowhitu, "Tackling Māori Masculinity: A Colonial Genealogy of Savagery and Sport," *The Contemporary Pacific*, 16:2 (2004), 259–84.
10. Said, *Orientalism.*
11. Hall, "The West and the Rest."

12. Smith, *Decolonizing Methodologies*; J. Archibald, "Coyote Learns to Make a Storybasket: The Place of First Nations Stories in Education," unpub. PhD diss. (Vancouver: Simon Fraser University, 1997); Kovach, "Emerging from the Margins."
13. Kovach, "Emerging from the Margins."
14. A "cleansing smudge" is a practice wherein a mix of Indigenous medicines like sage and sweet grass are burned in a smudge pot and the pot is passed around the people who cleanse their mind, heart, spirit and body by slowly, with their hands open, throwing the smoke generated by the medicines over their face, head, and (for some) the whole body. This practice is done in order to start any discussion, meeting or event with a clean and open mind, bringing in positive energy and cleansing oneself of any bad energy.
15. For a further discussion of the relationship between Indigenous women and feminisms, I refer readers to my dissertation which has an extensive review of the subject: M Sunseri, "Theorizing Nationalisms: Intersections of Gender, Nation, Culture and Colonialism in the Case of Oneida's Decolonizing Nationalist Movement," unpub. diss. (Toronto: York University, 2005), in particular chapters 1 and 6. See also A. Moreton-Robinson, *Talkin' up to the White Woman: Indigenous Women and Feminism* (Brisbane: University of Queensland Press, 2000).
16. Brown and Strega, *Research as Resistance.*
17. C.K Banks and J.M. Mangan, *The Company of Neighbours: Revitalizing Community Through Action-research* (Toronto: University of Toronto Press, 1999); B. Berg, *Qualitative Methods for the Social Sciences*, 6th ed. (Boston: Pearson Education, 2007); M.M. Fonow and J.A. Cook, *Beyond Methodology: Feminist Scholarship as Lived Research* (Bloomington: Indiana University Press, 1991); B. Harrison, *Collaborative Programs in Indigenous Communities: From Fieldwork to Practice* (Walnut Creek: Altamira Press, 2001); P. Lather, *Getting Smart: Feminist Research and Pedagogy with/in the Postmodern* (New York: Routledge, 1991); K. Strand, *Community-based Research and Higher Education* (San Francisco: Jossey-Bass, 2003).
18. See www.pre.ethics.gc.ca/english/policystatement/section6.cfm#6 as last accessed on 20 October 2007.
19. D. Martin-Hill, "What Is Community for the Purpose of Research?," keynote address presented at the Canadian Indigenous and Native Studies Association Annual Meeting at the 76th Congress of the Humanities and Social Sciences, Saskatoon, 2007.
20. K. Hodgson-Smith, "Seeking Good and Right Relations: Aboriginal Student Perspectives on the Pedagogy of Joe Duquette High School," unpub. MA diss. (Saskatoon: University of Saskatchewan, 1997).
21. Ibid., 38.
22. Louis, "Can You Hear Us Now?;" M. Stewart-Harawira, "Cultural Studies, Indigenous Knowledge, and Pedagogies of Hope," *Policy Futures in Education*, 3 (2005), 153–63.
23. Louis, "Can You Hear Us Now?"
24. Kovach, "Emerging from the Margins."
25. Smith, *Decolonizing Methodologies.*
26. Ibid.,7.

27. Louis, "Can You Hear Us Now?"
28. Ibid., 120.
29. Hodgson-Smith, *Seeking Good and Right Relations*, 48.
30. Smith, *Decolonizing Methodologies.*
31. B. Hokowhitu, "Te Mana Māori – Te Tatari i Nga Korero Parau," unpub. PhD diss. (Dunedin, NZ: University of Otago, 2002).
32. Bishop, *Collaborative Research Stories.*
33. Ibid.
34. Ibid., 232.

18

Rebellion of the Truth

TAIAIAKE ALFRED

> A Warrior is the one who can use words so that everyone knows they are part of the same family. A Warrior says what is in the people's hearts, talks about what the land means to them, brings them together to fight for it.[1]
>
> —Bighorse, Diné

Onkwehonwe existences in all their diverse expressions and experiences are rooted in the recognition and respect of sensitivity to one's place in creation and awareness of one's place in a circle of integrity. Our goal, regenerating authentic Onkwehonwe lives, means finding ways to restore the connections that define indigenous consciousness and ways of being; it means individuals and communities seeking the re-achievement of the elements of integrity: strength, clarity, and commitment. There are many pathways to the achievement. The freedom and power that come with understanding and living a life of indigenous integrity are experienced by people in many different ways, and respect must be shown to the need for individuals to find their way according to their own vision. We who seek to bring about change in others and in society can only offer guidance out of shared concern and reflection based on our own experience, so that others anxious for the journey can listen and then embark on the challenge for themselves. Awakened to their own freedom and power-generating potential, people living colonial lives will rise to the challenge and move toward achieving integrity.

We cannot give people definitive answers to their problems, or hope to lay out predetermined steps for them to recover indigenous identities; colonialism has affected people in too many varied and complex ways for simple answers to suffice. There is no redemptive teaching or easy answers on how to be happy, only directions towards the truth and rough pathways to freedom. This is what we have to offer.

PATHS OF LEAST RESISTANCE

Before we speak of restrengthening, we should begin with an acknowledgement of how weakened Onkwehonwe have become: as a whole we have been dispossessed of land and culture and disempowered as peoples. The main implication of this disintegration of important aspects of indigenous integrity is the spiritual defeat of our people on an individual level. The culture of being colonized takes away a peoples' ability to resist the racist aggression and political, economic, and cultural pressures of the colonial state and Settler society to surrender remaining land and rights and to further assimilate culturally. This relation of culture and politics is the fact of our existence. There is no longer a culture that supports resistance in most Onkwehonwe communities; it has long been defeated by force, co-optation, and a relentless grinding-down of the strength that was manifested in the resistance of the leaders and families who held on to Onkwehonwe ideas and identities. Political culture, if it exists at all in Onkwehonwe communities, is centred around the accommodation of oppression and modes of cynical survival. Political life in Onkwehonwe communities today is a vast hypocrisy unhinged and cut off from the cultural roots of peoples' lives, and Onkwehonwe have become subjects of the Settler society's power and colonialist manipulation on all levels of our existence.

The most obvious sign of Onkwehonwe weakness is the fact that fear, corruption, and greed have become normalized as governing principles in so many of our communities. In too many First Nations, traditional ceremonies and practices have become nothing more than a cover for the cynical manipulation of our peoples' weakness by the state in collaboration with our own indigenous politicians. One recent high-profile example serves to illustrate the continuing trend toward a complete abandonment of any form of resistance on the part of First Nations politicians. In 2003, two First Nations band councils agreed to pay the province of British Columbia $25,000 per year for the use of four acres of land in a resort town located in the First Nations' shared traditional territory. This was the first publicized case of Onkwehonwe paying white people for their own lands, of actually giving money to the people who stole their land in exchange for permission to use the very property that was stolen from them in the first place and which remains in legal limbo as far as its legal title goes, even in colonial courts. It would be easy to mock the band chiefs' cowardice, or shed tears over the loss of pride and culture that this represents, but this is the present reality of First Nations politics.[2]

This particular land deal was part of a band economic development scheme agreed to by the province. In exchange for the two band councils' participation in the City of Vancouver's successful bid to host the 2010 Winter Olympics, which is to take place in part in the resort town of Whistler, they were awarded the privilege of operating a cultural centre in the Olympic Village as a tourist attraction—to market their melanin and play the role of Friendly Indian for the otherwise white-washed Vancouver Olympic bid. The bid is a real estate developer's $850-million bonanza, bringing more jaded tourists, generic buildings, pollution-spewing resorts, and $387-million worth of paved roads and rail lines into the territory. The village of Whistler itself is one of the fastest growing winter playgrounds of the rich. Rather than dealing with land rights questions, the colonial government sought out Onkwehonwe politicians willing to speak in favour of the Olympic bid and offered them a partnership, making the deal one which money-minded band council chiefs could not refuse.

There are many questions being asked by Onkwehonwe communities about whether or not the band council chiefs should have accepted such a blatant pay-off for their support of the Olympic bid. Also, given the ongoing fight of Onkwehonwe in the area to protect their lands from predatory ski resort developers, and the fact that the planned Olympic facilities are to be situated on the pristine and unsurrendered lands of the Stl'atl'imx people (as is the resort town itself), there is much questioning of the sense and ethics of the band council's thrust toward an economic development agenda based on tourism. In justifying his council's actions, one of the band chiefs said that paying money to the provincial government in order to get involved in the tourist trade was the "path of least resistance." We can only suspect he believes this is a good thing.

Mimicking the masters of the money game, the chief informed the public that, "If you want to get into business, you've got to have the ability to go with the flow." How can anyone question that statement? But Onkwehonwe in the communities are probably asking themselves deeper questions than the chief is and wondering if "going with the flow" is the right thing to do when the mainstream is heading straight down the drain.

In colonized First Nations politics, principles don't mean a thing. Even moderate activism, the tame and constrained legalism of the current generation of leaders—futile court action to gain recognition of their rights and land title—is being rejected by the band council businessmen as too militant an approach. Further comments of the chief quoted above testify to a rejection of the responsibility to defend the peoples' ancestral rights in favour of embracing a capitalist ethic: "you know, we could spend a lot of money litigating, which would eat into the development dollars and profits." Going far beyond adopting the mentality of their colonizer, the new Onkwehonwe capitalists have taken to emulating the crass greed of their new partners in colonial crime. Said the chief of the deal, "this thing's going to give us the ability to shake all the dollars we can out of the tourists." Nice.

This deal is nothing more than a sell-out designed to benefit elite politicians on both sides of the colonial divide. It may be that some creative and (almost) credible arguments could be put forward if the package included skilled job training and long-term employment opportunities, as well as guaranteeing executive and management positions for band members. That at least would be justifiable in a materialist logic. But when questioned about the lack of experience and current capacity of people in their own communities to run even the cultural tourism centre, the chief responded, "That's the thing about having money, you can buy expertise, right?" Right. With that, it is clear that this deal is just another feeding frenzy for white consultants and band council politicians, who are collaborating to bury their faces in the government funding trough (which, it should be noted, provides the cash for the whole scheme). So much for the community development argument.

Aside from the common workaday corruption that continues to embed itself deeper in the lives of our communities through these types of deals, the most disturbing thing about the whole "cashing out" approach to decolonization—trying to buy freedom—is that the economic development agenda is founded on a basic concession to white power and the willing Onkwehonwe surrender of fundamental rights. In order to gain the benefit they expect from economic development partnerships, band council politicians have to accept the promise of money in exchange for protecting their people, lands, and cultures.

Situations such as this leave Onkwehonwe asking: Who is left to defend the land?

In the Whistler development deal, the band councils have sacrificed pristine valleys full of cultural memory, medicine, and spiritual power so that white people can build more ski runs for the Olympics and, later, turn them into tourist resorts. And they are promoting the land speculation that is already underway in the northwest corridor between the City of Vancouver and the resort area. The natural ecological balance will be disrupted and animal habitat will be lost as a result of huge increases in population and urban infrastructure. This will make it nearly impossible for people to use their lands for hunting, medicines, and ceremonial purposes. It is a sad claim to being Onkwehonwe when the identity being constructed is so closely associated with the destruction and not the love of nature.

What does it mean to be Onkwehonwe? The performance of dance shows and tourist "art" is not real culture. These are commodified artifices designed, packaged, and practised to entertain the rich and jaded bourgeois of the world, satisfying their craving for an "authentic" cultural experience—something completely out of reach in their own lives. Worse than just a pseudo-cultural sham, the tourist performance presents a false face of

Onkwehonwe life to the world, one that is completely different from the politics, spirituality, and truth of lived Onkewhonwe existences. The messy complexity and aching conflicts of our real lives don't market well to people on vacation, so the First Nations businessmen market romantic lies.

The chief who promoted the village deal was also quoted as saying that "we're setting a trail for the other First Nations. That's how I see it anyways." He sees the path to the future as a forest trail freshly paved over with money. It's plain and obvious by their actions that his band councillors also see it this way. From a true Onkwehonwe perspective, it's difficult to grasp the crude materialism and total disregard for our heritage and cultural values exhibited by these colonized band council chiefs—it's as if they're looking at the world through big 1970s-era sunglasses with dollar signs on the lenses that make everything look moneyed! Yet, I must grudgingly admit that these unfortunate glasses are in vogue; the weak-kneed "least resistance" mentality is, in fact, the dominant ethic in First Nations political circles today.

What about taking an honest and principled stand for what is right, working together and sacrificing to force change on the path of maximum resistance? Those words are a joke to most leaders in the band council system, who believe that any alternative to selling-out (or "paying-out," in this case) has been discredited. At political meetings and conferences, you can hear chiefs saying it all the time: "Nobody wants to fight anymore." Or, "We tried being adversarial in the 1960s and look where it got us!" So, was the short-lived radical phase of the American Indian and First Nations organizations responsible for poisoning the waters of integrity? It is not that simple. The path of least resistance I am pointing to is corrupted by greed. The majority of band chiefs don't care about community accountability and questions of integrity because the colonial gravy train keeps dropping loads of cash into their coffers. As a result, they continue to play their designated and essential role in the colonial system. Even when the endemic corruption of the Indian Affairs system is exposed and broadcast in the media, as has repeatedly occurred in both Canada and the United States, nothing of consequence happens to those involved.[3]

What kind of game are these band council chiefs playing? Onkwehonwe are seriously thinking about the future and the costs of continuing the charade that giving more authority and money to band and tribal councils—the very instruments put in place by the Settlers to divide and conquer their nations—will solve the problems and change the injustices faced by the masses of indigenous people on reserves and in the cities. Only people who have become dependent upon the state for their survival can possibly see a brighter future for their children in the bureaucratic notion of decolonization; enhanced governmental accountability mechanisms coupled with increased jurisdictional authority and more diverse governmental revenue streams are hardly an inspiring vision. When people who are *not* working for Indian Affairs hear that kind of language, they wonder what it means to anybody but government agents and colonial cooperators.

We must cut through the political rhetoric and legal double-talk. People need to think hard about the problems in their community and what is at stake in their children's lives and then ask themselves if the economic development and self-government agenda promoted by colonial authorities and cooperators is at all relevant to reality. The problems faced by Onkwehonwe have very little to do with the jurisdiction and financing of band councils or even with high unemployment rates. The real problems are the disunity of our people, the alienation of our youth, our men disrespecting our women, the deculturing of our societies, epidemic mental and physical sicknesses, the lack of employment in meaningful and self-determining indigenous ways of working, the widespread corruption of our governments, and the exploitation of our lands and peoples—all of which most of our current leaders participate in, rather than resist. Thinking this way, most Onkwehonwe will agree with me that against the fearsome enormity of this spiritual crisis, the chiefs' money-minded "path of least resistance" makes these men seem pale, small, and weak.

THE ETHICS OF COURAGE

Not all of us have been conquered. There are still strong Onkwehonwe who persevere in their struggle for an authentic existence and who are capable of redefining, regenerating, and reimagining our collective existences. If we are willing to put our words into action and transform our rhetoric into practice, we too can achieve the fundamental goal of the indigenous warrior: to live life as an act of indigeneity, to move across life's landscapes in an indigenous way, as my people say, *Onkwehonweneha*. A warrior confronts colonialism with the truth in order to regenerate authenticity and recreate a life worth living and principles worth dying for. The struggle is to restore connections severed by the colonial machine. The victory is an integrated personality, a cohesive community, and the restoration of respectful and harmonious relationships.

Translating this ethical sense and idea on a way of being into a concise political philosophy is difficult, for it resists institutionalization. I might suggest, as a starting point, conceptualizing *anarcho-indigenism*. Why? And why this term? Conveyance of the indigenous warrior ethic will require its codification in some form—a creed and an ethical framework for thinking through challenges. To take root in people's minds the new ethic will have to capture the spirit of a warrior in battle and bring it to politics. How might this spirit be described in contemporary terms relating to political thought and movement? The two elements that come to my mind are *indigenous*, evoking cultural and spiritual rootedness in this land and the Onkwehonwe struggle for justice and freedom, and the political philosophy and movement that is fundamentally anti-institutional, radically democratic, and committed to taking action to force change: *anarchism*.

This philosophical outlook is close to what Vaclav Havel described as his utopia—in his terms, a decentralized economy, local decision-making, government based on direct election of political leaders, and the elimination of political parties as governing institutions, a sort of a spiritual socialism:

> It's hard to imagine the kind of system I've tried to describe here coming about unless man, as I've said, "comes to his senses." This is something no revolutionary or reformer can bring about; it can only be the natural expression of a more general state of mind, the state of mind in which man can see beyond the tip of his own nose and prove capable of taking on—under the aspect of eternity—responsibility even for the things that don't immediately concern him, and relinquish something of his private interest in favor of the

> interest of the community, the general interest. Without such a mentality, even the most carefully considered project aimed at altering systems will be for naught.[4]

There are philosophical connections between indigenous and some strains of anarchist thought on the spirit of freedom and the ideals of a good society. Parallel critical ideas and visions of post-imperial futures have been noted by a few thinkers, but something that may be called anarcho-indigenism has yet to develop into a coherent philosophy.[5] There are also important strategic commonalities between indigenous and arnarchist ways of seeing and being in the world: a rejection of alliances with legalized systems of oppression, non-participation in the institutions that structure the colonial relationship, and a belief in bringing about change through direct action, physical resistance, and confrontations with state power. It is on this last point that connections have already been made between Onkwehonwe groups and non-indigenous activist groups, especially in collaborations between anarchists and Onkwehonwe in the anti-globalization movement.

But even before this, and without explicit linkages in theory or politically, resurgences of Onkwehonwe self-determination have been seen by the state in the same way as anarchist challenges to state authority: direct defence of rights and freedoms in a physical sense has been met with extremes of repression by the state.

The so-called "Oka Stand-off" in 1990 saw a surge of indigenous power in the resistance of the Kanien'kehaka communities located around the city of Montréal to the Canadian state's attempt to expropriate lands and impose its police authority on them. The determination and disciplined tactics of the *Kahnawakero*:non—people of Kahnawake, women and men alike and together—stymied the Canadian army's efforts to occupy the Kanien'kehaka village, and their ferocious but non-lethal defence of their lands and homes forced the army, trained and equipped only to confront other military forces in conventional armed combat, to withdraw after a prolonged effort. This incident, which happened at the end of a wearying 78-day stand-off between the Kanien'kehaka and Canadian police and military forces, is an unappreciated benchmark of indigenous resistance struggles.[6] Its lessons were reinforced by the so-called "Gustafsen Lake stand-off," where a serious paramilitary force was brought to bear by the Canadian state (this time in interior British Columbia) against a small group of Onkwehonwe who had occupied a sacred ceremonial site and refused to vacate the premises when ordered to do so by the Settler who held legal title to the land. Splitting-the-Sky, one of the defiant Onkwehonwe leaders in the conflict, describes what happened:

> We were unarmed. But then when they brought in arms, we weren't too far out to get some rifles, because everybody up there in the Cariboo hunts. I mean, everybody has shotguns and it's no big secret. So we ended up with about 17 shotguns to defend ourselves. I was asked at that time, "What should be the stance we make?" "Well," I said, "According to Canadian law, if somebody puts a gun in your face, you have a right to pick up a gun and defend yourself from being attacked or killed." That is Canadian law. Every citizen has that right according to due process of law—the Canadian law. And so it was a question of self-defense.[7]

These two incidents illustrate how immediate the issue of violence and self-defence is to any serious conception of resurgent indigenous power. The logic of contending with state power is inescapable. If contention is necessary to make change, if contention leads necessarily to confrontation, and if confrontation has an inherent element of potential or real violence, as the experiences in Oka and Gustafsen Lake demonstrate, then we must be prepared to accept violence and to deal with it. To continue advancing, the intelligent course of action is contention. Dogmatically pacifist movements have only succeeded in making change when they are backed by the support of the threat of violence—either explicit in the form of organized armed resistance movements subsumed within the larger non-violent movement or implicit in the fear the state and Settler society have of the potential of unorganized violence coming out of frustration. Thus we must contend, and we must confront, and we must be prepared to shoulder the burden of conflict. But how?

Governments will always use violence, and it is a responsibility to recognize this. How do we resist the power of violence and prevent it from becoming a way of life as it has become in places like Israel, Northern Ireland, and Sri Lanka? In a sense, the question could be framed as: What is our theory of violence?

There are arguments in any movement both for and against violence that make sense. Western notions of non-violence are rooted in a counter-historical reading of Jesus' teachings (given that Judeo-Christian societies are among the most violent the world has ever known) and are advocated as a moral choice. But, in fact, this reading is unnecessary; non-violent action coupled with a capacity for physical self-defence is a strategic choice, not a moral choice. It is simply the best way of prevailing in a struggle. In a state context, rather than attempting to destroy or displace state authority, non-violence offers a sound strategic vision that will mobilize a movement to deny rulers control of community life and will undermine the legitimacy of the state both domestically and internationally. Because the practices and theories of politics today are so permeated with the logic of armed force, my argument here will seem counterintuitive to most people. But in fact, non-violent resistance, as the foundation of a movement made up of many different tactics responding to the demands of circumstance, has been historically widespread and effective against all types of repressive regimes.[8]

It is a warrior's definition of courage that most concerns us, as it will be individuals who will contend against the state. Willpower and determination are the elements of courage. They are not a finite reserve and must be nurtured, fed, and developed if Onkwehonwe are to be able to stand up to state authority in any way—whether it be protest, contention, or more aggressive assertions. People who engage in battle in whatever form are not "fearless." Ninety-eight per cent of combat soldiers break down mentally in wartime situations, and for military commanders, the question is not if, but when will the men's well of courage run dry?[9] Onkwehonwe, like all warriors in battle, will realize our collective courage from sharing in others' wells of strength and determination and building up our collective store of mental and emotional strength by supporting each other in struggle and achieving victories along the way to our goals. And it is the warrior's question that is our challenge: How to shore up courage?

The answer is that we, like any warrior in battle, need to realize that our collective fortitude consists in sharing in others' courage, as leaders inspire and motivate us to persevere when we feel like quitting the fight, and by building up our collective store of mental and emotional strength through the uplifting and cumulative effects that victory provides a people.

Yet it is the bold and unchallenged white arrogance and racial prejudice against indigenous peoples that is the first and most important target of action. The personal and mundane maintenance of colonialism and colonial power relations through words and behaviours on a one-to-one level, conversationally and socially, must be stopped. Psychological research has conclusively proven that the racism of white society manifest in its most basic form, hostility and aggression, more than anything else assaults the sense of self-control and affects the health and well-being of people who are discriminated against.[10] The constant hostility of white people and lack of acceptance on a societal level has been proven in studies of African-Americans in the United States to have a direct effect on the rates of high blood pressure in that population, for example. So racism is expressed on all levels in many forms as a personal relationship, and it is embedded systemically in colonial relations. This must be acted against if our people are to survive and restrengthen. This is the orienting first goal of the Onkwehonwe struggle for freedom.

There is a real question here on how we would inoculate people from fear to allow them to act with courage against this root cause of Onkwehonwe stress, suffering, and premature death. I believe it must be a primary belief in all of our leaders and peoples' minds that the racism of white society *can* be overcome; our people must be reoriented culturally and politically so that their conscious and subconscious minds learn that stress and hate can be defeated and that they have a responsibility to act against racism. In doing so, they will be truly living the ethic of courage as warriors for the next generations.

REGARDLESS OF THE CONSEQUENCES

We turn now to the strategic and tactical questions of how to carry the fight to the adversary, by taking a long view on the struggles of people who have sought to free themselves from colonial oppression. All successful anti-colonial wars have been based on the existence of three factors favouring the indigenous people. First, the conditions must be ripe, meaning that domination was a present and onerous fact and that there was a focused colonial assault on the national and social aspects of the indigenous people's existence. Second, anti-colonial forces were organized to endure over the long term and to grow. And, third, the anti-colonial forces had disciplined organizations that were connected to other resisting organizations. The defeat of colonialism has always been essentially political and caused by a few key factors being made into realities by the indigenous people: the continuing organizational strength of the indigenous people as distinct from any battlefield or other tactical successes or failures; the growth and maintenance of the support of the people; the imposition of what eventually becomes unbearable costs on the colonial government and economic systems; internal divisions in Settler society; spiritual and financial weariness among the Settlers over time; and attacks on the position and power of co-opted comprador politicians who work with the colonial regime to ensure, or to participate directly in, their own people's oppression.[11]

It cannot be denied that these factors apply in the description of Onkwehonwe situations. But it is rare, due to widespread intellectual and moral cowardice among aboriginal elites and supposedly progressive Settlers, to apply the remedy that is so obviously applicable as well to the situation of Onkwehonwe.

But there may be cause for caution in following the anti-colonial line all the way to its logical conclusion. The pattern of revolutionary struggles has historically been one of a cycle of futility. As peoples have sought the long-term objective of achieving their freedom, their movements have been paired with the rhetoric and politics of anti-imperialism. This has led to the development of nationalism, an ideology of anti-colonialism geared toward regaining control of resources and imposing socially and culturally conservative cultures as a counter to colonial corruptions. In fact, this has been the experience of the limited resurgences undertaken by Onkwehonwe thus far, except for the notable exception of the Zapatista movement in Mexico. Even in the context of the small-scale localized struggles of the past generation on Turtle Island, *Anówara Kawennote,* consider how the radical potential has been co-opted by the inherent conservatism of nationalist ideologies and agendas. Out of the Oka Crisis and the military stand-off in 1990 between the Kanien'kehaka and the governments of Québec and Canada came the entrenchment and growth of the co-optive band councils in the Kanien'kehaka communities that had fought to protect their lands from further encroachment by Settlers. Out of the Burnt Church resistance in the mid-1990s to defend inherent and treaty rights to fish came the co-optive fishing agreements recognizing colonial authority signed by all of the Mi'kmaq band council governments. Our experience thus far demonstrates that "revolutionary" conditions as described above do exist in our communities, but that we completely lack the structural capacity to carry our movement beyond actions that leverage political advantage to comprador politicians operating within colonial systems. There is no indigenous organizational capacity for taking advantage of the oppressive conditions that do exist and for channeling Onkwehonwe stresses into an effective action strategy against the colonial system itself. Action in our current frame leads, ironically, to further entrenchment of the colonial system. As the state represses radical direct action by funding conservative, legal negotiation accommodations, nations and the movement are divided, thus ensuring the continuing role of colonial law, which is supported by the ideology of "Aboriginal" nationalism constructed within the frame of colonialism. Such a nationalism is actually a cover for accommodation and surrender.

The most difficult question facing us is: How do we break from the cycle of assertion/co-optation and become a credible threat to the colonial state? How do we end passivity and natural human inertia and get people to activate and engage in organization and direct action in the social, cultural, and political spheres of society? In this, it is important to grasp the insight, pointed out by the political scientist Sidney Tarrow, that the causes of any political action are not the *conditions* endured by people, who can always find ways to live with much deprivation, but the *opportunities* to and limits placed on their collective action. This is even more so when the people believe that doing nothing is even more dangerous than taking action. So, in getting Onkwehonwe to become involved and take on the challenges and sacrifices needed to build a movement for change, on a collective basis, three important things must be realized in concrete ways, as Tarrow discovered in his own research:[12]

- *self-sufficiency*: people must have access to the resources that will allow them to defy the state and the control of colonial institutions;
- *reorganization*: new channels for people's energy must be created for them to take part in contentious action against state structures and colonial power; and,
- *reculturation*: people and communities must come to understand that cooperating with colonial authorities is wrong and must be acted against.

People like Sakej and the warrior societies of Anówara don't do submission. And a growing number of Onkwehonwe youth are following their lead. The Chinese classic Taoist teaching, I *Ching*, advances that, "In contention there is sincerity."[13] Trouble doesn't start without reason, contention arises because of a need for change. Contention flows from a manifested unwavering commitment to the truth: sincerity. The Chinese hexagram for "contention" shows both internal desire and outward strength; the etymology of the Chinese hexagram shows how, in Chinese philosophy as in Onkwehonwe teachings, contention is natural and organic to human relations. The wisdom here is deep: the symbol evolved from divinations traced as far back as 1300 BCE, through narrative descriptions, or stories, of observed changes in the natural world; for at least 2,000 years, the hexagrams have been explicit and specific in this teaching. Philosophically, contention can be seen as the dialectical unity of polar energies bringing together opposed forces that need to and must be reconciled if life is to continue. It is not something to be feared or avoided—people seeking balance and harmony *must* embrace the process of contention. The *I Ching* also teaches that contention is related to the concept of impermanence, that struggle is constant and that it is only the form of contention that changes over time. This is an obvious truth, too, in our experience as Onkwehonwe. We can choose to live with colonial conflict inside of ourselves and our families, or we can redirect that negative force through contention toward its source, which is the injustice of the state towards our people.

Contention is always justified when it challenges oppression as an opposition to suffering caused by others' greed and aggression. There is no valid argument for complacency or tolerance of colonial injustice that is not based on moral failure or sheer cowardice. As becomes clear in my discussion with Sakej, the only real question is *how* to fight.

We need to take our direction from the achievements of our Onkwehonwe ancestors in creating balanced and respectful relationships. In advocating a strategy of contention, what I am acknowledging is that peace and harmony are only possible if we take the possibility of contention to its limit. That limit is reached by developing a renewed sense of pride in bold and serious disruptions of the status quo. Reaching that limit is only possible if we discipline ourselves to reject the promotion of offensive violence as the means of advancing our struggle; if we commit to using words, symbols, and direct non-violent action as the offensive weapons of our fight on a battlefield that is the critical juncture of contention and conflict; if we push disruptive direct action tactics right up to the point that they will become a means of violent attack on our adversaries. Arms are certainly necessary, only because we must protect ourselves from violent attack and survive in a physical sense, but we should have faith in the power of our ideas and in our abilities to communicate our ideas without resorting to the mute force of violence to bring our message to people. We should seek to contend, to inform our agitating direct actions with ideas, and to use the effects of this contention to defeat colonialism by convincing people of the need to abandon the cycle of subjugation and conflict and join us in a relationship of respect and sharing.

SACRED PROTECTORS

Reconnection to our heritage and the sources of our strength as Onkwehonwe will not happen just because we need it or wish it to be so. In laying out an essentially non-violent theory of resistance, I should make clear again that I am not saying that we should avoid engaging the adversary in battle. Make no mistake: Onkwehonwe will have to fight to recover our dignity, and we will need to sacrifice and struggle to make the connections that are crucial to our survival.

We need to become warriors again. When we think of those people who take on the responsibility to act against threats to the people, we think of the word, "warrior." But, obviously, the way that word is understood is just one of the meanings of the term. It is European in origin and quite a male-gendered and soldierly image in most people's minds; it doesn't reflect real Onkwehonwe notions from any of our cultures, especially that of the ideal we are seeking to understand and apply here, of men and women involved in a spiritually rooted resurgence of Onkwehonwe strength. Trying to gain a deeper understanding of the various Onkwehonwe senses of being a warrior, let us consider the different ways of thinking about being a warrior in the framework of our own languages. My friend Thohahoken helped me understand the Kanien'kehaka meaning and concept of a warrior. Right away we noticed how interesting it was that the English word "warriors" and the self-referent Kanienkeha word for someone who is a fighter, *wateriyos*, sounds almost the same![14]

The most common English-Kanienkeha translation for the word "warrior" is *rotiskenhrakete*, which literally means, "carrying the burden of peace." The word is made up of *roti*, connoting "he"; *sken* in relation to *skennen*, or "peace"; and *hrakete*, which is a suffix that combines the connotations of a burden and carrying. Many of our people say that the concepts built into the word *rotiskenhrakete* form the contemporary Kanien'kehaka cultural basis for the militant action in our communities for the last couple of generations. It was one of the founding ideas of the Mohawk Warrior Societies that emerged in Kanien'kehaka communities in the 1970s and 1980s and a strong link between the traditional teachings and the contemporary movement ideology. But the word has a much broader usage in Rotinoshonni traditional culture relating to warriors: in the Condolence ceremony, in which grief is assuaged and new chiefs are raised up by clans, young men of the nation are referred to as *Rotiskenhraketakwa;* in the traditional Thanksgiving Address, in which gratitude for our place in creation is expressed, the sun is called *Rotiskenhraketekowa*. So the word has a deeper meaning than the simple equivalent to "warrior" in English.

In fact, Thohahoken told me that he had spoken with elders about this question many times. They said that the concept of a warrior in our "real old" traditional culture has no relation to *rotiskenhrakete* and that the name for warriors that more closely reflects the notion of warriors of the people living the values and principles of the *Kaienerekowa*, the Great Law of Peace, is *Oyenko:ohntoh*, which means "the tobacco hanging." As Thohahoken put it:

> *Rotiskenhraketakwa* are like conscript fighters, men who would normally not be fighting except when conscripted to defend the peace, *Oyenko:ohntoh* are more akin to the Japanese samurai. One of our more sacred protection medicines is tobacco, *oyenkwehonwe*, and in the old days it was cured by hanging it up in the rafters of the longhouse, *arhenton*, "in the shadows." Thus, hanging tobacco in the longhouse rafters protects the house. *Oyenko:ohntoh* are not conscripts, but sacred protectors; they are anonymous shadow warriors in a secret society whose duty it is to protect the house.

This is the depth of understanding and appreciation of the warrior's role that is lacking from less culturally grounded notions. Thohahoken's teaching on the traditional Rotinoshonni ideas of the warrior as sacred protector is a good place to begin a further exploration of the modern concept of the warrior in Onkwehonwe societies. The great Lakota scholar Vine Deloria, Jr. has written on the qualities of warriors like Tecumseh, Crazy Horse, and other old war chiefs:

> they had a sense of personal worth, of a mission to be accomplished, and of a relationship with the life forces of the greater cosmos in a measure that we have not seen since. Fighting overwhelming odds, suffering the loneliness of knowing the situation was hopeless, and maintaining their sense of person was an achievement few of us can conceive and none of us can match.[15]

This description of the warrior spirit resonates with the response I received from a long-time Onkwehonwe activist, who commented on the theme of restoration that I developed in my last book, *Peace, Power, Righteousness*.[16] The comments reflect the spiritual need for this kind of inspirational and courageous individual among Onkwehonwe today:

> Condolence cannot happen if we are all in grief. The healthy ones, the bright-eyed ones, must accept their responsibility to restore those in grief, temporarily in dysfunction, so to speak, to health, to accept, recognize, restore, ameliorate, admonish, and provide the new mentor, model and inspiration. In today's context, that is a primary task and responsibility of the warrior. We need a new statement, a creed, almost, which sets out some near-universal (at least North American) concept of the traditional person, the traditional life, the traditional society—what it is we are aspiring to be.[17]

From ancient teachings embedded in our indigenous languages, to the most thoughtful of modern reflections, there seems to be a great consistency among Onkwehonwe about the idea of the warrior.[18] To get at this notion beyond my own people's cultural heritage, I asked a number of people from different nations to tell me what the *warrior* word was in their own language:[19]

- for the Kuna: there is no literal equivalent for the English term (it should be noted that they interact mostly with Panamanians, Spanish-speaking Settlers), and the closest approximation is *napa-sapgued*, meaning "One who protects or guards the land, or nature."
- for the Dakota: *akicita* generally refers to those who have engaged in war combat, though linguistically the word is related to *akita*, which means "to seek" (as in something lost) or to hunt for something, generally in reference to hunting animals.
- for the Wsanec: *stomish* means those who protect the territory and defend the names with honour and discipline.
- for the Dakelh: the word for soldier is *lhudughan*, meaning "he who kills" and includes the notion that he is in the midst of killing (the word for war is *lhadughan*, meaning "they kill one another"). I was told that these words came about only recently, as until recent years most Dakelh spoke only their own language, and that they had to formulate such words to talk about what was happening in Europe.
- for the Pawnee: there is no particular word for warrior, although *Ootakissh*, the phonetic spelling of "young man," is a word mentioned frequently in war dance songs. A more familiar word is *heluska*: the warrior, the war dance, the war, battle, struggle.

A Pawnee friend summarized beautifully for me the unified Onkwehonwe concept of warrior and the struggle of our peoples when she further explained her people's idea of *heluska*:

> One of my most favourite sayings is *tu-da-he*, "the war, the battle, the struggle is good, sacred, right." In other words, life and the everyday struggles of living, good or bad, is the epitome of life. It is how you know you are living. Nothing is easy, and because it isn't easy, one should truly value the blessings. Of course, in a warrior society the warrior ideal is how life is lived. It is what you do, it is who you are—you fight. Defeat is painful, but it is only temporary because you still live to get yourself up and see the dawn.

After hearing her teaching, my earlier framing of the question of how to fight seemed superficial. Faced with the thought of struggle as an identity, the more appropriate question becomes now *how to* fight but, *what is* the fight? We often talk in radical political circles of the need to take action. But what is "action"? Is it movement with purpose? Decision put into practice? Is it constituted in moral, physical, or cultural acts? This is difficult to grasp.

We can be sure that action is not reaction. It is not declarations or statements or rhetoric by which people are affected and then decide to plan to take action. Action is spirit and energy made into a driving force for change. Action is the manifestation in physicality of the spiritual energy of the warrior. It is behaviours, methods, goals, desires, and beliefs, all expressed in real ways in relationships with other people and forces.

In the context of colonialism and the Onkwehonwe struggle, proactive and protective actions are the elements of a resurgence. Actions—thinking, feeling, and behaving indigenously—are the things that make our movement Onkwehonwe, not simply what we might name it. Our indigeneity is observable as a personal quality. Are we living culturally as Onkwehonwe? This kind of authenticity is a powerful melding of renewal and continuity. It can be figured with reference to history, drawing on indigenous values and teachings, and from recent cultural developments that respect indigenous principles. Such a combination creates a cultural foundation for contemporary forms of resistance, making in effect new cultural practices to shape authentically indigenous movements which are both outgrowths from historic forms and organic expressions of timeless indigenous values. There are five main characteristics, both authentically indigenous and effective as a means of confronting colonial dominion, which are evident in Onkwehonwe movements from the Mohawk Warrior Societies to the Zapatistas to the recent mass movements against governmental corruption in Ecuador:

- They depend on and are led by women.
- They protect communities and defend land.
- They seek freedom and self-sufficiency.
- They are founded on unity and mutual support.
- They are continuous.

It is a much easier task to describe the qualities of a real indigenous movement than it is to point out examples of their emergence in Onkwehonwe societies. An authentic movement for change that is true to Onkwehonwe values and effective at engaging unjust power is very rare because there are so few Onkwehonwe who have remained connected to their traditional cultures and fewer still who have preserved within their culture and philosophies the ethic and practice of the sacred protector—the warrior spirit. There are no movements for change among indigenous peoples generally because the sad fact is that there are hardly any more warriors, sacred or otherwise. Are we fearful of the consequences of contentious action? This may be so, but the fact is that, whether we realize it or not, as Onkwehonwe nations today we have a stark choice: organize or die.

As we seek restrengthening and embark on pathways to recover the personal and collective ethic of the warrior, we should hearken back to the spirit of the teachings that do survive, like that of the Kiowa nation's Dog Warriors, whose founder had a dream in which warriors were led into battle by a dog singing the *Ka-itsenko* song. Ten men only, the bravest in the nation, wore a ceremonial sash and carried a sacred arrow, and, as described famously by the writer N. Scott Momaday, "In times of battle he must by means of this arrow impale the end of his sash to the earth and stand his ground to the death.[20]

In most people's minds, the words "North American Indian warrior" invoke images of futile angry violence or of noble sacrifice in the face of the white march of triumph over this continent. To Euroamericans, the descendents and beneficiaries of conquest, "Indian warriors" are artefacts of the past; they are icons of colonization, that version of history in which the original people of this land have been defiant but defeated by Euroamerican land greed and the unstoppable advance of Euroamerican civilization. But history has not ended. There are still Onkwehonwe lands, souls, and minds that have not been conquered. For them, a warrior is what a warrior has always been: one who protects the people, who stands with dignity and courage in the face of danger. When lies rule, a warrior creates new truths for the people to believe.

NOTES

1. Tiana Bighorse, *Bighorse the Warrior,* ed. Noel Bennett (Tucson, AZ: University of Arizona Press, 1990) xxiv.
2. This section was previously published as an article entitled, "The Path of Least Resistance," archived on my Website, <http://www.taiaiake.com>. The article was based on information originally reported in Joan Taillon, "First Nations Team Up for Economic Opportunities," *Raven's Eye* (July 7, 2003), from which all of the direct quotes attributed to the band council chief in this section are taken.
3. The best sources of information on these issues are on the Internet. On management problems with the Canadian Department of Indian Affairs, see Chapter 8 of the 2003 Auditor General of Canada report <htpp://www.oagbvg.gc.ca/domino/reports.nsf/html/20031108ce.html>; on the Samson Cree Nation's breach of trust suit against the Government of Canada in relation to the Department of Indian Affairs' mismanagement of the Nation's oil gas revenues, *Buffalo v. The Queen*, see the Website set up to provide information on the case <http://www.samsoncree.org>; and on the suit against the United States for Bureau of Indian Affairs trust fund misappropriations, *Cobell v. Norton*, see an overview report from the *Boulder Weekly*, online at <http://www.monitor.net/monitor>.
4. Vaclav Havel, *Distrubing the Peace*, trans. Paul Wilson (New York, NY: Vintage, 1991) 17.
5. See Hakim Bey, *T.A.Z.* (San Francisco, CA: AK Press, 1985) and *Immediatism* (San Francisco, CA: 1994). See also Richard Day, "Who is This We That Gives the Gift? Native American Political Theory and *The Western Tradition*," *Critical Horizons* 2:2 (2001): 173–201; Richard Day and Tonio Sadik, "The BC Land Questions, Liberal Multiculturalism, and the Spectre of Aboriginal Nationhood," *BC Studies* 143 (Summer 2002): 5–34.
6. Sean M. Maloney, "Domestic Operations: The Canadian Approach," *Parameters*, US Army War College Quarterly (Autumn 1997). See also Sandra Lambertus, *Wartime Images, Peacetime Wounds: The Media and the Gustafsen Lake Stand-off* (Toronto, ON: University of Toronto Press, 2004).
7. *Monday Brownbagger*, 25 March 2002, Co-Op Radio, CFRO, Vancouver, BC.

8. Peter Ackerman and Jack DuVall, *A Force More Powerful: A Century of Nonviolent Conflict* (New York, NY: Palgrave, 2000) 5. Also, for solid arguments for non-violence as a strategic option based on the success of Latin American social movements in forcing change in those societies, see Philip McManus and Gerald Schlabach, eds., *Relentless Persistence: Nonviolent Action in Latin America* (Santa Cruz, CA: New Society Publishers, 1991). And for an argument against the notion that violence is the most powerful means of political action, see Gene Sharp's "Non-violent Struggle Today," *Unarmed Forces: Non-violent Action in Central America and the Middle East,* ed. Graeme MacQueen, Canadian Papers in Peace Studies 1 (Toronto, ON: Science for Peace, 1992).
9. Dave Grossman, *On Killing: The Psychological Cost of Learning to Kill in War and Society* (Toronto, ON: Little, Brown and company, 1995) 83–85.
10. Grossman, *On Killing* 77–84.
11. Gérard Chaliand, *Revolution in the Third World,* rev. ed. (New York, NY: Penguin, 1989) 37–49.
12. Tarrow, *Power in Movement* 71.
13. Thomas Cleary, *Classics of Strategy and Counsel* (Boston, MA: Shambhala, 2000) 194.
14. Similarities in the concept of the warrior as Sacred Protector as explained by Thohahoken can be found in Tibetan Buddhism as explained in Chögyam Trungpa, *Shambhala: The Sacred Path of the Warrior* (Boston, MA: Shambhala, 1998).
15. Frank Waters, *Brave are My People: Indian Heroes Not Forgotten* (Athens, OH: Ohio University Press, 1993) xiv.
16. Taiaiake Alfred, *Peace, Power, Righteousness: An Indigenous Manifesto* (Don Mills, ON: Oxford University Press, 1999).
17. Ratihokwats to Taiaiake, personal communication, 2003.
18. For an additional appreciation of the traditional warrior's view of the meaning of resistance, see Jon W. Parmenter, "Dragging Canoe: Chickamauga Cherokee Patriot," *The Human Tradition in the American Revolution,* ed. Nancy Rhoden and Ian Steele (Lanham, MD: SR Books, 2000) 117–37.
19. The definitions and explanations of the various Onkwehonwe words were provided by Brock Pitawanakwat (Spanish and Kuna), Nick Claxton (Sencoten), Angela Wilson (Dakota), Patty Loew (Ojibway), and Lyana Patrick (Dakelh/Carrier).
20. N. Scott Momaday, *The Way to Rainy Mountain* (Albuquerque, NM: University of New Mexico Press, 1969) 21.

BIBLIOGRAPHY

Ackerman, Peter, and Jack Du Vall. *A Force More Powerful: A Century of Nonviolent Conflict.* New York, NY: Palgrave, 2000.

Alfred, Taiaiake (Gerald). *Peace, Power, Righteousness: An Indigenous Manifesto.* Don Mills, ON: Oxford University Press, 1999.

Bey, Hakim. *Immediatism.* San Francisco, CA: AK Press, 1994.

——. *T.A.Z.* San Francisco, CA: AK Press, 1985.

Bighorse, Tiana. *Bighorse the Warrior*. Ed. Noel Bennett. Tucson, AZ: University of Arizona Press, 1990.

Chaliand, Gérard. *Revolution in the Third World: Currents and Conflicts in Asia, Africa, and Latin America*. Rev. ed. New York, NY: Penguin, 1989.

Cleary, Thomas. *Classics of Strategy and Counsel. Vol. I: The Collected Translations of Thomas Cleary*. Boston, MA: Shambhala, 2000.

Day, Richard. "Who is This We That Gives the Gift? Native American Political Theory and *The Western Tradition*." *Critical Horizons* 2:2 (2001): 173–201.

Day, Richard, and Tonio Sadik. "The BC Land Questions, Liberal Multiculturalism, and the Spectre of Aboriginal Nationhood." *BC Studies* 143 (Summer 2002): 5–34.

Grossman, Dave. *On Killing: The Psychological Cost of Learning to Kill in War and Society*. Toronto, ON: Little, Brown and Company, 1995.

Havel, Václav. *Disturbing the Peace*. Trans. Paul Wilson. New York, NY: Vintage, 1991.

Lambertus, Sandra. *Wartime Images, Peacetime Wounds: The Media and the Gustafsen Lake Stand-off.* Toronto, ON: University of Toronto Press, 2004.

Maloney, Sean M. "Domestic Operations: The Canadian Approach." *Parameters*, US Army War College Quarterly (Autumn 1997).

McManus, Philip, and Gerald Schlabach, eds. *Relentless Persistence: Nonviolent Action in Latin America*. Santa Cruz, CA: New Society Publishers, 1991.

Momaday, N. Scott. *The Way to Rainy Mountain*. Albuquerque, NM: University of New Mexico Press, 1969.

Parmenter, Jon W. "Dragging Canoe: Chickamauga Cherokee Patriot." *The Human Tradition in the American Revolution*. Ed. Nancy Rhoden and Ian Steele. Lanham, MD: SR Books, 2000.

Sharp, Gene. "Non-violent Struggle Today." *Unarmed Forces: Non-violent Action in Central America and the Middle East*, ed. Graeme MacQueen. Canadian Papers in Peace Studies 1. Toronto, ON: Science for Peace, 1992.

Tarrow, Sidney. *Power in Movement: Social Movements and Contentious Politics*. 2nd ed. Cambridge, UK: Cambridge University Press, 1998.

Trungpa, Chögyam. *Shambhala: The Sacred Path of the Warrior*. Boston, MA: Shambhala, 1988.

Waters, Frank. *Brave are My People: Indian Heroes Not Forgotten*. Athens, OH: Ohio University Press, 1993.

Conclusion

FREEDOM DREAMS: BUILDING OUR COMMUNITIES AND ADDRESSING SOCIAL INJUSTICE

In this book, we trace the continuity of colonialism and racism from Canadian history to contemporary events. The processes of racialized domination, marginalization, and exploitation, manifested in different historical moments, but also occurring at various levels of the Canadian State, principally through its early "white supremacist inspired" nation-building project that became codified through discriminatory policies and legislation, continue to be evident in everyday cultural practices that range from racial profiling, job discrimination, misrepresentations in the media to the construction of the "racialized other" as outside the Canadian norm. Canada's colonial legacy persists in its contemporary institutionalized forms of internal colonization. In this conclusion, we explore the current patterns of internal colonization and provide a critical historical analysis of their impact on contemporary Canadian society. We connect these socio-political and cultural practices to theoretical interpretations and reflections to ask the questions: How do we move forward beyond this calcified reality? Indeed, what is to be done?

A contemporary example of the continuity of colonized relations is Canada's preparation for the Winter Olympics in Vancouver in 2010. Many Indigenous communities, especially those located in the BC area, have been protesting the extensive destruction of traditional homelands of local Indigenous peoples.[1] More than 75 percent of British Columbia is unceded native land. This land belongs to the Indigenous peoples and not the Canadian government. A formal complaint made to the International Olympic Committee (IOC) informed them of violations of Indigenous rights that were also violations of human rights. The IOC has an official policy to not hold events in countries where human rights abuses occur; however, Indigenous peoples' complaints were ignored. The Native Youth Movement is actively resisting the building of highways, resorts, and condos—the construction of infrastructure for the 2010 Olympics: "By them opening up our land, our sacred sites, our medicine grounds. . . . We want investors to know our land is not for sale."[2]

Another issue that highlights the ongoing internal colonization of Indigenous peoples is documented in the 2004 Amnesty International Report, *Stolen Sisters: Discrimination and Violence Against Indigenous Women in Canada*. It is important to point out that it was Amnesty International, an international non-governmental organization, and not the Canadian government, that raised public awareness of the violence against Indigenous women. According to this report, "A shocking 1996 Canadian government statistic reveals that Indigenous women between the ages of 25 and 44, with status under the Indian Act, were five times more likely than all other women of the same age to die as the result of violence."[3] This report notes the indifference of various levels of Canadian authorities to treat these murders and the outstanding cases of missing women with the seriousness and attention they deserve. This report also makes the case of how Indigenous women have been systematically violated due to government policies in the 1870s to mid-1980s when

Indigenous women lost their status as Indigenous people under the federal Indian Act. The AI Report concludes: "It is Amnesty International's view that the role of discrimination in fuelling this violence, in denying Indigenous women the protection they deserve or in allowing the perpetrators to escape justice is a critical part of the threat faced by Indigenous women.[4]

The disregard for certain groups' rights and concerns is also evident in the aftermath of the January 2008 decision by the Toronto School Board approving an Africentric Alternative School as a pilot project. Dalton McGuinty, the Premier of Ontario, responded to the outcome of this democratic process by seeking to rouse mob action against a minority group, saying: "The good news for Torontonians is that they now have 18 months . . . to put a stop to this."[5] In addition, some school trustees attempted to hold an emergency meeting to reverse the decision immediately after the Premier's intervention. This intervention showed both a contempt for the democratic process and a complete disregard of the concerns of the affected communities that have been advocating for this action for decades as a way of creating a supportive learning environment in which the cultural, social, and life experiences of black youth would matter as reference points for their educational experience. The communities had long argued that contrary to the Premier's assertions that this was "segregation," the objective of the pilot was to successfully integrate the disengaged students into the education system by giving them an education that did not marginalize their history and social experiences.[6] Moreover, the school was the outcome of an application process duly prescribed by the Toronto District Alternative Schools Policy and Procedures, the basis for thirty-six other schools in the system, including the Native School Program and the Triangle School Program (part of the OASIS Alternative School), that aimed to provide a safe and culturally relevant learning environment for Aboriginal and LGBTT (Lesbian, Gay, Bisexual, Transgendered, and Transsexual) students. Toronto Star columnist Jim Coyle aptly asked the following questions: "What's the Premier saying? That votes we don't like are merely the starter's pistol to let lobbying of elected representatives begin and are to be retaken until we get the result we want?"[7] The debate over the Toronto District School Board's decision about Africentric schools, the provincial government's decision not to fund this pilot project, and the federal government's decision to withdraw from Durban Two (discussed in our introduction) have in common a "do-nothing"strategy combined with a destruction of alternatives that others are putting forward. Neither the federal nor provincial governments are proposing pro-active measures to address what groups of people are saying is important to them.

Another opportunity to glance at Canadian society's processes of internal colonization is provided by the excitement over the United States' Democratic primaries and the battle between Hillary Clinton and Barack Obama for the Democratic election. On January 10 2008, Haroon Siddiqui, columnist for the *Toronto Star*, asked a critical question: "Can a Black man become our PM?" Siddiqui quotes, among two others, Professor Alan Hutchinson of York University's Osgoode Law School: "The chances are close to zero. . . . We may be somewhat better than the States but not much when it comes to dealing with race. Visible minorities remain woefully under-represented in not only politics but in civic and business leadership . . . I look at my legal profession and it's mostly white men on the higher courts. It's still news when we get a woman but that's still white women. No aboriginals on the Supreme Court, no visible minorities."[8] Why is it that the chances are close to zero of a Black man, or a racialized woman, becoming Canada's Prime Minister? Why isn't there an Indigenous presence on the Supreme Court? What does this say about Canadian society?

Social scientists have analyzed social inequality in Western societies and have largely discredited individual-based causal explanations for social problems as convenient victim blaming. For our purposes here, we will highlight Charles Tilly's work published in a book titled *Durable Inequality* in 1998.[9] Tilly argued that durable inequality is created and maintained by "social closure" among categories—male/female, black/white, citizen/foreigner, etc.: "People who control access to value—producing resources to solve pressing organizational problems by means of categorical distinctions. Inadvertently or otherwise, those people set up systems of social closure, exclusion, and control." (1998:7–8). This is another way of commenting on white-settler assumptions, behaviours, and social relations that continue to retain and reproduce white supremacy in contemporary Canadian society. These relations of domination are maintained at a cost to society because organizations that run this way do not necessarily operate more efficiently. Durable inequality persists because "categorical inequality . . . facilitates exploitation and opportunity hoarding by more favored members of a given organization, who have the means of maintaining their advantage even at the expense of overall inefficiency" (1998:81). Charles Tilly suggests that his research on durable inequality has three strong implications for intervention:

- First, changes in attitudes will have weak and indirect effects on existing patterns of inequality. Education for tolerance and understanding may ease the way, but it will not attack the root causes of discrimination.
- Second, alteration of categorical differences in human capital through education, on-the-job training, or transformation of social environments will affect categorical inequality chiefly through their impact on the organization of opportunity rather than their improvement of individual capacities.
- Finally, reorganization of workplaces and other sites of differential rewards with respect to the location and character of categorical boundaries can produce rapid, far-reaching changes in categorical inequality. Breaking connections between interior categories and widely adopted exterior categories, diminishing cross-boundary reward differentials, or facilitating mobility across boundaries should produce significant reductions in overall inequality.(1998:244)

Accountability and rewards need to be implemented for making inroads and progress in terms of diversity, representation, and the transformation of education, politics, healthcare, policing, etc. Yet as we have shown, several levels of authority in Canada have actively resisted structural transformations.

This resistance resonates with the range and depth of what George Lewis (2006) described as "Massive Resistance" to the Civil Rights Movement and specifically desegregation in the United States. In *Massive Resistance: The White Response to the Civil Rights Movement, 2006*, George Lewis writes: "As a number of historians have cogently argued, white supremacy in the post-Emancipation South was not merely the product of historical happenstance, but was created, crafted and carefully managed" (2006:13).[10]

White supremacy was entrenched in all aspects of life not just the political, judicial, and education systems. As a result, Blacks were denied access to institutions and resources to seek redress. In interpreting the "deep structures" he was observing, Lewis came to the following conclusion: "It is . . . essential to envision massive resistance not as a single homogeneous movement, but as a conglomeration of concomitant conversations of resistance. At its central, binding core was the preservation and continued maintenance of a segregated, white-dominated society." (2006:185) The similarities in the resistances of the pre-Civil Rights Movement and in Canada's response to Indigenous and racialized peoples' struggles today are striking. It raises a critical question: How do we move forward?

For Indigenous Peoples of Canada, the answer has been clear and consistent. As Phil Fontaine, National Chief of the Assembly of First Nations said as recently as January 25, 2008: ". . . The solutions must always start with Canadian governments effectively and meaningfully engaging First Nation governments."[11] The priority is Land Claims: "First of all, it is imperative that governments negotiate and resolve all of the outstanding land claims."[12] The natural resources sector of Canada's economy operates on First Nations' traditional territories, so Fontaine has called on governments to "act responsibly to fulfill the duty to consult and negotiate resource revenue sharing agreements with First Nations."[13] Fontaine emphasizes that the "solid plan laid out in the report of the Royal Commission on Aboriginal Peoples remains the basis of the First Nations plan for action." The Right to decide their future as a Nation is the consistent action called for by First Nations peoples in response to calls for initiatives to break down the legacy of historical and contemporary patterns of colonization. Fontaine concludes: "The Assembly of First Nations, which is the only national organization with political legitimacy to speak for First Nations, has advocated for self-reliance for a very long time. Our answer is to recognize and respect First Nations' ability to make our own decisions. That is real accountability. Investing in First Nations is investing in Canada's future."[14]

While Fontaine speaks on behalf of the Assembly of First Nations, a body that is made of elected band chiefs at the community level, many other Indigenous peoples struggle on a daily basis against oppressive conditions and confront a State that countenances these conditions. Many of these people act outside of the leadership model of the AFN attempting to have traditional ways of governing acknowledged and respected by outsiders. We see this happening currently in the Caledonia example (Ontario), where the traditional Confederacy government has been at the forefront of the land reclamation struggle. What the community there demands is very clear; as Hazel Hill, one of the community members who has been quite outspoken during the ordeal, said: "we are left with trying to get the Crown to uphold its end of this Treaty Relationship. The task has never been more in the forefront as it has been since the land reclamation of Kanonhstaton (former Douglas Creek Estates) began in February 2006 by the Onkwehonweh of what is now known as the Six Nations. Since this negotiation process has started, the Haudenosaunee representatives have continually reminded the Crown representatives of Canada that their obligation to uphold and protect this relationship is just as binding today as it was in the early 1600's when the first friendship treaty was established. No law or any unilateral acts of the British Crown in the formation of what is now known as Canada or the United States can take away from those responsibilities, nor can any other law supercede these Treaties."[15]

What Hazel Hill refers to is the Two-Row Wampum Belt, the Friendship Belt agreed upon by the Six Nations peoples and the early European settlers that stipulated that the two sides would co-exist in this land while respecting each other's cultures and laws and not interfering with each other. Yet, as history has taught us and has been demonstrated in this book, the Canadian State has repeatedly ignored the treaty, but Six Nations peoples have always remembered it and called upon it when needed.

The experience of exclusion extends beyond Indigenous populations, although in different ways. For Muslims and Arab communities in the post-September 11, 2001 era, the war on terror and the intensification of the focus on Islamic and Arab identity means that these identities are more pronounced for their bearers and that processes of racialization that historically structure state–society relations are being deployed in contemporary contexts to determine a particular norm of Canadian-ness that is racially and culturally distinctive in such a way that even de-emphasizing one's cultural and religious affinities cannot protect against differential experiences of citizenship. The intensification of a moral panic in the post-September 11 period, which arose from the ubiquitous discourses of national security, anti-terrorism, and community safety, has had negative implications for the exercise of equal citizenship, particularly for Arab and Muslim populations in Canada. While the discourses and practices of religious and racial profiling have set community against community, drawing boundaries around and within communities, the political and social salience of Islamophobia has become a basis for the reassertion of a narrow concept of citizenship—one that emphasizes dominant society's socio-cultural forms as Canadian values while challenging the validity of multiculturalism and difference as dangerous and a threat to a Pan-Canadian idea based on assumed Anglo–Franco, Judeo–Christian values. Sedef Arat-Koc has recently suggested that Arab and Muslim communities in Canada are under siege "not only to safely *cross* borders, but also to live in safety and as equal citizens *within* borders" (2006:216).[16] Arat-Koc argues that "since the 1990s in Canada, but more specifically since 11 September 2001 . . . Anti-Arab racism and Islamophobia have not only intensified, but also, and more important, been legitimated through public discourse and mainstream institutions" (2006:216).

Thus Muslims and Arab communities are fixed as the "dangerous other," even as a growing proportion of the communities are Canadian-born. This imposition of a very ethno-specific concept of Canadian-ness means that space for cultural and religious difference is narrowing at the expense of pluralism during a time when Canadian society should be embracing diversity because of the momentum of South-North immigration and globalization. The disproportionate focus on the power of right-wing Muslim voices, presented as representing the Muslim community and caricatured as medieval, misogynistic, and homophobic, clearly constitutes a form of moral panic that transcends the political spectrum from left to right as a key element in the racialization of Islam in the Canadian context. In this environment, is it possible to assert citizenship claims by those who have been demonized within Canadian society as the "jihad generation," although many are Canadian-born?

The condition of white supremacy in Canadian society should be addressed on a number of levels—national, provincial, local, and community. The Canadian government has developed a framework to address racism titled: Canada's Action Plan Against Racism. Its short-comings have necessitated community action. Currently, the Canadian Race Relations Foundation (CRRF) is organizing an effort to draft a national anti-racism policy framework to propose to the federal government.[17] There are various community initiatives across the country that seek to undo the damage caused by the persistent relations of colonization and structural racism. But social change requires a commitment to holding those with the power to make change accountable. It also requires building people power to confront governments, professionals, and institutions of education, health, and the judiciary in Canada. Key institutional arrangements that sustain relations of colonization and hierarchies of racism have to be challenged. The state–society relations that reproduce such discourses as "reasonable accommodation," which seek to legitimate the Eurocentric character of Canadian society, have to be challenged. The unequal processes and outcomes experienced by racialized citizens undermine citizenship and represent a crisis of legitimacy. Canadian society needs to address communities' concerns about disproportionate poverty, unemployment, suicides, and depression. These represent structural issues that are perpetuating durable inequality. In the face of all the research and the need for intervention, a do-nothing strategy that uses the rhetoric of treating everyone equally is just another step toward the continuing crisis of legitimacy. When durable inequality disadvantages people and groups of people at the starting line, what is needed is leadership to educate people that equity and substantive equality require treating different groups of people differently to make them truly equal.

Finally, Indigenous and racialized communities need to recognize the classic divide-and-conquer strategies those in power use to divide them. There are a number of groups focusing on their own specific issues: land claims for Indigenous peoples, anti-black racism highlighted by various Black communities, racial and religious profiling against Muslim and other populations, the devaluation of foreign-trained professionals in Asian, South Asian, Arab, African, Caribbean, Latin American and other communities, redress for the Head Tax victims in East Asian communities, and various forms of anti-Asian racism against South and South East Asian communities etc. This work is important, and can be advanced with a strategy for building alliances across the various groups and communities. The Ontario Premier's destructive interference in the debate around Afrocentric schools and the federal government's refusal to endorse the UN Declaration of the Rights of Indigenous peoples are not issues specific to the Black communities and the Indigenous peoples. The struggle for structural changes that will transform the lives of specific groups of people in Canada effects all people and groups that are marginalized and devalued. Even as groups focus on strengthening their own communities, it is imperative to build alliances with other communities to present a united front. The issues of white supremacy and the utter lack of democratic transparency and accountability effects each and every Canadian.

The task of documenting and being a witness to these resistances and struggles for equity and justice is the responsibility of all. As the crisis of legitimacy unfolds, it should be documented to ensure that communities struggling with growing rates of poverty and sense of hopelessness can be given voice. There is the added responsibility to develop hope and a desire for our children's full development of their human potential. This is a responsibility to create what Robin Kelley (2002) has called "freedom dreams." In *Freedom Dreams: The Black Radical Imagination, 2002*, Robin Kelley makes a case for an alternative vision: "Unfortunately, too often our standards for evaluating social movements pivot around whether or not they "succeeded" in realizing their visions rather than on the merits or power of the visions themselves. By such a measure, virtually every radical movement failed because the basic power relations they sought to change remain pretty much intact. And yet it is precisely these alternative visions and dreams that inspire new generations to continue to struggle for change" (2002: Preface).[18]

Today, internationally, Canada is at war in Afghanistan. There are suicide bombings in the Middle East and elsewhere, and massacres in Kenya, Darfur, and beyond. At home, there is growing racialization of poverty and poverty among women and children; the expansion of precarious employment; the disappearance of full-time quality jobs; the neglect of violence against women; the persistent dearth of racialized women as decision-makers—politicians, CEOs, presidents of public and private institutions—in critical public spaces; and push-back against indigenous claims for control over their land and resources. In a world with so much inequality, oppression, violence, hate, greed, and destruction, how can we consider ourselves "free"? What we have are distortions of our fundamental moral values. According to Kelley: ". . . the pervasive consumerism and materialism and the stark inequalities that have come to characterize modern life under global capitalism could not possibly represent freedom. And yet, freedom today is practically a synonym for free enterprise." (2002:ix) We need to be vigilant against such distortions of our values of freedom, justice, and democracy as we build our movements for authentic liberation. Our challenge is to be vigilant against the distortions of our values of freedom, equality, justice, and democracy, while we guard against despair. We can draw inspiration from Dr. Martin Luther King, Jr., who, in a similar time of challenge said these words:

> "We Negroes have long dreamed of freedom, but still we are confined in an oppressive prison of segregation and discrimination. Must we respond with bitterness and cynicism? Certainly not, for this will destroy and poison our personalities . . . To guard ourselves from bitterness, we need the vision to see in this generation's ordeals the opportunity to transfigure both ourselves and American society. Our present suffering and our nonviolent struggle to be free may well offer to Western civilization the kind of spiritual dynamic so desperately needed for survival.
>
> [quoted in Kelley, 2002: x]

The vision to see in this generation's ordeals the opportunity to transfigure both ourselves and Canadian society is our primary responsibility. We must, and will create such a transcendant vision.

NOTES

1. "Two Aboriginal Youths Speak Out Against Vancouver Olympics," *Timmins Daily Press.* http://www.timminspress.com/articledisplay.aspx?e=883594
2. "No Olympics on Stolen Land: Great Lakes and East Coast Speaking Tour Details." http://www.rabble.ca/whats_up_details.shtml?x=66739
3. Amnesty International Canada, *Stolen Sisters—Discrimination and Violence Against Indigenous Women in Canada* (A Summary of Amnesty International's concerns), 2004:2.
4. Ibid, 2004:5.
5. Jim Coyle's column "McGuinty's Big Step Backward," *Toronto Star,* February 5, 2008.
6. James Bradshaw, "Africentric Schools Not Segregation, Advocates Argue," *Globe and Mail,* A11, February 8, 2008; Keith Leslie, "Aim of Afrocentric School Is Rentention, Not Segregation: Advocates," *Canadian Press,* February, 8, 2008; Louise Brown, "Black-focused School Endorsed: Community Representatives Say They're 'United' on Controversial Plan," *Toronto Star,* A07, February, 8, 2008; Jack Boland, "Black-focused School Pushed at Legislature: Supporters Say It Would Not Segregate Blacks, But Would Trim a High Drop Out Rate," *London Free Press,* B4, February 8, 2008; G.E. Galabuzi, "Making the Case for an Africentric Alternative School," *Our Schools OurSelves,* Vol 17, #3 Spring 2008 (Canadian Centre for Policy Alternatives).
7. Jim Coyle's column, February 5, 2008.
8. See Haroon Siddiqui's column "Can A Black Man Become Our PM?", *Toronto Star,* January 10, 2008.
9. Charles Tilly, *Durable Inequality* (California: University of California Press, 1998).
10. George Lewis, *Massive Resistance: The White Response to the Civil Rights Movement* (Great Britain: Hodder Education, member of the Hodder Headline Group, 2006).
11. Phil Fontaine, "Deal with Us As First Nations," *National Post,* January 15, 2008.
12. Ibid.
13. Ibid.
14. Ibid.
15. www.honoursixnations.com/Sept25.07.htm
16. Sedef Arat-Koc, "Whose Transnationalism? Canada, 'Clash of Civilizations' Discourse, and Arab and Muslim Canadians" in Vic Satzewich and Lloyd Wong, eds., *Transnational Identities and Practices in Canada* (Vancouver: UBC Press, 2006).
17. In 2005, the federal government announced the creation of Canada's Action Plan Against Racism as a federal initiative aimed at addressing racism in the federal government and within Canadian society. It fell short of the expectations of a national Pan-Canadian anti-racism strategy. Durban II in 2009 would have been the place for the Government of Canada to report on its process regarding Canada's Action Plan and to make itself accountable to all Canadian citizens.
18. Robin D. Kelley, *Freedom Dreams: The Black Radical Imagination* (Boston: Beacon Press, 2002).

Plain Language Version of the United Nations Draft Declaration on the Rights of Indigenous Peoples

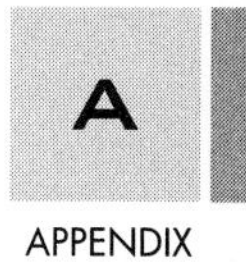

APPENDIX

PREAMBLE

The Preamble lists some of the reasons that led the United Nations to develop a declaration on Indigenous peoples' rights.

The Declaration sets out the rights of indigenous peoples. The language of "peoples" is important. Indigenous peoples do not want to be treated simply as numerical "populations". They want recognition of their rights as distinct peoples, including the right to self-determination and the right to control development of their societies.

RACISM

Rejection of the view that some peoples are better than others. This view is racist and wrong.

COLONIZATION

Recognition that indigenous peoples have been deprived of their human rights and freedoms and that this has led to their colonization and the taking of their land.

RESPECT

Recognition of the urgent need to respect the rights of indigenous peoples, particularly their rights to their land and resources.

INDIGENOUS ORGANIZATIONS

Recognition that indigenous peoples are getting together to end discrimination and oppression.

ENVIRONMENT

Recognition that respect for indigenous peoples' knowledge can contribute to fair and lasting development and better management of the environment.

PART I - FUNDAMENTAL RIGHTS

Article 1: Human Rights

Indigenous peoples have the right to all the human rights and freedoms recognized in international law.

Article 2: Equality

Indigenous peoples are equal to all other peoples. They must be free from discrimination.

Article 3: Self-determination

Indigenous peoples have the right of self-determination.

This means they can choose their political status and the way they want to develop.

Article 4: Distinct Characteristics

Indigenous peoples have the right to keep and develop their distinct characteristics and systems of law.

They also have the right, if they want, to take part in the life of the rest of the country.

Article 5: Citizenship

Every indigenous person has the right to be a citizen of a country.

Part I sets out some very fundamental rights of indigenous peoples. It states that indigenous peoples have the same rights as other peoples and must be treated like everybody else. Part I recognizes that indigenous peoples have the right of self-determination and the right to keep their distinct characteristics.

There is much disagreement over what self-determination means. Indigenous peoples base their claims to self-determination on the fact that they were the first peoples in their territories. Self-determination means the right of indigenous peoples to choose their political status and to make decisions about their own development. Self-determination can take a variety of forms.

Some governments reject the right of indigenous peoples to self-determination or they try to limit its scope. They are fearful of independence movements and the possibility of national disintegration.

Indigenous representatives at the United Nations consider this view to be racist and discriminatory. They point to the fact that the United Nations Charter and the main human rights instruments state self-determination as a right of all peoples.

PART II - LIFE AND SECURITY

Article 6: Existence

Indigenous peoples have the right to live in freedom, peace, and security.

They must be free from genocide and other acts of violence.

Their children must not be removed from their families and communities for any reason at all.

Article 7: Cultural Integrity

Indigenous peoples shall be free from cultural genocide.

Governments shall prevent:

a) actions that take away their distinct cultures and identities;
b) the taking of their land and resources;

c) their removal from their land;
d) measures of assimilation;
e) propaganda against them.

Article 8: Identity

Indigenous peoples have the right to their distinct identities.

This includes the right to identify themselves as indigenous.

Article 9: Communities and Nations

Indigenous peoples have the right to belong to indigenous communities and nations in accordance with their traditions and customs.

Article 10: Removal and Relocations

Indigenous peoples shall not be removed from their land by force.

They shall not be relocated without their agreement. Where they agree, they should be given compensation and the possibility to return.

Article 11: Time of War

Indigenous peoples shall have protection in time of war.

Governments shall respect international law and must not:
a) force indigenous peoples to enter the army, particularly if they have to fight against other indigenous peoples;
b) allow indigenous children to join the army;
c) force indigenous peoples to leave their land;
d) force indigenous peoples to work for the army under discriminatory conditions.

Part II sets out the right of indigenous peoples to exist as distinct peoples. Indigenous peoples are to be free from genocide and their children must not be removed from their culture and identity.

They have the right to stay on their land and must be specially protected in time of war.

Genocide means the physical destruction of a people, including through the removal of children.

Cultural genocide refers to the destruction of a people's culture.

PART III - CULTURE, RELIGION, AND LANGUAGE

Article 12: Culture

Indigenous peoples have the right to their cultural traditions and customs.

This includes aspects of their culture such as sacred sites, designs, ceremonies, technologies, and performances.

Their cultural property shall be returned to them if it was taken without their permission.

Article 13: Spiritual and Religious Traditions

Indigenous peoples have the right to their spiritual and religious traditions, their customs, and their ceremonies.

They have the right to their sacred sites, ceremonial objects, and the remains of their ancestors.

Governments shall assist indigenous peoples to preserve and protect their sacred places.

Article 14: Language

Indigenous peoples have the right to their histories, languages, oral traditions, stories, writings, and their own names for people and places.

Governments shall ensure that in courts and other proceedings indigenous peoples can understand and be understood through interpreters and other appropriate ways.

PART IV - EDUCATION, MEDIA, AND EMPLOYMENT

Article 15: Education

Indigenous children have the right to the same education as all other children.

Indigenous peoples also have the right to their own schools and to provide education in their own languages.

Indigenous children who do not live in indigenous communities shall be able to learn their own culture and language.

Article 16: Information

All forms of education and public information shall reflect the dignity and diversity of indigenous cultures, traditions, and aspirations.

In consultation with indigenous peoples, governments shall take measures to promote tolerance and good relations between indigenous and other peoples.

Article 17: Media

Indigenous peoples have the right to their own media in their own languages.

They shall also have equal access to non-indigenous media. Government-owned media must reflect indigenous culture.

Article 18: Employment

Indigenous peoples have rights under international labour law and under national laws.

Indigenous peoples must not be discriminated against in matters connected with employment.

Part IV sets out the rights of indigenous peoples in the areas of education, the media, and employment. Indigenous children have the right to education, including education in their own languages and culture. They have the right to use mainstream media and to establish their own media. They have the right to be treated fairly in all matters relating to employment.

PART V - PARTICIPATION AND DEVELOPMENT

Article 19: Decision-Making

Indigenous peoples have the right to participate in decisions that affect them.

They can choose their own representatives and use their own decision-making procedures.

Article 20: Law and Policy-Making

Indigenous peoples have the right to participate in law and policy-making that affects them.

Governments must obtain the consent of indigenous peoples before adopting these laws and policies.

Article 21: Economic Activities

Indigenous peoples have the right to their own economic and social systems and to pursue their own traditional and other activities.

Where indigenous peoples have been deprived of their means of subsistence, they are entitled to compensation.

Article 22: Special Measures

Indigenous peoples have the right to special measures to improve their economic and social conditions.

This includes in the areas of employment, education, housing, health, and social security.

Article 23: Economic and Social Development

Indigenous peoples have the right to determine priorities and strategies for their development.

They should determine health, housing, and other economic and social programs and, as far as possible, deliver these programs through their own organizations.

Article 24: Health

Indigenous peoples have the right to their traditional medicines and health practices. The plants, animals, and minerals used in medicines shall be protected.

Indigenous peoples shall have access to all medical institutions and health services without discrimination.

Part V sets out the rights of indigenous peoples to participate in decisions and developments that affect them. Indigenous peoples must participate in, and give their consent to, decisions on law-making that affect them. They have the right to their own economic activities and to special measures to improve their economic and social conditions.

PART VI - LAND AND RESOURCES

Article 25: Distinctive Relationship

Indigenous peoples have the right to keep and strengthen their distinctive spiritual relationship with their land and waters.

Article 26: Ownership

Indigenous peoples have the right to own and control the use of their land, waters, and other resources.

Indigenous laws and customs shall be recognized.

Article 27: Restitution

Indigenous peoples have the right to return of land and resources taken without their consent.

Where this is not possible, they shall receive just compensation in the form of other land and resources.

Article 28: Environment

Indigenous peoples shall receive assistance in order to restore and protect the environment of their land and resources.

Army activities shall not take place on the land of indigenous peoples without their consent.

Hazardous material shall not be stored or disposed of on the land of indigenous peoples. Governments shall take measures to assist indigenous peoples whose health has been affected by such material.

Article 29: Cultural and Intellectual Property

Indigenous peoples have the right to own and control their cultural and intellectual property.

They have the right to special measures to control and develop their sciences, technologies, seeds, medicines, knowledge of flora and fauna, oral traditions, designs, art, and performances.

Article 30: Resource Development

Indigenous peoples have the right to determine strategies for the development of their land and resources.

Governments must obtain the consent of indigenous peoples before giving approval to activities affecting their land and resources, particularly the development of mineral, water, and other resources.

Fair compensation must be paid for such activities.

Part VI sets out the right of indigenous peoples to their land. They have the right to maintain their distinctive spiritual relationship with their land, waters, and resources. They have the right to own and develop their land, waters, and resources and to the return of land taken without their consent. Their environment and their cultural and intellectual property must be protected. Indigenous peoples have the right to control development of their land.

Cultural Property

There is high demand for indigenous artwork and cultural artefacts. Through theft and unauthorized use and sale, indigenous peoples have been robbed of their cultural heritage. Therefore, indigenous peoples are seeking protection of their "cultural property". This includes sites, human remains, oral traditions, designs, arts, and ceremonies.

Intellectual Property

Indigenous knowledge is a valuable resource; however, the profits are rarely shared with indigenous peoples. Therefore, indigenous peoples are seeking protection of their "intellectual property". This means indigenous knowledge in areas such as medicinal plants, agricultural biodiversity, and environmental management.

PART VII - SELF-GOVERNMENT AND INDIGENOUS LAWS

Article 31: Self-Government

As a form of self-determination, indigenous peoples have the right to self-government in relation to their own affairs.

These include culture, religion, education, media, health, housing, employment, social security, economic activities, land and resources management, environment, and entry by non-members.

Article 32: Indigenous Citizenship

Indigenous peoples have the right to determine who are their own citizens.

They have the right to decide upon the structures and membership of their organizations.

Article 33: Indigenous Laws and Customs

Indigenous peoples have the right to their own legal customs and traditions as long as they accord with international human rights law.

Article 34: Responsibilities

Indigenous peoples can decide the responsibilities of individuals to their communities.

Article 35: Borders

Indigenous peoples separated by international borders have the right to maintain relations and undertake activities with one another.

Article 36: Treaties and Agreements

Governments shall respect treaties and agreements entered into with indigenous peoples.

Disputes should be resolved by international bodies.

Part VII sets out guidelines for situations in which indigenous peoples exercise their right of self-determination through self-government. It recognizes the right of indigenous peoples to determine their citizenship, to their own laws and customs, to relations with other peoples across borders, and to treaties and agreements with governments.

Treaties and Agreements

In the past, many indigenous peoples have reached treaties and agreements with governments. For indigenous peoples, treaties have great spiritual meaning, and they provide recognition of their status as self-governing peoples and their right to self-determination.

Indigenous peoples have struggled for the recognition of their treaty rights in domestic and international law. Treaty rights can also be a way to regain control over indigenous land and resources.

PART VIII - IMPLEMENTATION

Article 37: National Law

In consultation with indigenous peoples, governments shall take measures to effect this Declaration. This includes making the rights recognized in the Declaration into national law so that they can be enforced by indigenous peoples.

Article 38: Financial Assistance

Indigenous peoples have the right to financial and other assistance from governments and international organizations in order to exercise the rights recognized in this Declaration.

Article 39: Disputes

Governments shall establish fair procedures for resolution of disputes with indigenous peoples. These procedures must take account of indigenous customs and traditions.

Article 40: United Nations

The United Nations and other international organizations shall provide financial and other assistance in order to effect the rights recognized in this Declaration.

Article 41: Special International Body

The United Nations shall create a special international body in order effect this Declaration. Indigenous peoples shall participate directly in this body.

Part VIII sets out what governments and the United Nations must do in order to put the Declaration into practise.

PART IX - UNDERSTANDING THE DECLARATION

Article 42: Minimum Standards

This Declaration contains only minimum standards for indigenous peoples.

Article 43: Men and Women

The rights recognized in this Declaration apply equally to indigenous men and women.

Article 44: Other Indigenous Rights

Nothing in this Declaration affects other rights indigenous peoples presently hold or may obtain in the future.

Article 45: United Nations Charter

Nothing in this Declaration allows any action against the Charter of the United Nations.

Part IX provides some guidance as to how to understand the Declaration.

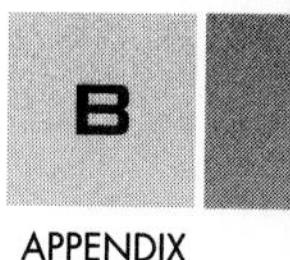

International Convention on the Elimination of All Forms of Racial Discrimination

Adopted and Opened for Signature and Ratification by General Assembly Resolution 2106 (XX) of 21 December 1965

entry into force 4 January 1969, in accordance with Article 19

The States Parties to this Convention,

Considering that the Charter of the United Nations is based on the principles of the dignity and equality inherent in all human beings, and that all Member States have pledged themselves to take joint and separate action, in co-operation with the Organization, for the achievement of one of the purposes of the United Nations which is to promote and encourage universal respect for and observance of human rights and fundamental freedoms for all, without distinction as to race, sex, language or religion,

Considering that the Universal Declaration of Human Rights proclaims that all human beings are born free and equal in dignity and rights and that everyone is entitled to all the rights and freedoms set out therein, without distinction of any kind, in particular as to race, colour or national origin,

Considering that all human beings are equal before the law and are entitled to equal protection of the law against any discrimination and against any incitement to discrimination,

Considering that the United Nations has condemned colonialism and all practices of segregation and discrimination associated therewith, in whatever form and wherever they exist, and that the Declaration on the Granting of Independence to Colonial Countries and Peoples of 14 December 1960 (General Assembly resolution 1514 (XV)) has affirmed and solemnly proclaimed the necessity of bringing them to a speedy and unconditional end,

Considering that the United Nations Declaration on the Elimination of All Forms of Racial Discrimination of 20 November 1963 (General Assembly resolution 1904 (XVIII)) solemnly affirms the necessity of speedily eliminating racial discrimination throughout the world in all its forms and manifestations and of securing understanding of and respect for the dignity of the human person,

Convinced that any doctrine of superiority based on racial differentiation is scientifically false, morally condemnable, socially unjust and dangerous, and that there is no justification for racial discrimination, in theory or in practice, anywhere,

Reaffirming that discrimination between human beings on the grounds of race, colour or ethnic origin is an obstacle to friendly and peaceful relations among nations and is capable of disturbing peace and security among peoples and the harmony of persons living side by side even within one and the same State,

Convinced that the existence of racial barriers is repugnant to the ideals of any human society,

Alarmed by manifestations of racial discrimination still in evidence in some areas of the world and by governmental policies based on racial superiority or hatred, such as policies of apartheid, segregation or separation,

Resolved to adopt all necessary measures for speedily eliminating racial discrimination in all its forms and manifestations, and to prevent and combat racist doctrines and practices in order to promote understanding between races and to build an international community free from all forms of racial segregation and racial discrimination,

Bearing in mind the Convention concerning Discrimination in respect of Employment and Occupation adopted by the International Labour Organisation in 1958, and the Convention against Discrimination in Education adopted by the United Nations Educational, Scientific and Cultural Organization in 1960,

Desiring to implement the principles embodied in the United Nations Declaration on the Elimination of Al l Forms of Racial Discrimination and to secure the earliest adoption of practical measures to that end,

Have agreed as follows: . . .

Articles 1 – 25

(http://www.unhcr.ch/html/menu3/b/d_icerd.htm)

CREDITS

Chapter 1: Mackey, Eva. 1956. *The House of Difference: Cultural Politics and National Identity in Canada.* © University of Toronto Press Incorporated, 2002. Reproduced by permission of Taylor & Francis Books, UK.
Chapter 2: © Sherene H. Razack, *Race, Space, and the Law: Unmapping a White Settler Society* (Toronto: Between the Lines, 2002).
Chapter 3: Copyright © Sylvia Hamilton.
Chapter 4: *The Dark Side of the Nation: Essays on Multiculturalism, Nationalism and Gender* by Himani Bannerji. © 2000. Reprinted by permission of Canadian Scholars' Press Inc.
Chapter 5: Excerpt by Patricia Monture-Angus, "Theoretical Foundations and the Challenge of Aboriginal Rights" from *Journeying Forward: Dreaming First Nations' Independence.* © 1999 by Patricia A. Monture-Angus. Reprinted by permission of Fernwood Books.
Chapter 6: From Henry, *The Colour of Democracy: Racism in Canadian Society.* © 2006 Nelson Education Ltd. Reproduced by permission. www.cengage.com/permissions
Chapter 7: Excerpt from *The Security of Freedom: Essays on Canada's Anti-Terrorism Bill.* © 2001. Edited by Ronald J. Daniels, Patrick Macklem, and Kent Roach. Reprinted with Permission of the Publisher, University of Toronto Press Inc.
Chapter 8: Excerpt from *Who Da Man? Black Masculinities and Sporting Cultures.* © 2005 by Gamal Abdel-Shehid. Reprinted by permission of Canadian Scholars' Press Inc.
Chapter 9: Excerpt from *Claiming Space: Racialization in Canadian Cities.* © 2006. Edited by Cheryl Teelucksingh. Reprinted by permission of Canadian Scholars' Press Inc. and/ Women's Press.
Chapter 10: Excerpt from *Claiming Space: Racialization in Canadian Cities.* © 2006. Edited by Cheryl Teelucksingh. Wilfrid Laurier University Press, Waterloo, Canada N2L 3C5 www.wlupress.wlu.ca ISBN 0-88920-499-3. Reprinted with Permission.
Chapter 11: Excerpt from "Canadian Woman Studies," Summer 2000, Volume 20, Number 2, August 2000, Publications Mail Registration, Number 08045. © Angela Aujla. Reproduced with Permission.
Chapter 12: Excerpt from *Canadian Journal of Political Science* XXXIII:3 (September 2000) 465–497. © Canadian Political Science Association. Reproduced with Permission.
Chapter 13: Excerpt from *Social Determinants of Health: Canadian Prespectives.* © 2004. Edited by Dennis Raphael. Reprinted by permission of Canadian Scholars' Press Inc.
Chapter 14: Excerpt from *The Canadian Review of Sociology of Anthropology;* Nov 2001; 38, 4; Research Library pg. 415. © Blackwell Publishing Ltd.
Chapter 15: This excerpt is reprinted with permission of the Publisher from Transnational Identities and Practices in Canada. Edited by Vic Satzewich and Lloyd Wong. © University of British Columbia Press, 2006. All rights reserved by the Publisher.
Chapter 16: Excerpt from *Anti-Racist Feminism: Critical Race and Gender Studies.* © 2000. Edited by Agnes Calliste and George J. Sefa Dei. Reprinted by permission of Fernwood Books.

INDEX